Biological Individuality

Biological Individuality

Integrating Scientific, Philosophical,
and Historical Perspectives

EDITED BY
SCOTT LIDGARD AND
LYNN K. NYHART

The University of Chicago Press
Chicago and London

The University of Chicago Press, Chicago 60637
The University of Chicago Press, Ltd., London
© 2017 by The University of Chicago
Published 2017
Printed in the United States of America

26 25 24 23 22 21 20 19 18 17   1 2 3 4 5

ISBN-13: 978-0-226-44631-8 (cloth)
ISBN-13: 978-0-226-44645-5 (paper)
ISBN-13: 978-0-226-44659-2 (e-book)
DOI: 10.7208/chicago/9780226446592.001.0001

Library of Congress Cataloging-in-Publication Data

Names: Lidgard, Scott, editor. | Nyhart, Lynn K., editor.
Title : Biological individuality : integrating scientific, philosophical, and historical
   perspectives / Scott Lidgard and Lynn K. Nyhart.
Description: Chicago ; London : The University of Chicago Press, 2017. | Includes
   bibliographical references and index.
Identifiers: LCCN 2016049002 | ISBN 9780226446318 (cloth : alk. paper) |
   ISBN 9780226446455 (pbk. : alk. paper) | ISBN 9780226446592 (e-book)
Subjects: LCSH: Biology—Philosophy. | Variation (Biology).
Classification: LCC QH331 .B495 2017 | DDC 570.1—dc23 LC record available at
   https://lccn.loc.gov/2016049002

Contents

Introduction: Working Together on Individuality

LYNN K. NYHART AND
SCOTT LIDGARD

This volume is premised on the idea that biologists, historians of biology, and philosophers of biology can learn more about individuals and individuality by working in concert than by working separately. Biologists and historians and philosophers of biology have a long history of working productively together; indeed, one of the hallmarks of the discipline of philosophy of biology as it has developed since the 1960s has been that it didn't treat the science and scientists at arm's length the way philosophy of physics often did (Hull 2000, 69). The successful meetings of the International Society for the History, Philosophy, and Social Studies of Biology since the late 1980s have further demonstrated the fruitfulness of conversation and collaboration across these disciplines. The problem of individuality is a natural locus for such collaboration, as it is a fundamentally *philosophical* problem with a long *history* that impinges directly on how *biologists* do their work.

As this volume shows, philosophical and historical perspectives can illuminate concepts of biological individuality and help open up productive avenues for biologists. In addition, we argue that a truly cross-disciplinary approach should contribute to more even-handed reciprocity among these fields, in which interdisciplinary examinations of biological individuality would also enhance the history and philosophy of biology. How might such mutual enlightenment work?

To answer that question, we must first acknowledge that biologists, historians, and philosophers all have different goals, and their areas of interest overlap only partially. A crude but useful formulation might be as follows. Biologists want to understand how living nature works. Historians of biology want to understand how *scientists think* about how nature works, and how scientific construals of nature have changed over time. Philosophers of biology want to understand and improve upon the conceptual structure of biology. They might partake in the biologists' or the historians' kinds of questions, depending on whether their focus is more ontological (concerning the basic categories of what is real in nature), which tilts them toward the biologists, or epistemological (how we understand nature), which tilts them toward the historians as well as the biologists.

Historians and philosophers can help biologists by clarifying hidden assumptions and reviving older questions and perspectives that have fallen into the shadows, by imagining new approaches to solving existing problems and identifying where seemingly new approaches have in fact been tried before, thus inviting scientists (notoriously forward-looking as a group) to look to their forebears for insights. In addition, because the research of historians and philosophers is not tied to particular biological systems or levels, and tends to be more attuned to broader contextual factors and underlying philosophical questions, they may provide insights deriving from those more distant perspectives on how biology works. These broad claims to the utility of philosophy and history for biology are not novel, but they bear repeating. We think that their application to the more proximate, specific problems that arise when biologists put individuality concepts to work do in fact yield novel insights (see Lidgard and Nyhart, this volume).

But what is the role of the biologist in enhancing the history and philosophy of biology? This question is asked less frequently. For both historians and philosophers, biologists and their confrontations with nature provide the fodder for analysis, but they tend to do so in different ways and for different ends. Present-day biologists—who are usually well acquainted with how they themselves work—provide "reality checks" to historians and philosophers (Waters 2014). "No, that's not actually how I think; I think about it like *this*" is one sort of help a biologist can offer to a historian or philosopher. Another is to clarify, based on deep familiarity, how particular biological systems work, according to the present state of observation and theory. The work of past biologists—especially dead ones who can't talk back—provides further material upon which historians and some philosophers of biology might perform overlapping kinds of analyses. Both groups often begin by seeking to understand how past biologists thought and what they knew. What were the fundamental underlying commitments that shaped how they thought about nature? By what processes did their ideas about living nature change? How did they gain new knowledge? At this point, historians and philosophers of biology diverge.

Historians will tend to seek plausible explanations for the process of change from the particulars of historical contexts—whether those particulars have to do with the biographies of the individuals involved, their scientific and technical practices, or the institutional, economic, religious, political, or cultural circumstances under which they operate. Because the world of historical events is unimaginably complex and its record is always incomplete, historians typically don't argue that the trajectories they trace through the historical events, or even their causal explanations, are the only "true" ones. Rather, they seek to reconstruct what happened based on the available evidence, and then provide a plausible and persuasive account of why it happened in a particular way. In recent decades, such accounts in the history of science have tended to highlight the social and material features of doing science that have contributed to structuring the directions of new knowledge, with a heightened sense that those directions are context-dependent. Even historians of science who are not "strict" social constructionists (and most aren't) usually accept this basic premise and are thus more likely to adopt a pluralist account of nature: historical and cross-cultural analyses emphasize that humans have developed many different ways to make sense of nature, and they are not all commensurable with one another.

Philosophers of biology with a historical bent (often part of or influenced by a long Continental tradition that unites the history and philosophy of science) tend to focus more exclusively on intellectual history, unpacking the ideas of past biologists and philosophers. This can resemble the work of historians of science with strong philosophical interests. But the historical work that characterizes philosophers usually has different goals from those of historians. Such work may seek to generalize about the nature of conceptual change in science, rather than to give an account of that change tied strictly to the particulars of history (Kuhn 1962; Lakatos and Musgrave 1970; Hull 1988; Griesemer and Wimsatt 1989; LaPorte 2004; Brigandt 2011, esp. 251–54). Or their work may address examples or texts from the past to analyze or critique a still-current philosophical problem. For example, Kant's notion of organicism or Leibniz's monads are relevant to arguments about part-whole relations that continue today, and even recent philosophers sometimes argue with positions represented by long-dead people in a way that may strike the historian as weirdly anachronistic (e.g., Bennett 1974). Because philosophers of biology operate within a discipline in which seeking truth is a constitutive feature, the question of whether we should adopt a single (monistic) approach to truth or a pluralistic one is much more fraught—and much more central—than it currently is among historians. Thus philosophers may offer a check on historians' ready (and possibly sloppy) pluralism, while historians are likely to push back against unitary accounts of nature's truths. The diversity of research problems and plethora of biological individuality concepts that biologists have appealed to just add urgency to the questions: Is there a single way to "carve nature at its joints"? If so, what's needed to do it? If not, what are we left with?

To begin with, we're left with prefatory questions: "What is biological individuality, and why does anyone care?" When we talk about "individuals" or "individuality" in casual conversation, we tend to mean something about identity—that each of us is unique, different from everyone else (perhaps by virtue of our personality). Identity is also part of some biological and biomedical discourses of individuality, such as epigenetic and phenotypic discordance between twins despite shared genetic identity (Bell and Saffery 2012; Yet et al. 2015), or rethinking immunology's notion of "self and other" in light of pervasive symbioses and autoimmune disorders (Pradeu 2012; Tauber 2015a and b). However, biological individuality can refer not only to identity but also to unity or wholeness. Examined still more closely, biological individuality may entail a host of related criteria, among them boundedness, integration, the nature of interactions among parts and wholes, agency or governance of parts, propagation by a variety of means, continuity over time, comprising or being part of a biological hierarchy, being a potential unit of selection, contributing to theoretical evolutionary fitness, and, yes, identity or autonomy (Lidgard and Nyhart; Elwick; Love and Brigandt; all this volume). Further complicating matters, we might notice that whereas many of the above-listed features direct our attention to *structural* aspects of individuality (such as hierarchy or boundedness), others address biological individuality in terms of dynamic *processes* (such as part-part interactions or the maintenance of continuity over time) and even *agency* (such as governance or functional capacity; Wilson and Barker 2013). Working out the relations among structural and processual approaches to individuality is a longstanding problem (Brigandt, this volume).

In both biology and the philosophy of biology today, then, individuality is a complex topic, hot with contestation. Levels of selection, major transitions in evolution, relationships among parasites or symbionts and their hosts, the nature of coloniality, the control of cells and communication among them, immunological recognition and tolerance, genetic and developmental modularity, even the essential question "What is a gene?"—all these specific problem areas require concepts of individuality before they can even be posed, much less answered. Biologists recognize this; hence the interest in individuality. Yet at the same time, these very problem areas—not to mention the wealth of possible criteria listed above—defy a single consistent, unified definition of biological individuality.

Despite the wide range of problems associated with biological individuality, its analysis in recent years has mostly been framed within evolutionary theory, especially in relation to the units or levels of natural selection, or with respect to the major transitions in evolution (for leading recent discussions, see esp. Godfrey-Smith 2009; Clarke 2010; Bouchard and Huneman 2013; Sterner 2015). And yet, many biological processes may be productively studied independently of their relation to Darwinian evolution. The historians, philosophers, and biologists contributing to this volume seek to deepen and broaden the contexts within which biological individuality is currently discussed. At the same time, we call attention to the range of philosophical and historical contexts and questions in which biological individuality concepts play a role.

The Chapters

In bringing together the contributors to this volume, we sought out people who not only are open to different disciplinary perspectives but also have been willing to think and write amphibiously about biological individuality, blending perspectives from biology, philosophy, and history. While most chapters are readily identified as drawing more dominantly from one perspective or another, individually and collectively they signal the importance and productivity of bringing together multiple perspectives. An almost natural consequence of this is a stance of epistemological modesty, an awareness that what we know is limited, and that the way toward bigger truths—if perhaps never "the" truth—is to view knowledge as multifaceted, and to see the outcome of this interdisciplinary endeavor as being to enhance our understanding of biological individuality itself as a problem agenda (Love 2014, forthcoming) that belongs not only to biology but also to the history and philosophy of science.

After an opening chapter that offers a problem-centered framework for analyzing concepts of biological individuality and their contexts, the book offers nine research-based chapters, clustered in three groups of three, followed by three commentaries. Chapters 2–4 share an interest in single-celled and multicellular individuals, and how we might think about their relations in evolutionary, developmental, and functional terms. Chapters 5–7 analyze historical cases of scientists in the nineteenth and early twentieth centuries who struggled to conceptualize complex biological individuality across multiple temporal scales, often from non-Darwinian evolutionary perspectives. And chapters 8–10, on parasitism, metabolism, and functional parts, share an emphasis on physiological or functional considerations in part-part and part-whole relations. Different chapters also speak to one another in alliances beyond these groupings: for instance, emphasizing the important interactions between theorizing about biology and theorizing about the sociopolitical realm, or grappling with questions of agency. The volume ends with three commentaries emphasizing historical, biological, and philosophical takes on biological individuality and the research chapters. In this way, we hope to offer fresh perspectives on old questions and open up new avenues of research.

In chapter 1, Lidgard and Nyhart begin by addressing the work that individuality concepts do for biologists. Taking off from an idea first proposed by Allen Newell and Herbert Simon (1972), they develop the idea of a "problem space" used to work toward a solution to a given biological problem. The problem space may incorporate ideas, hypotheses, experimental information, and other elements with which the biologist is working, and may itself change over time as its external environment changes. That is, it is a semipermeable domain of thinking, one that excludes that which is unimportant to the problem but that can change to incorporate new elements if they come to be perceived as important. This general model, they suggest, is useful for thinking about what gets put inside the space for different biological problems, by different biologists, and also for thinking about how historians and philosophers construct problem spaces for biological individuality differently than do biologists. Then, after enumerating a wide range of proposed definitional criteria for biological individuality, past and present, they suggest four broad kinds of problems that confront most discussions of individuality: individuation, or the delineation of one entity from another, and relationships that unify and are not shared by anything else; hierarchy, or the idea of individuals nested into levels; constitution, or the constituent parts, their functional relations, and their interactions, as well as the interpenetration of entities and their external environments; and temporality, or the notion of individuality expressing itself over time. They then broaden the scope of inquiry still further, exploring how network, system, and process theories in biology open new perspectives on established concepts and definitional criteria of biological individuality. Rather than seeing biological individuality as a scatter of highly proximate, specific problems that are insoluble if approached as instances of an undifferentiable singularity, Lidgard and Nyhart reenvision it as a problem agenda (Love 2014, forthcoming) that provides a positive, constructive way forward. They argue that this problem agenda is a stable, broad domain that has persisted through much of biology's history, and that it abstractly corresponds to a constellation of specific problems whose explanatory demands differ somewhat in particular contexts.

In recent years, hierarchies of individuality have been addressed almost exclusively within two overlapping discussions: one on levels of selection (first in the "levels of selection" debate and more recently with the development of multilevel selection theory), and the other on evolutionary transitions in individuality. A key focus of attention in these discussions is the transition from unicellularity to multicellularity (Maynard Smith and Szathmáry 1995). Chapters 2 through 4 expand our perspective on these discussions of the relations of cells to one another, as they take different approaches to asking how individuality is maintained or suppressed in groups of cells where the group or the cell has the greater degree of autonomy. Such "groups" might include loose aggregates of single cells of the same kind, more tightly bound cell colonies, multicellular organisms, or biofilms (comprising microorganisms that can be of different species). In these chapters, these different kinds of groups are put to work to examine evolutionary transitions in individuality (from single-celled to multicellular organisms), the regulation of life-cycles and development, and even cell death. To address such questions, Matthew Herron (chapter 2) and Beckett Sterner (chapter 3) take as their key biological exemplars the volvocine algae, for this group shifts between unicellularity and multicellularity in ways that make it an especially apt model. These authors also share with Andrew Reynolds (chapter 4) an interest in the role of the extracellular matrix in binding cells together, though their analyses contextualize it quite differently. Yet together they highlight the fruitfulness of attending to the functional roles of material properties in producing complex individuals.

In chapter 2, Herron illuminates questions of hierarchy and emergence through a careful study of development among the volvocine green algae, in which three kinds of biological units—the cells, the colonies, and the clones—have a justifiable claim to a degree of biological individuality. He then asks how much individuality is present in each one, but the answer is hardly simple. Volvocines produce new generations of cells asexually that can themselves form new colonies, and in a few lineages, some cells within colonies become structurally and functionally differentiated. Herron employs a phylogenetic framework to show how different volvocine lineages with different kinds and degrees of individuality are related to one another through evolutionary descent with modification, which he uses to ground different scenarios by which single-celled organisms might have evolved into multicellular ones. In doing so, he makes a compelling case that degrees of individuality can and do vary among different biological individuality concepts and their respective criteria. Of equal or greater importance is the implication that degrees of individuality depend in part on a variety of determinants, such as developmental programs, mutation rates, and ecological and demographic circumstances, that operate through population structure changes in ecological time as well as changes over evolutionary time.

In chapter 3, Beckett Sterner views evolutionary transitions in individuality with a philosopher's concern for the frequently neglected roles of material and causal structure. He points out that key explanatory challenges to such transitions involve how and why transitions happened at each level of a compositional hierarchy, why the higher level individual persists or changes, and how to determine whether an entity is an individual or not, or to what degree. He argues that multilevel selection theory (Okasha 2006; Clarke 2014) is inadequate by itself to stand as a complete explanatory theory: it cannot initially choose candidate evolutionary units, offer criteria to decide where and how in a complex life cycle conflict and cooperation affect an evolutionary transition, or take into account the material structure in a given case that specifies or realizes a multilevel selection model (see also Winther 2006, 2011; Love and Brigandt, this volume). Sterner devises a complementary theory of individuality based not on fitness, but on other fundamental capacities inherent in the idea of a "demarcator." The demarcator may be either developmental or ecological, either a substantive material object or a processual sequence of events, connected by causal relations. Complementing Herron's biological analysis, Sterner bolsters the demarcator idea with promising examples from cell differentiation in volvocine algae and group-level evolution in bacterial biofilms. Yet the potential of this new theory may be more far-reaching, helping biologists and philosophers to investigate degrees of individuality at different hierarchical levels, and to form a more complete and more representative mechanistic understanding of specific evolutionary transitions in individuality.

Andrew Reynolds (chapter 4) draws on history and the analysis of scientists' language to introduce yet other considerations into the discussion of cell-cell relations and the transition to multicellularity. He sketches a history of the key nineteenth-century tropes of the cell state and the division of physiological labor, which infused political and economic language into biology. Then he charts the emergence in the twentieth century of the language of communication—a major novelty in thinking about cells. Seeing cells as "communicative agents" rather than the body's "building blocks" views them not merely as architectural parts or even as political or economic individuals but as *social* ones, which collectively display group-level properties. Reynolds analyzes the work of cell sociologists since the 1970s to elucidate an intertwined vocabulary or discursive field of cell-cell and cell-tissue communication that employs such terms as *signal, cell,* and *coercion* to describe how cells are held together in a multicellular organism, and how cells' interactions (represented as social and agential) lead to their differentiation. Communication is thus essential to the cooperation of cells in producing a more complex individual, and must feature importantly in evolutionary transitions in individuality.

Whereas these three chapters focus on cellular and multicellular individuals to move beyond current understandings of evolutionary transitions in individuality, the next three take us further away from Darwinian evolution. They examine how biologists from the nineteenth and early twentieth centuries created non-Darwinian frameworks to address problems combining hierarchy at multiple scalar levels (both below and above the organism) with temporal considerations—how individuality is expressed or produced through time. Lynn Nyhart and Scott Lidgard (chapter 5) examine problematics surrounding temporal expressions of individuality developed in frameworks that were pre-Darwinian, but in some cases evolutionary; Snait Gissis (chapter 6) traces out key aspects of the views on individuality of Herbert Spencer, who began developing his evolutionary philosophy before Darwin published the *Origin* and continued long thereafter in his own mode; and Olivier Rieppel (chapter 7) considers the lasting influence of a non-Darwinian vision of nested hierarchy developed in the late nineteenth century that spread into several different biological areas by the 1930s, including a temporal version in paleontology. Together, these chapters remind us that individuality has long been intertwined with large-scale questions about the organization of nature and its changes over time, questions that could be fruitfully addressed without putting selection at their center.

Nyhart and Lidgard's chapter on alternation of generations takes as its central puzzle how biologists have thought about complex life cycles that extend temporally across generations. In the first half of the nineteenth century European biologists established the existence of this broad cyclical pattern. Controversies over its meaning and generality extended across classification, reproduction, development, parasitism, and evolution. The authors show how formal, functional, and temporal criteria for individuation were intimately embedded in these problems. Despite many differences, the historical biologists under study tended to share certain common commitments: to a developmental perspective, yielding a new emphasis on temporal and cyclical definitions of individual wholeness; to the search for generalizations across the animal and plant kingdoms in pursuit of biological laws; and to nested hierarchies of individuality as a way to reconcile the diversity of relations they saw among form, function, and development (a practice continued into the twentieth century, as Rieppel's and Reynolds's chapters in this volume delineate). Nyhart and Lidgard's chapter thus shows that discussions treating the roles of hierarchy, complexity, emergence, reproduction, and even evolution in defining individuals were already well developed in the pre-Darwinian period. Rather than dismissing them because of their non-Darwinian perspective, we might consider what ways of thinking about individuality from this earlier period might provide inspiration and new directions for thinking by disentangling individuality from Darwinism.

The same is true of Gissis's chapter 6, which offers a surprising new take on the nineteenth-century philosopher Herbert Spencer. Spencer's work has been taken to epitomize a nineteenth-century version of nested individuality stretching from cells to societies (see, e.g., Elwick 2015). Yet in Gissis's nuanced reinterpretation, he was less interested in asserting levels of individuality or any nested relation among them than in detailing the gradual emergence of increasingly complex individuals. Gissis presents a Spencer working to articulate a form of evolutionary emergence based on a continuous loop of interactions between organism and environment. The process itself grew more complex over time, as the organism, the environment, and their interaction gained in complexity. In this view, individuals are not an ontologically prior category of nature, but emerge (with different degrees of complexity) as the result of evolution. For Gissis's Spencer, stable interaction with the environment, not demarcation from it, characterized the individual—to the extent that we can identify it. (For a modern definition of individual that sounds remarkably similar, but stands within a Darwinian framework, see Bouchard 2010). Like today's scientists and philosophers who posit degrees of individuality, Spencer engaged centrally with the *process* of individuation. Gissis shows how, in seeking to capture moments of that process, he struggled to articulate a concept of individuality right at the shifting edge of the collectivity—an assemblage of individuals—and a more tightly integrated complex individual. Gissis's novel treatment invites reassessment of the historical Spencer and his times. At the same time, as the chapters by Herron and Sterner show, that tension between seeing a hierarchy and describing its emergence—between structure and process—remains central to the biological and philosophical problematics surrounding individuality today.

In chapter 7, Olivier Rieppel analyzes the synthetic concept of *Enkapsis*, developed in the early twentieth century by the Tübingen anatomist Martin Heidenhain to characterize levels within the multicellular organism. This was applied in the 1930s by August Thienemann and Karl Friederichs to ecology, and by the paleontologist Karl Beurlin to phylogeny (where, in Beurlin's conception, the tree of life as a whole directs the development of its subordinate branched lineages, and Darwinian selection is nowhere to be seen). These enkaptic hierarchies sat well with Nazi ideology, which emphasized the subordinated character of the "parts" of the "national community" to the whole German *Volk.* Rieppel thus shows how arguments about biology can intersect and interact with political arguments—a central concern for historians of biology. Yet as a biologist-philosopher, he also argues that enkapsis remains useful to systems-oriented biologists today. Rieppel's paper is philosophically useful in arguing for a fundamental distinction between logical individuals and biological ones. It challenges historians to think again through the problem of "Nazi taint" in science, and more generally the shaping of science by politics and ideology. And it invites biologists to look across scales and realms for fundamental conceptual similarities in concepts of hierarchy applied to the body, ecological units, and phylogenetic ones.

Up to this point in the book, the essays have largely placed relations between hierarchy and temporality in the foreground, framed in good part by questions about evolution (if not necessarily Darwinian evolution). The last three research chapters, by contrast, focus more on aspects of individuality structured around constitution—the maintenance of a whole by its parts (or its abrogation via parasitism and symbiosis)—as a key aspect of individuation, and on constitution's corollary, function (understood as activity). They also offer further directions for work in the history and philosophy of biology that connects to broader themes in these disciplines. In particular, for historians of science, ideas about the agency, governance, and regulation of parts within a whole illuminated by Osborne (chapter 8) and Landecker (chapter 9) connect to a significant historiographic theme: the fruitful borrowing and interchange of language between the biological and social sciences (also treated by Reynolds, Gissis, and Rieppel, this volume). This focus on figurative language, analogy, and metaphor is one of the most enduring themes in the historiography of biological individuality (especially in the French-language tradition, from Perrier [1881] to the present). For philosophers, these themes as well as questions of robustness and structure-function relations (Brigandt, chapter 10) offer new takes on classic philosophical questions about individuals, parts, and wholes.

Many chapters already discussed address phenomena that call into question clear boundaries of individuals and suggest instead degrees of individuality (Herron, Sterner, Gissis). Parasitism and symbiosis blur the boundaries still more, as in these cases, basic physiological processes are achieved through participation across different species. Michael Osborne's chapter 8, on parasitology and ideas of individuality in late nineteenth- and early twentieth-century France, addresses a classic instance of the problematics of constitution. In biomedical definitions, parasitism undermined a key feature of individuality, namely autonomy, since a parasite depended on an organism of another kind to sustain its life-functions. But this was never a "pure" matter of scientific investigation. Osborne shows how medical parasitology developed as both discipline and discourse in the French colonial state: the "body" of Greater France would be healthy only if the tropical, largely parasite-borne diseases of North Africa and other French colonies could be vanquished. At the same time, the inhabitants of those colonies were sometimes seen as socioeconomic parasites on the body of France. Parasitological tropes extended further into the political and sociological realms via the Solidarists, in whose philosophy the individual citizen voluntarily limited his autonomy by subordinating himself to the greater good. Like the social parasite, the Solidarist citizen could not function on his own; unlike the social parasite, the good citizen's contribution benefited the state. Osborne's chapter presents new material for historians of science, medicine, and the state, sketching out the conditions under which different levels of historical action—linguistic, disciplinary, and political—layered over one another to create a powerful discursive structure uniting biomedical science and politics around part-whole relations.

Both Reynolds (chapter 4) and Osborne (chapter 8) are interested in questions about autonomy and dependence and how the agency of the individual has been understood to be mediated in hierarchical social wholes—states or organisms. Hannah Landecker's chapter 9, on metabolism and individuality, takes the question of agency still further. In outlining the development of ideas of metabolism since the mid-nineteenth century French scientist Claude Bernard (who influenced a number of scientists discussed by Osborne), she uncovers a "logic of total conversion" in which the prime work of metabolism is to produce and maintain the autonomous, perduring whole individual in the face of a fluctuating environment. In consuming and metabolizing food, the eater has complete agency, while the thing eaten has none. However, Landecker contends, new research on horizontal gene transfer, ingestion of commensals and symbionts, and metabolites as signals or cues, is challenging the assumptions underlying this logic and returning agency to the eaten (see also Grote and Keuck 2015). This in turn calls into question the basic logic of conversion itself. For biology, this represents a challenge to assumptions about what metabolism *is*, and its role in maintaining the (multicellular) individual as an autonomous entity. For history, this challenge makes visible the logic as a historically specific and contingent one, rather than a truth of nature. Landecker's essay thus suggests new possibilities for both biological and historical analysis and their fruitful interactions.

The final chapter is squarely philosophical. Here Ingo Brigandt argues that, to understand part-whole relations in complex organisms, we should go beyond considering structural parts and incorporate functions (understood as activities) as equally legitimate agents of part-whole relations. Brigandt contends, first, that individuation of a structure can depend on its activity-functions, on either ontological or epistemological grounds. More controversially, he argues that functions can themselves be considered parts; these involve a temporal aspect (because they are activities), can be organized in multilevel hierarchies that also include structures, and can be homologized across species. Finally, he offers cases to show how considering activity-functions as parts reduces tensions that have built up in biology from the usual treatment of structure and function as opposed or antagonistic considerations. Brigandt's argument thus contributes to a broader move among some philosophers to integrate structure and function, or at least to show them as complementary (Griesemer 2003, 2014; Pigliucci 2003; Breuker et al. 2006; Booth 2014).

Though this is one of the most purely philosophical chapters in the volume, parts of Brigandt's argument are readily put in conversation with other chapters in this section, and indeed in the book as a whole. As Nyhart and Lidgard show, already in the nineteenth century Schleiden explicitly conceptualized temporal parts, and Braun treated life cycles as operating in a nested hierarchy that was simultaneously temporal and taxonomic. Brigandt's focus elsewhere in the chapter on robustness (the ability of a structure to persist in the face of a fluctuating environment) as resulting from developmental and functional processes recalls Landecker on Bernard's view of metabolism as creating an internal milieu separate from the external one, which in turn allowed the organism to maintain its autonomy from the environment. One suspects, based on Gissis's chapter on Spencer, that the nineteenth-century philosopher, too, would have been sympathetic to Brigandt's desire to treat structure and function as both involved, in concert and in interaction, in producing complex organisms. In this way, history provides examples for contemporary philosophy, while philosophy offers relevance to historical research. Both expand the framework for biological inquiry by lifting our gaze above particular problems in the present moment, and inviting all of us to think more broadly and deeply about the nature of our inquiries.

The volume closes with three commentaries. Their authors were given the challenging task of reflecting on the problem of biological individuality, and reflecting on the chapters in this volume, from the perspectives of history, biology, and philosophy, respectively. Each of these commentaries offers a broad overview of biological individuality in its respective discipline, together with a particular argument for where scholars might head next—syntheses that can profitably be read by those both within and outside these disciplines.

Historian James Elwick helps us understand why intuitive notions of biological individuality have persisted despite being contested by biologists for over two and a half centuries, and suggests how a history organized around this contestation might yield new insights into the history of biology, extending even into such canonical areas as Darwin's own understanding of nature. Biologist Scott Gilbert defends a new level of biological individuality called the "holobiont," in which organisms of different species joined into symbiotic associations may be usefully considered "individuals"—with consequences for how we think about development, genetics, immunity, and selection. And philosophers Alan Love and Ingo Brigandt argue for pursuing an understanding of biological individuality that begins from epistemology rather than ontology. In particular, this pursuit starts from the practical perspective of what sorts of problems a biologist wants to address. Descending from universals to the level of problem-appropriate criteria of individuality, they argue, is more likely both to help biologists solve their problems *and* to provide philosophically robust accounts of nature—even ontological ones, in the end. The price is a pluralist account, but as they demonstrate, such an account actually enhances the work that philosophy can do.

While remaining sensitive to the interactions among all three disciplines, each commentary articulates difficulties facing existing individuality concepts from within the contexts characteristic of its respective discipline, and opens up new avenues of thought. Collectively, the plethora of concepts and problems of biological individuality enumerated by biologists, philosophers, and historians puts pressure on a claim that there is a single overarching definition of individuality to be agreed upon.

In closing, we would be remiss not to acknowledge other disciplinary realms and problem areas of biological individuality not treated in this volume. Biological individuality concepts play important roles in studies of eusociality, immunology, physiology, neurobiology, psychology, and philosophy of mind—areas that we have only touched upon or haven't mentioned here and in the chapters that follow. Nor do we situate problems of biological individuality in relation to other sciences that also deal with individuality, without special reference to life. (For a comparative approach to problems of individuality, see Guay and Pradeu 2016.) All of these deserve more attention. It is our hope that biological individuality will become a focus of interdisciplinary study in its own right. We offer this volume toward that end.

References for Introduction

Bell, Jordana T., and Richard Saffery. 2012. "The Value of Twins in Epigenetic Epidemiology." *International Journal of Epidemiology* 41 (1): 140–50.

Bennett, Jonathan. 1974. *Kant's Dialectic.* Cambridge: Cambridge University Press.

Booth, Austin. 2014. "Populations and Individuals in Heterokaryotic Fungi: A Multilevel Perspective." *Philosophy of Science* 81 (4): 612–32.

Bouchard, Frédéric. 2010. "Symbiosis, Lateral Function Transfer and the (Many) Saplings of Life." *Biology & Philosophy* 25: 623–41.

Bouchard, Frédéric, and Philippe Huneman, eds. 2013. *From Groups to Individuals: Evolution and Emerging Individuality.* Cambridge, MA: MIT Press.

Breuker, Casper J., Vincent Debat, and Christian Peter Klingenberg. 2006. "Functional Evo-Devo." *Trends in Ecology & Evolution* 21 (9): 488–92.

Brigandt, Ingo. 2011. "Philosophy of Biology." In *The Continuum Companion to the Philosophy of Science*, edited by Steven French and Juha Satsi, 246–67. London: Continuum.

Clarke, Ellen. 2010. "The Problem of Biological Individuality." *Biological Theory* 5 (4): 312–25.

———. 2014. "Origins of Evolutionary Transitions." *Journal of Biosciences* 39 (2): 303–17.

Elwick, James. 2015. "Containing Multitudes: Herbert Spencer, Organisms Social and Orders of Individuality." In *Herbert Spencer: Legacies*, edited by Mark Francis and Michael W. Taylor, 89–110. Abingdon, Oxon: Routledge.

Godfrey-Smith, Peter. 2009. *Darwinian Populations and Natural Selection.* Oxford: Oxford University Press.

Griesemer, James R. 2003. "The Philosophical Significance of Gánti's Work." In *The Principles of Life* by Tibor Gánti, 169–86. New York: Oxford University Press.

———. 2014. "Reproduction and the Scaffolded Development of Hybrids." In *Developing Scaffolds in Evolution, Culture, and Cognition*, edited by Linnda R. Caporael, James R. Griesemer, and William C. Wimsatt, 23–55. Cambridge, MA: MIT Press.

Griesemer, James R., and William Wimsatt. 1989. "Picturing Weismannism: A Case Study of Conceptual Evolution." In *What the Philosophy of Biology Is: Essays Dedicated to David Hull*, edited by Michael Ruse, 75–138. Dordrecht: Kluwer Academic.

Grote, Mathias, and Lara Keuck. 2015. "Conference Report 'Stoffwechsel. Histories of Metabolism,' Workshop Organized by Mathias Grote at Technische Universität Berlin, Nov. 28–29th, 2014." *History and Philosophy of the Life Sciences* 37 (2): 210–218.

Guay, Alexandre, and Thomas Pradeu. 2016. *Individuals Across the Sciences.* Oxford: Oxford University Press.

Hull, David L. 1988. *Science as a Process: An Evolutionary Account of the Social and Conceptual Development of Science.* Chicago: University of Chicago Press.

———. 2000. "The Professionalization of Science Studies: Cutting Some Slack." *Biology & Philosophy* 15 (1): 61–91.

Kuhn, Thomas S. 1962. *The Structure of Scientific Revolutions.* Chicago: University of Chicago Press.

Lakatos, Imre, and Alan Musgrave, eds. 1970. *Criticism and the Growth of Knowledge.* Proceedings of the International Colloquium in the Philosophy of Science, London, 1965, Vol. 4. Cambridge: Cambridge University Press.

LaPorte, Joseph. 2004. *Natural Kinds and Conceptual Change.* Cambridge: Cambridge University Press.

Love, Alan C. 2014. "The Erotetic Organization of Developmental Biology." In *Towards a Theory of Development*, edited by Alessandro Minelli and Thomas Pradeu, 33–55. Oxford: Oxford University Press.

———. Forthcoming. "Individuation, Individuality, and Experimental Practice in Developmental Biology." In *Individuation across Experimental and Theoretical Sciences*, edited by O. Bueno, R.-L. Chen, and M. B. Fagan. Oxford: Oxford University Press.

Newell, Allen, and Herbert A. Simon. 1972. *Human Problem Solving*. Englewood Cliffs, NJ: Prentice-Hall.

Maynard Smith, J., and E. Szathmáry. 1995. *The Major Transitions in Evolution*. Oxford: Oxford University Press.

Okasha, Samir. 2006. *Evolution and the Levels of Selection*. Oxford: Oxford University Press.

Perrier, Edmond. 1881. *Les Colonies Animales et La Formation Des Organismes*. Paris: G. Masson.

Pigliucci, Massimo. 2003. "From Molecules to Phenotypes? The Promise and Limits of Integrative Biology." *Basic and Applied Ecology* 4 (4): 297–306.

Pradeu, Thomas. 2012. *The Limits of the Self: Immunology and Biological Identity*. Oxford: Oxford University Press.

Sterner, Beckett. 2015. "Pathways to Pluralism about Biological Individuality." *Biology & Philosophy* 30 (5): 609–28.

Tauber, Alfred I. 2015a. "The Biological Notion of Self and Non-self." *Stanford Encyclopedia of Philosophy*. http://plato.stanford.edu /archives/sum2015/entries/biology-self/.

———. 2015b. "Reconceiving Autoimmunity: An Overview." *Journal of Theoretical Biology* 375 (June): 52–60.

Waters, C. Kenneth. 2014. "Shifting Attention from Theory to Practice in Philosophy of Biology." In *New Directions in the Philosophy of Science*, edited by Maria Carla Galavotti, Dennis Dieks, Wenceslao J. Gonzalez, Stephan Hartmann, Thomas Uebel, and Marcel Weber, 121–39. The Philosophy of Science in a European Perspective 5. Cham: Springer.

Wilson, Robert A., and Matthew Barker. 2013. "The Biological Notion of Individual." *Stanford Encyclopedia of Philosophy*. http://plato.stanford.edu /archives/spr2013/entries/biology -individual/.

Winther, Rasmus G. 2006. "Parts and Theories in Compositional Biology." *Biology & Philosophy* 21 (4): 471–99.

———. 2011. "Part-Whole Science." *Synthese* 178 (3): 397–427.

Yet, Idil, Pei-Chien Tsai, Juan E. Castillo-Fernandez, Elena Carnero-Montoro, and Jordana T. Bell. 2015. "Genetic and Environmental Impacts on DNA Methylation Levels in Twins." *Epigenomics* 8 (1): 105–17.

1

The Work of Biological Individuality: Concepts and Contexts

SCOTT LIDGARD AND LYNN K. NYHART

> The whole question seems to turn upon the meaning
> of the word "individual."
> THOMAS HENRY HUXLEY (1852, 185)

> There is, indeed, as already implied, no definition of
> individuality that is unobjectionable.
> HERBERT SPENCER (1864, 206)

Between them, Huxley and Spencer got it right. Individuality is a basic concept in biology, yet there is no broad agreement on its meaning. Not unlike other wide-ranging concepts, biological individuality is a nomadic chameleon, "subject to constant reification and change while crossing and turning across disciplines and non-scientific domains" (Surman et al. 2014, 127). Not only biologists, but also historians and philosophers of biology adopt and reanalyze the concept. Why is it so difficult to isolate a meaning for biological individuality, yet so useful to biologists that they continually put different versions of an individuality concept to work?

Put to work on specific biological problems, individuality concepts provide structure and buttress generality, predictability, and explanatory power —though seldom all together (J. Wilson 1999). They do their work in the contexts of biologists' questions and epistemic practices, in a historically contingent yet continually changing landscape of biological knowledge and theoretical interpretation, and in philosophical views biologists draw upon. We suggest that biologists, historians, and philosophers need to appreciate all of these. In this chapter, we first show that a broad spectrum of definitional criteria of individuality has been integral to biological studies for at least 170 years. We then argue that this work is necessarily contextualized by specific problems, and lay out the notion of a problem space as a way to think about such contextualization and the place of individuality concepts within it. Our next section shows how many (if not most) *specific* problems concerning individuality can be seen as distributed across four stable and continuous *kinds* of problems: individuation, hierarchy, temporality, and constitution. We then briefly consider individuality from perspectives of relationality and interaction—systems, networks, and processes. In closing, we advocate a re-thinking of what biological individuality concepts are meant to do in biology. Rather than arguing over whether this or that specific concept and defini-tional criterion is right or best, a more promising way forward is to recast biological individuality as a broad, stable domain of problems, a "problem agenda" (Love 2014, forthcoming).

Definitions and Criteria of Biological Individuality

There is now a sprawling definitional diversity of biological individuality concepts, constituted by dozens of old and new criteria defining "individ-ual" or its contained subset "organism." These definitions and criteria don't always agree in picking out the same biological objects (Clarke 2010), and perhaps agree even less when some of their theory-driven underpinnings are compared. This section presents some of the definitional criteria that have been proposed for either biological "individuals" or "organisms" (Table 1.1), before moving on to the four kinds of problems we introduced earlier. Our compilation includes some 146 publications from 1800 to the present that have offered definitions or criteria for either one or both of these terms. While the terms are not equivalent, they have been used interchangeably in many publications, precluding a simple separation here. We have attempted to parse the references into groups by approximate criteria in order to facili-tate comparisons.

It is important to recognize that a number of these criteria overlap, and that a reference may endorse a definition using either a single criterion or several. Also, many definitional criteria allow for degrees of individuality and degrees of organismality (Verworn 1899; Bergson 1911; Conklin 1916; J. Huxley 1926, 1949; Sober 1991; Godfrey-Smith 2009; Queller and Strassmann 2009; Strassmann and Queller 2010; Herron et al. 2013; Wolfe 2014), although some references don't endorse this view. A few authors have combined multiple criteria into higher definitional categories or factors that abstract away from some of the more material criteria in Table 1.1. For instance, the criterion of high cooperation and low conflict could be considered to subsume germ-soma separation, policing, and bottlenecks (Herron et al. 2013). Godfrey-Smith (2011) discusses two categories that may overlap in different senses of individuality, "Darwinian individuals" and (in a non-Darwinian sense) or-ganisms. Clarke (2012) subsumes a range of different criteria within each of

TABLE 1.1. Some definitional criteria for "biological individuals" or "organisms" in biology

Definitional Criteria	References
Propagation and capacity for reiterating a qualified sameness, irrespective of means of propagation	E. Darwin 1800; C. Darwin 1839; Gaudichaud 1841; Steenstrup 1845; Owen 1849, 1851; Leuckart 1851; Braun 1853, 1855–56; Hofmeister 1862; Spencer 1864; Haeckel 1866; Weismann 1889, 1893; Bailey 1906; Bergson 1911; Geddes and Mitchell 1911; W. Wheeler 1911; Child 1912, 1915; J. Huxley 1912, 1949; Thomson 1920; Arber 1930, 1941; Agar 1948; Bell 1982; Vuorisalo and Tuomi 1986; Tuomi and Vuorisalo 1989; Margulis 1993; Gould and Lloyd 1999; Griesemer 2000; R. Wilson 2005; Dupré and O'Malley 2009; Godfrey-Smith 2009, 2011, 2013; A. Hamilton et al. 2009; Wilson and Barker 2013; Hamilton and Fewell 2013; Booth 2014b; Moreno and Mossio 2015a and b; L. Ma et al. 2016
Sex or sexually demarcated life cycle, excluding other means of propagation	Gallesio 1814; Carpenter 1848; T. H. Huxley 1852; Burnett 1854; Cobbold 1869; Janzen 1977; de Sousa 2005
Demarcated life cycle or phases of life cycle, irrespective of means of propagation	Steenstrup 1845; Owen 1849, 1851; Leuckart 1851; Braun 1853, 1855–56; Hofmeister 1862; Haeckel 1866; Bonner 1974; Dawkins 1982; R. Wilson 2005; Rainey and Kerr 2011; Wilson and Barker 2013
Causal integration, cohesion, collaboration, or agency of parts (often functional, sometimes metabolic, developmental, or related to division of labor)	Milne-Edwards 1827; Leuckart 1851; Spencer 1864; Bergson 1911; Geddes and Mitchell 1911; W. Wheeler 1911; J. Huxley 1912, 1949; Montgomery 1880; Child 1915; Thomson 1920; Fisher 1934; Emerson 1939; Agar 1948; Jeuken 1952; Vuorisalo and Tuomi 1986; Tuomi and Vuorisalo 1989; Wilson and Sober 1989; Margulis and Guerrero 1991; Sober 1991; Zylstra 1992; Margulis 1993; Baum 1998; Bolker 2000; Korn 2002; R. Wilson 2005; Dupré and O'Malley 2009; Gardner and Grafen 2009; Godfrey-Smith 2009, 2011, 2013; A. Hamilton et al. 2009; Folse and Roughgarden 2010; Wolfe 2010; Gilbert et al. 2012; Bouchard 2013; Haber 2013; Hamilton and Fewell 2013; Turner 2013; Wilson and Barker 2013; Arnellos et al. 2014; Booth 2014b; Gilbert 2014; Huneman 2014a and b; O'Malley 2014; Schneider and Winslow 2014; Boon et al. 2015; Bordenstein and Theis 2015; Moreno and Mossio 2015a and b; West et al. 2015; Guay and Pradeu 2016
Constituting or a part of a biological hierarchy (generally formed by a process at one level and contributing to another process forming an object at a higher level)	Gaudichaud 1841; Owen 1849; Leuckart 1851; Braun 1853, 1855–56; Haeckel 1866; Geddes and Mitchell 1911; Child 1915; Arber 1930, 1941; Tuomi and Vuorisalo 1989; Wilson and Sober 1989; Zylstra 1992; Korn 2002; Huneman 2014a and b; Schneider and Winslow 2014; Moreno and Mossio 2015a and b
Exhibiting adaptation(s)	Spencer 1864; Gardner and Grafen 2009; Queller and Strassmann 2009; West and Kiers 2009; Folse and Roughgarden 2010
Partition of germ and somatic cell lineages (and division of labor in some views)	Weismann 1889, 1893; Buss 1987; Michod and Roze 2001; Michod and Nedelcu 2003; de Sousa 2005; Michod 2007; Godfrey-Smith 2009, 2011, 2013; Folse and Roughgarden 2010

(continued)

TABLE 1.1. (*continued*)

Definitional Criteria	References
Genetic homogeneity or near-homogeneity, and genetic distinctness or uniqueness	Weismann 1889, 1893; Todd and Rayner 1980; Dawkins 1982; Vuorisalo and Tuomi 1986; Smith et al. 1992; G. Williams 1992; Santelices 1999; de Sousa 2005
Immunological or allorecognition system integration (including tolerance and symbiotic microorganism integration in some views)	Ehrlich and Morgenroth 1900; Metchnikoff 1905; Loeb 1921, 1937; Todd 1930; Medawar 1957; Burnet 1959, 1969; Todd and Rayner 1980; Gilbert et al. 2012; Pradeu 2012, 2013; Brusini et al. 2013; Anderson and McKay 2014; Tauber 2015a and b; Guay and Pradeu 2016
Autonomy or discreteness: functional, metabolic, physiological, or otherwise	Sutton 1902; Bailey 1906; W. Wheeler 1911; J. Huxley 1912, 1949; Jeuken 1952; Buss 1987; G. Williams 1992; Santelices 1999; R. Wilson 2005; Wilson and Barker 2013; Arnellos et al. 2014; Booth 2014b
Spatial and temporal continuity	Sutton 1902; J. Huxley 1912, 1949; Hull 1978, 1980, 1992; Bell 1982; Zylstra 1992; Gould and Lloyd 1999; de Sousa 2005; A. Hamilton et al. 2009; Haber 2013; Hamilton and Fewell 2013; Guay and Pradeu 2016
Spatial and temporal boundedness	J. Huxley 1912, 1949; Emerson 1939; Jeuken 1952; Ghiselin 1974; Hull 1978, 1980, 1992; Todd and Rayner 1980; Bell 1982; Margulis and Guerrero 1991; Zylstra 1992; Baum 1998; Gould and Lloyd 1999; de Sousa 2005; A. Hamilton et al. 2009; Haber 2013; Hamilton and Fewell 2013
Indivisible without losing character or function	Montgomery 1880; J. Huxley 1912; Michod and Roze 2001; Michod 2007, 2011
Bottleneck in life cycle, or accompanying some forms of propagation (a narrowing of material constituents between generations, typically unicellular)	J. Huxley 1912; Dawkins 1982; Fagerström 1992; Grosberg and Strathmann 1998; Godfrey-Smith 2009, 2013
Unit of selection or capacity to undergo selection (often with a common evolutionary fate)	Emerson 1939; Tuomi and Vuorisalo 1989; Wilson and Sober 1989; Gould and Lloyd 1999; Michod and Roze 2001; Michod and Nedelcu 2003; Michod 2007, 2011; Bouchard 2008, 2013; A. Hamilton et al. 2009; Leigh 2010; Gilbert et al. 2012; Clarke 2012, 2013; Goodnight 2013; Hamilton and Fewell 2013; Booth 2014b; Gilbert 2014; Boon et al. 2015; Bordenstein and Theis 2015; L. Ma et al. 2016
Fitness maximization or alignment	Janzen 1977; Grafen 2006; Gardner 2009; Gardner and Grafen 2009; Folse and Roughgarden 2010; Leigh 2010; Niklas and Kutschera 2014; West et al. 2015
High cooperation and low or restrained conflict among cell lineages or other contituents	Wilson and Sober 1989; Michod and Roze 2001; Michod and Nedelcu 2003; Michod and Herron 2006; Michod 2007, 2011; Queller and Strassmann 2009; West and Kiers 2009; Strassmann and Queller 2010; West et al. 2015

(*continued*)

TABLE 1.1. (*continued*)

Definitional Criteria	References
Particular genetic/epigenetic regulatory interaction and integration that generates and retains phenotype structure (often as genetic or, more broadly, developmental modules)	Wagner 1989a and b, 2014; Gass and Hall 2007; Arnellos et al. 2014
Persistence	Thomson 1920; Van Valen 1989; Bouchard 2008, 2013; Haber 2013; Turner 2013
Mechanisms of policing or diminishing cell lineage conflict or cheating (sometimes stipulating reduced within-object selection)	Grosberg and Strathmann 1998; Michod and Roze 2001; Michod and Nedelcu 2003; Godfrey-Smith 2009; Clarke 2013
Non-genetic microbial interaction or integration, and microbial phenotypic heterogeneity among cells or among groups of like cells, in either isoclonal or mixed populations	Avery 2006; Davidson and Surette 2008; Huang 2009; Ackerman 2013; Doolittle 2013; Martins and Locke 2015; Van Gestel et al. 2015
Factors that provide heritable variation in fitness in populations of objects, together with factors that mediate interactions among parts of objects, constraining variation at that level	Clarke 2012, 2013
Cognition or self-awareness	Turner 2013
High relative interaction strengths among groups of living entities (as in ecological communities or sub-communities)	Huneman 2014a and b

Note: The terms "biological individuals" and "organisms" are not distinguished in this compilation. References from the beginning of the nineteenth century onward are listed adjacent to the respective criteria. Some criteria overlap, and the same reference may specifiy more than one criterion. See text for explanation and discussion.

two factors, one that ensures variation among biological objects in a population and another that acts to constrain variation within each object. These factors attempt to combine the effects of certain other criteria listed in Table 1.1 that could influence heritable variation in fitness, her overriding concern. Goodnight (2013) offers three definitions of individuality in terms of fitness, with an informed discussion of models, criteria, or fitness proxies that could actualize them: units of fitness at a chosen organizational level, objects at the lowest level experiencing natural selection, and objects at the lowest level displaying an evolutionary response to natural selection. He states clearly that it is the observer who imposes an abstract concept of individuality. This is hardly a new observation: "in general the concept of individual being, in the various special sciences, as well as in inquiry, shows many characters that cannot be derived from mere perception in any case" (Royce 1901, 535) Yet these sorts of abstractions, some with strong theoretical underpinnings, may lead to an epistemological or decidedly empirical question when framed in terms of the practice of biologists. What sorts of assumptions or observations or workflows are needed to actualize a given criterion? This last question has garnered far too little attention.

Numerous comparative accounts of biological individuality and organism concepts exist as narratives, surveys, or compilations of definitional criteria (Geddes and Mitchell 1911; Mumford 1925; Bell 1982; Hull 1992; Keller and Ewing 1993; J. Wilson 1999; de Sousa 2005; R. Wilson 2005; Clarke 2010; Strassmann and Queller 2010; Boulter 2013; Herron et al. 2013; Wilson and Barker 2013). Neither our compilation nor any one of these are necessarily the "right" one, all the more so as criteria often remain implicit in definitions. Thus the criteria as we have specified them are heuristic approximations. They might be stated or grouped differently by another worker, and certain references might be reassigned relative to the arrangement in Table 1.1. Lastly, while our compilation is more comprehensive than others, it is hardly complete. A fuller exploration would undoubtedly add more references for various criteria, and extend some criteria further back in history.

Setting all these provisos aside for the moment, what can we say about "individuality" and "organism"? Figure 1.1 shows definitional criteria of "biological individuals" or "organisms" in historical view. It confirms that they have been topics of research in biology continuously for at least 170 years and possibly more. Criteria of propagation and life cycles, and of causal integration of biological objects as individuals or organisms, have held steadfast. Similarly, the criterion of comprising or being part of a biological hierarchy has a history showing the persistence of "individuality at different organizational levels." Looking more closely at the substance of publications in Table 1.1 reveals that individuality of species, solitary and modular organisms, and cells all received attention in the nineteenth century. So too did problems that involved continuity and change over time, from reproduction and life cycles to evolutionary lineages. Soon after, chromosomes, genes, immunological units, developmental modules, and symbiotic entities garnered attention. Near the end of the twentieth century through today, more abstract criteria related to Darwinian evolution, especially selection and theoretical fitness, assumed a central role. We suggest that this last pattern has drawn from what Gould

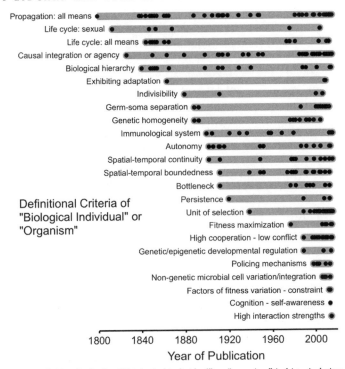

FIGURE 1.1. Definitional criteria of "biological individual" or "organism" in historical view. Refer to Table 1 for fuller statements of the criteria (abbreviated here) and for references. The years of publication for references are shown as black points.

(1983) termed the hardening of the modern synthesis, from units of selection (Lewontin 1970) and subsequent multilevel selection theory (Okasha 2006), and from the evolution of individuality (Buss 1987) and major evolutionary transitions (Maynard Smith and Szathmáry 1995). Yet despite the brash prominence of these newer criteria, they have not replaced their predecessors in the discourse over biological individuality—far from it. Many earlier criteria and problems of biological individuality in philosophy and science did not lose their meaning, significance, or influence with the invention and diversification of their most recent counterparts. (Individuality criteria that are no longer used at all, such as those reflecting extreme goal-directedness or divine providence, have been omitted here.) The enduring character of definitional criteria as an assemblage becomes even more apparent as we turn to situating biological individuality concepts within different problem-contexts. We begin by considering the nature of those problem-contexts.

Concepts—Moving through Changing
Problem Spaces and across Contexts

Historians, philosophers, and biologists interested in biological individuality share certain crucial areas of concern. The largest and most obvious are matters of epistemology and scientific theory, and involve *concepts* of biological individuality. Concepts perform a wide variety of work in biological problem-solving and theorizing (Brandon 1996; Hall and Olson 2007; Schwarz 2011; Feest and Steinle 2012a). We argue here that the work of biological individuality concepts has to be understood contextually. While "context" can be taken at many levels, here we are most concerned with the roles of individuality concepts in the relation to biologists' immediate activities: forming problems within a given domain, classifying, reasoning, investigating, experimenting, modeling, explaining, and communicating (Feest and Steinle 2012b; Sterner 2015).

We focus on three senses of contextuality, without excluding the prospect of others. The first is the practice-driven approach of biologists themselves, involving the interaction of concepts and problem spaces (e.g., Bechtel 1984; Brigandt and Love 2012). Here it may be useful to have a model to think with. Allen Newell and Herbert Simon (1972) devised the notion of a "problem space" in which a problem solver, say a biologist, confronts a specific problem or task and determines a task environment in terms of an initial problem state (Fig. 1.2). This environment supplies but is not limited to a breadth of information about the specific system under consideration, associated knowledge, and theory. The problem space within it is constrained by the problem-solver by selecting certain relevant information in the task envi-

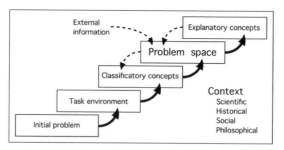

FIGURE 1.2. A greatly simplified schema of concepts and a problem space. Solid arrows indicate prospective roles of an initial problem, the task environment relevant to the system under consideration, concepts doing classificatory or explanatory work, which in turn structure and allow exploration of the problem space, all in the course of a problem-solver attempting a solution to the problem. Dashed arrows indicate prospective modifications or revisions of the problem space and concepts. Note especially the potential for circulation of concepts (in different roles) around and within the problem space.

ronment, the state of the problem in its initial form, alternate or intermediate problem states considered along the way toward the goal state, and the goal or solution state. The problem space itself changes as the problem is studied, as information and alternate problem states are investigated. As originally proposed, its multiple dimensions are explored mathematically and algorithmically in attempting a solution. A key insight is that "a great many of the characteristics of the problem solving systems . . . can be inferred, at least in broad outline, from the structure of the environments in which they operate. . . . The forms this information takes condition the possible modes of its exploitation" (Newell and Simon 1972, 834). In other words, the problem context conditions the terms for solving the problem. Kovaka (2015) defends a roughly similar claim of contextuality with respect to biological individuality and scientific practice.

Here we step away from Newell and Simon's precise mathematical statement. Instead, we use a less formal dynamic problem space in which preexisting as well as newly developing concepts can move through permeable boundaries, and historical, social, and philosophical facets are also present. For any problem-solver, each distinct problem has its own problem space (cf. Nersessian 2006; Love 2014, 2015a, forthcoming). Consider a solving process that begins with a scientific problem involving biological individuality, one that needs to individuate the entities under consideration, whether in a physiological, ecological, evolutionary, or other area (designated by the full interior of Figure 1.2). A biologist's initial problem helps circumscribe a task environment. Individuation of key entities can draw on one or more existing (or new) individuality concepts and definitional criteria, which are contextualized by their reference to the specific problem and its system. A problem involving the individuality of a consumed organism converted metabolically in its consumer (Landecker, this volume) calls for an individuality concept that will classify entities differently from a problem involving evolutionary transitions to multicellularity in green algae (Herron, this volume). The two problems will also support radically different explanatory inferences.

While biological individuality concepts can perform work in a classificatory role (identifying kinds of individuals or criteria of individuality), they can also do so in explanatory or other roles. Thus, the "epistemic goal" of a research problem (Brigandt 2010a, 2012) is another potentially important part of a concept's work. Is the problem more empirical or theoretical? Is it synchronic, in a short moment, or diachronic, extending over time? Is it on one level of a biological hierarchy or many? Does it focus on pattern or process, structure or function? These and other sorts of questions reflect epistemic goals that shape the problem space and further contextualize the work done by a given individuality concept (or organism concept [Wolfe 2014]). A concept may work in a distinctly explanatory role as our biologist works through her problem space toward a solution; it may be part of the goal or even the goal state itself. For example, concepts that frame biological individuals principally or entirely in terms of natural selection or fitness are often used in explanatory roles in the study of major evolutionary transitions. Yet these concepts may bracket off material or structural elements that are needed for a fuller understanding, and might necessarily employ a broader or complementary concept of biological individuality (Sterner, this volume; Love and Brigandt, this volume). Also, many problems are initially "ill-structured" (Simon 1973), resulting in modification of the problem space, drawing new information from the task environment or externally (Fig. 1.2). Here again, the concepts can change with a revised context.

A second sense of the contextuality of biological individuality concepts exposes the different perspectives that biologists, philosophers, and historians bring to the table. The schema for our biologist (Fig. 1.2) could apply as well for a historian or a philosopher, but the concepts and problem space would differ for their specific problems, reflecting what they consider germane to their task environments. An experimental developmental biologist analyzing functional integration, for example, might organize a problem space largely in terms of structural objects, spatial boundaries, and morphogenetic interactions to individuate molecular or morphological objects and understand changes during ontogeny. A philosopher studying the same system from an evolutionary viewpoint might organize a problem space to develop an axiomatic or otherwise formal account of individuality, relying on a "fundamental" theory of natural selection and functional properties. While the two contextual perspectives may indeed individuate the same objects, the structural properties approach of the developmental biologist is not generally directed by fundamental philosophical theory (Love forthcoming).

More generally, philosophers are likely to contextualize individuality with respect to fundamental categories of their field, such as metaphysics and epistemology. The variety of ways they do so is shown in several recent volumes (R. Wilson 2005; Okasha 2006; Godfrey-Smith 2009; Bouchard and Huneman 2013a), which, however, tend not to give serious attention to history and historical change (but see Tauber [1997] and Pradeu [2012]). Metaphysics and more precisely ontology—the study of the nature of being and what is real in the world—might not enter into problem spaces of working biologists and historians. But more specific metaphysical considerations of mereology, or formal analysis of the part-whole relation, have been incorporated into some philosophical analyses of biological individuality. David Hull defended

an abstract logical view of biological individuals as (minimally) spatiotemporally continuous localized entities, yet he warned us, "real examples tend
to be much more detailed and bizarre than those made up by philosophers"
(Hull 1978, 344). Where philosophy of biology traditionally focused more on
fundamentalist ontology and epistemology, as in history, here too there has
been a shift in attention over the past several decades from theory-centered
toward practice-centered views (Waters 2014), and new exploration of the
warrant for monism with respect to biological individuality as opposed to
a pluralism of biological individuality concepts (see also Kellert et al. 2006;
Brigandt 2010b; Love forthcoming).

The problem space of historians concerned with biological individuality
is yet different. It includes considerations such as how scientific knowledge
is constructed in a given context (Winsor 1976), how knowledge circulates
within and across communities (Secord 2004), and how contexts and questions change over time. They may call attention to which objects are attached
to the label "individual" (Elwick 2007; Nyhart and Lidgard, this volume), and
to the role of figurative language, especially metaphor and analogy, in analyses of how biological concepts are generalized from one kind of individual to
another, and from one kind of problem to another. Calling two different levels of biological objects—say, cells and organisms—individuals involves at
least implicit attribution of the same features to both kinds, whether treated
explicitly as analogous or more casually similar. But the historian's problem
space around language is often bigger, encompassing how the movement of
political, economic, and social vocabulary into and out of the sphere of biology influences part-whole conceptions (Gissis, Osborne, Reynolds, Rieppel,
all this volume). Whereas biologists might be especially interested in enumerating properties and levels for the purposes of understanding different
kinds of living things as well as for general theoretical purposes, historians
tend to be most interested in the scientists who talked about those levels, the
intellectual and social circumstances surrounding that discussion, and the
ways such language reflects continuities and discontinuities over time.

This brings us to a third sense of contextuality, that biological and philosophical problems involving concepts of biological individuality aren't static.
They change over time, creating an ebb and flow in which biologists, historians, and philosophers all participate. An example suggests one way in which
all three perspectives might be combined.

On a spring evening in 1852, Thomas Henry Huxley addressed the Royal
Institution of London on individuality. He identified different kinds. One
philosophically inconsequential kind exists as someone arbitrarily points to
that "individual" in everyday life. A less capricious kind involves material

properties of a thing that if altered or divided cause a loss of individuality. A third kind is Huxley's authoritative biological individuality, a law of succession that is "the total result of the independent development of a single impregnated ovum" (T. H. Huxley 1852, 14). He partitioned off difficulties or exceptions as well as other structural and functional concepts of biological individuality (Causey 1935; Elwick 2007), for which he invented the term "zooids," that "are like individuals, and yet are not individuals, in the sense that one of the higher animals is an individual" (T. H. Huxley 1852, 14). His problem space, at least for this talk, seems to be designed starting from an organism's development understood in terms of von Baerian embryology. Development ramifies from an unspecialized homogeneous state to a more specialized heterogeneous one, irrespective of the independent body forms or parts and their functions that emerge from developmental stages over the life cycle (Elwick 2007). Huxley drew from the inductive philosophy of John Stuart Mill (Ellegård 1957). Huxley's method purported to begin with facts gathered directly from sensory experience and inductive inference from that experience, yielding both generalizations of pattern and causal explanation, and seldom if ever admitting preexisting theories or suppositions. He privileged sexual reproduction as *the* criterion of his biological individual. Something that could actually be observed, it was placed in a causal explanatory role, and became the dominant individuality concept for much of biology (Nyhart and Lidgard, this volume).

Thomas' grandson Julian Huxley wrote from a more pluralist perspective than his grandfather and developed a biological individuality concept spanning multiple levels of organization. His concept included sub-organismal individuals, and simultaneous or spatial individuals like ant colonies. It extended over time as *either* sexually or asexually produced generations of individuals, and endorsed gradients or degrees of individuality both within and among organisms. Thus Julian Huxley's problem space was far broader than his grandfather's, embracing interactive systems as individuals, "any one in any grade being able to combine with others like itself or with others unlike itself to form the beginnings of a new system, a new individual" (J. Huxley 1912, 153). This problem space was designed around ideas of division of labor and ever-continuing change in part-whole hierarchies, and not constrained only to a single organism's development. Julian Huxley took up the philosophical views of Mill's sometime opponent, Herbert Spencer (J. Huxley 1912, 1926, 1949; see also R. Wilson 2005). Spencer's far less restrictive empiricism tried to unite emerging complexity and the correspondence of organism-environment interaction with evolutionary change (Pearce 2014; Gissis, this volume). A biological individual could be "any concrete whole having a structure which enables it, when placed in appropriate conditions, to continuously adjust its internal relations to external relations, so as to maintain the equilibrium of its functions" (Spencer 1864, 207). Unity results not from subordination to a sole governing agent or function (as for T. H. Huxley's concept), but from diverse interdependency of biological entities at different levels of organization that assume functional agency and harmonious order from the bottom up (Elwick 2003, 2007). Moving from Huxley to Huxley, we see that science and philosophy repeatedly foreground some problems over others, as new discoveries and research domains move on and off center stage, and prevailing social and philosophical milieus ebb and flow.

Deeper transformations can be seen over an even longer timescale. "Individual" originally meant indivisible, rather than stressing a distinction from other entities as in common speech. Philosophical attitudes from Aristotle through the early seventeenth century stressed numerical singularity, "the same in kind"—views that Gottfried Leibniz challenged by proposing that living individuals contain other individuals in a hierarchy. Beginning late in the 1690s, Leibniz's pluralist stance sanctioned a nested individuality of self-sufficient and unique monads that nonetheless harmonize with one another (Nachtomy 2007). It was not until the nineteenth century in biology and social thought that "*an* individual" as a singularity in a group of things was overtaken by "*the* individual" as a fundamental order of being (Royce 1901; R. Williams 1985), as it became closely connected to the concept of the organism—a term with its own complex history (Cheung 2010; Pepper and Herron 2008; Wolfe 2014). Yet biological and philosophical ideas of the most fundamental level of biological individuality were challenged in the early nineteenth century by complex life cycles in "lower" animals and plants, and by the cell theory. Later on, the chromosome, chromatin, the classical gene of late Neodarwinism and the Modern Synthesis, resurgent organicism and the theory of emergent evolution, and the material gene of molecular biology inspired a succession of claims to ultimate individuality, with attendant philosophical considerations (e.g., Harwood 1994; Gilbert and Sarkar 2000; Delisle 2008; Gibson et al. 2012). Today, discoveries and challenges continue to expose crucial interchanges as well as intensely varied purposes among scientific and philosophical concepts of biological individuality in problem areas such as multiple levels of selection, the major evolutionary transitions, epigenetics and evo-devo, mosaicism and chimerism, symbioses, immune self and non-self, and holobionts (Buss 1987; Sapp 1994; Maynard Smith and Szathmáry 1995; J. Wilson 1999; R. Wilson 2005; Okasha 2006; Godfrey-Smith 2009; Calcott and Sterelny 2011; Pradeu 2012; Gilbert and Epel 2015; and many more). We can surely anticipate further changes in the contexts and concepts of biological individuality in the future. Indeed, one object of this paper is to help us rethink the uses of such concepts.

Problems in Practice

How, then, do biologists' individuality concepts operate in different problem spaces? Definitions and criteria both emerge from and are applied to biologists' specific problems, as they work through them. Problems relating to individuality may be clustered into four major groups, which we might call biologists' philosophical *kinds* of problems: individuation, hierarchy, temporality, and constitution. Variation exists within each kind, and the kinds may coincide with one another in practice, as relevant to a specific problem. We do not claim that these four kinds are all-inclusive, and as we discussed in relation to their different problem spaces, philosophers' and historians' notions and problems around individuality might be different. But these four recur often in the work of biologists and therefore invite further clarification. Here we call attention to the range of work associated with these different kinds in biological problem solving.

INDIVIDUATION

Individuation is among the most basic things that a biologist normally attends to when constructing a problem space. It involves decisions about the entities that are being investigated, and the classificatory role of a biological individuality concept (Fig. 1.2). Individuation is often thought of in two modes, identity and unity. *Identity* involves distinguishing one instance of a class from another instance, typically by some persistent exclusive property. *Unity* might be used to distinguish parts of an instance from everything else in the world; parts are held together by some unifying relationship not shared by anything else. Thus for a single sterile worker bee in a colony, its identity (a first mode of individuality) may be given by unique nuances of form, functional duties, spatial and temporal boundedness, and so on, that distinguish it from all other bees. Worker bees as a group might constitute a functional part of the larger whole that is the colony, but might not always be considered collectively as an individual. Yet the colony could be distinguished as an individual as well, its unity (a second mode) given by its reproductive and (imperfect) genetic continuity, or its autonomous function relative to other parts of the world. Child (1915) also describes dynamic unity, a sense of unity maintained throughout changes inherent in development, and as components are broken down and built up anew during life. Both the worker and the colony could be seen as natural wholes distinct from other wholes that have properties referring to identity, unity, or both.

Biologists need to identify specific instances of things as individuals and often to count things as individuals in order to identify agency. This requires making boundaries, at least epistemically. But those boundaries might be fuzzy, and might differ with different problems, epistemic goals, and consequent problem spaces. A conservation biologist or physiological ecologist studying habitat preferences or competition for resources between bracken and herbaceous shrubs in several meadows might individuate and count plants as individuals by their above-ground clumps to determine distribution and abundance; consider material growth form, soil type, shading, soil wetness, or evapotranspiration by leaf area as competitive mechanisms; and structure a problem space on such concrete information. An evolutionary biologist studying the same meadows might measure relative reproductive output of the bracken and shrubs as proxies for evolutionary fitness. But bracken multiplies rampantly by underground runners, so one habitat might contain dozens of genotypes, and another habitat only a single one that has grown sprawling for decades. The evolutionary biologist's problem space is partly structured by the function of reproductive output as a proxy for theoretical evolutionary fitness. The entire bracken clone would likely better represent the genetic individual than would the forms, size distribution, and edaphic factors of clumps. This contrast is the classic puzzle of individuation represented by genets (genetic individuals) and ramets (material individuals) known in modular plants and animals (Harper 1977; Jackson et al. 1986; N. Hamilton et al. 1987; Scrosati 2002; Clarke 2012), and reflecting identity and unity. Different purposes dictate the ways biologists carve things up, and workers taking a more theoretical evolutionary approach seldom explicitly discuss the practical requirements of broad-scale comparisons (Dupré 2010; Herron et al. 2013). We don't seek to resolve this contrast, but rather accept that different problems will invite the use of different individuality concepts.

We might also ask how organisms themselves use individuation, how they recognize or draw boundaries or distinctions. For example, since the late 1940s, some experimental population geneticists have seen sexual selection as a form of species individuation, demonstrated by female fruit flies preferring to mate with males of the same species over males of different species (Milam 2010). In another example, the immunological view of organisms, "self/ nonself discrimination," is a longstanding biological individuality concept focused on individuation, in which self equals individual. This view is being challenged by work on chimeric transplants, autoimmune disorders, and tolerated or symbiotic microbiotas. Thomas Pradeu (2012) argues that these studies give evidence precluding static and distinct functional boundaries of organismal individuality; yet there is still a basis for establishing individuation. Here, identity and unity intersect, reflecting "an expanded 'sensibility' to the environment, where flexible borders of the organism and changing parameters of individuality dispense with any characterizing essences— genetic, molecular, or immune" (Tauber 2015a, 3).

HIERARCHY

Hierarchy represents levels of organization in complex biological systems; hierarchical structure can pose problems for certain individuality concepts but may also aid in conceptual understanding. Thinkers concerned with the units of life have long debated whether biological individuals could be (or are) made up of smaller individuals, and in an evolutionary frame, the units of selection and their relative importance. Hierarchy often gives the impression of a paradox, contravening the historical definition of indivisibility of individuals or else challenging common-sense recognition of individuals. Common sense has generally lost out, though as James Elwick points out in his commentary to this volume, it keeps bouncing back. A familiar solution is the idea that individuality operates at multiple levels. There may be ultimate indivisible units, but then there are also higher units with their own distinct properties that make them individuals, too. Charles Darwin, early on recognizing polypides as individuals within colonial Bryozoa (1839), later wrestled with a specific problem of hierarchy confronting his theory of natural selection: sterile castes of workers in eusocial insects (Gibson 2013). In the sixth edition of the *Origin* Darwin wrote, "This difficulty, though appearing insuperable, is lessened, or, as I believe, disappears, when it is remembered that selection may be applied to the family, as well as to the individual, and may thus gain the desired end" (1872, 230).

Philosophers like Zylstra (1992), Korn (2002), and Wimsatt (1994, 2007) argue that the living world is constituted of levels—molecules, genes, cells, developmental primordia and tissues, organs, bodies, reproducing populations, species, and so on in various versions—wherein one level of entities differs qualitatively from another, yet is making or made up of entities at another level (nestedness). In much of biology, a whole is understood in terms of its parts, and mechanistic explanations often attempt to decompose the functions of the whole by allocating portions of these among meaningful parts and hierarchical levels (McShea and Venit 2001; Winther 2006, 2011). But whether or not such stratified levels are universal and discrete, and precisely what those parts and levels are, is another matter entirely.

A traditional view of organizational levels (e.g., Feibleman 1954) suggests that higher levels are entirely and solely dependent upon lower ones for continuance, higher levels have emergent properties that cannot be predicted and explained by lower-level properties, and organization at a given level is accomplished by mechanisms at a lower level. Potochnik and McGill (2012) dispute and provide evidence against all these claims. Wimsatt (2007) and Gillett (2010) argue that uniformity of composition or properties related by mechanistic explanations break down at higher levels. Phenomena like chimerism, epistatic interactions in development (primarily interactions among genes), horizontal gene exchange, functional dependencies in symbioses and microbiotas, immunological tolerance and response call into question the universality, autonomy and discreteness of levels: "It is certainly not the case that every whole is composed of only parts at the next lower level. Nor is it the case that each type of whole is composed of all and only the same types of parts" (Potochnik and McGill 2012, 127).

Biological individuality concepts used in explanatory roles may also differ in how they treat organizational hierarchies: more reductionist (as in how organisms are put together, and what controls assembly) or holistic (as in why they are put together a certain way and how they function). Some reductionist individuality concepts make a mechanistic claim that properties at upper levels, say structures or functions of material organ systems or whole organisms, are fully determined by establishing the properties of lower-level components, say genes or certain genetic regulatory networks, that are assumed to be more fundamental. This claim is not always justified, and may turn on different scientific interests, problems, and problem spaces, especially when considered in an epistemological frame (Love and Brigandt, this volume; Love forthcoming).

Nevertheless, hierarchy itself remains a component of many important biological questions about individuality. How do individuality concepts work to specify individuals at different levels? Are the hierarchical levels constituted of integral structural wholes or of parts, and how is this difference contextualized with respect to a given problem and problem space? For example, a more abstract problem such as "How does organismal variation arise?" might engage individuality concepts at multiple hierarchical levels, as opposed to a more concrete concept focused at a single level: "What gene or genes are signalling the onset of this genetic regulatory cascade for this structure?" (cf. Brigandt and Love 2012). How do entities at different levels interact: as natural wholes (an enkaptic hierarchy [Rieppel, this volume]) or as parts and wholes? If interaction exists, how do scientists account for agency when they specify individuals at different levels and use/develop criteria of individuality? In discussions of the units of selection, relative contributions to fitness from different levels, as well as contributions from level-to-level interactions, have been assigned fundamental importance; individuality at hierarchical levels is necessary for explanatory adequacy (Okasha 2006; Rainey and Kerr 2011). But we emphasize that hierarchy is useful beyond a narrowly Darwinian fitness-oriented perspective. Hierarchy, for example, has proven useful to biologists in understanding cell theory, developmental change, and life cycles, from its deep pre-Darwinian roots to the present (Owen 1849; Korn 1999). This last comment reminds us just how permeable our four kinds are in practice, for the next one is temporality.

TEMPORALITY

Temporality treats part-whole relations occurring over time—especially, how to accommodate changes in part-whole structure and function and changes in emergent individuality, or emergent hierarchy. Time can refer to scales ranging from the nearly instantaneous to the geological. Temporality includes topics as diverse as emergent complexity or novelty; Darwinian evolution; major evolutionary transitions such as unicellularity to multicellularity; the organism or its parts or a sub-organismal hierarchy of individuals in a life cycle (as opposed to a more static physiological or morphological unit); reproductive, developmental, and morphogenetic stages; and more.

Leo Buss' *Evolution of Individuality* (1987) provided a strong argument that units of selection could change over the grand scale of evolutionary time. This was developed in important ways by Maynard Smith and Szathmáry in *The Major Transitions in Evolution* (1995), and has since blossomed into a virtual industry investigating and theorizing on the origins of eukaryotes, multicellularity, and eusociality (e.g., Calcott and Sterelny 2011). Individuality concepts applied to major transitions confront many problems: explaining new mechanisms of heredity, say, fission and sex; how new levels of biological individuality are formed from groups of species or the same species; how previously independent entities reproduce together; how cooperation might be selected for; how conflict between distinct genetic or cellular lineages in a derived composite individual is mediated; and others. Different biological individuality concepts applied to evolutionary transitions expose tensions between structural and functional properties, eco-physiological properties and Darwinian fitness, monistic reductionism and pluralism (Michod 1999; Griesemer 2001; McShea and Anderson 2005; Michod and Herron 2006; Calcott 2011; Bouchard and Huneman 2013b; Pradeu 2013; Clarke 2014; Herron, this volume; Sterner, this volume).

On a developmental scale throughout an organism's life cycle, what criteria should define a biological individual (or individuals) as changes occur? This was a central question for our two Huxleys, and it is still relevant today for organisms with alternation of generations, haploid and diploid phases of varied lengths, sexual and asexual reproduction, and genets and ramets that may later become physically separated (Bell 1994; Minelli 2011, 2015; Clarke 2013; Herron et al. 2013; Niklas and Kutschera 2014). The term "development," suggesting change over time, challenges definitions of individuality. Maienschein (2011, 23) calls our attention to "when development begins or ends and whether it has defined stages along the way," and to spatial boundaries: "where does the developing object start and end?" (see also Minelli and Pradeu 2014). Accordingly, we may ask of biological individuality concepts put to work in developmental problems, "What is an individual (or a part), and what defines it (or the parts) as it changes through developmental stages over time?"

Problems of temporality are central to assessing homology. In the post-Darwinian era, homology is a hypothesis aligned with similarity due to common descent, but must also account for changes during organismal development and at an evolutionary scale. These problems often involve both classifying and explaining corresponding entities, "the same but different," that occur within different organisms and sometimes within the same organism. Individuality concepts here require some shared criterion or guarantor of "sameness," but those criteria and associated problem spaces have differed historically as well as today (Rieppel 1994; Brigandt 2003; Kleisner 2007). As far back as Étienne Geoffroy Saint-Hilaire in the 1820s (Appel 1987), a scientist would individuate homologous entities using criteria of invariant numbers, relative position, and connectivity to other recognized within-organism structural entities. Decomposition of the whole organism into constituent entities (often without reference to function) helps construct this problem space. Alternatively, a scientist might use a developmental individuality concept that is diachronic rather than synchronic, with shared, sequential reproductive, embryological, or morphogenetic changes as criteria (Nyhart and Lidgard 2011, this volume). Such was the basis of Ernst Haeckel's (1866) approach to comparative evolutionary morphology (which explicitly distinguished three kinds of individuality: morphological, physiological, and genealogical or developmental) and the broader late nineteenth-century German tradition (Nyhart 1995); and such remains at base the same today. While congruence of these two sets of criteria often supports a hypothesis of homology, they do sometimes clash. It is widely recognized that putatively homologous structures may originate and transform through different developmental mechanisms (Claus 1874; Brigandt 2003; Hall 2003b). Homologous entities might also be individuated separately at successive stages of change in order to separately evaluate ontogenetic identity and phylogenetic identity (Havstad et al. 2015).

Günter Wagner (2014) presents a remarkably innovative synthesis in support of a genetic theory of homology, focused in part on the reliable maintenance of characters on an evolutionary scale, yet retaining a capacity for variation and change. For him, a homologue is a character, and its different manifestations are character states such as the wings of birds versus bats as states of tetrapod limbs. Wagner's theory is reductionist in the sense that modifications of gene regulatory networks, especially those strongly conserved relations that he terms "character identity networks" (ChINs), result in morphological changes in evolution. What is most relevant for our discussion is that his account foregrounds conceptual criteria, particularly the temporal segments of development when the ChIN activates, and at the same time a particular character comes into being. Wagner's molecular-biological problem space includes homology of individual genes or proteins and would dictate an individuality concept with sequence similarity as one criterion. His version differs from most previous accounts, which employ a functional criterion of what counts as an organized individual (Love 2015b), and thus also involve functional-phenotypic elements. The strengths and weaknesses of these views are not our focus here. Rather, we wish to show that different concepts and problem spaces work to individuate and explain homologous entities, addressing problems related to temporal change at many scales.

CONSTITUTION

The most capacious of our four kinds is *constitution*; what constitutes an individual necessarily and sufficiently for a particular problem and its problem space. This kind has undergone the most rapid growth in recent decades, engaging problems of modularity, symbiosis and holobionts, superorganisms, microbial communities, and genetic mosaicism and chimerism. It refers to problems of compositional, interactive, or causal organization of parts in making a whole, or of wholes in supporting a larger enkaptic whole (Rieppel, this volume). Different approaches have been used for distinguishing parts and wholes—material structure, function, physiological or ecological processes, or foci of Darwinian selection—and these approaches are frequently used in the senses of binding organization and interactivity (Winther 2001, 2006; Bouchard and Huneman 2013b). Immunologically based individuality concepts also need to account for problems of constitution. These problems involve recognition, tolerance, or defense directed at allogeneic and xenogeneic cells and their metabolites, whether these originate "externally" or from the organism itself (Eberl 2010; Pradeu 2012; Brusini et al. 2013; Tauber 2015a and b).

One way scientists and philosophers have discussed constitution is in terms of modularity. Though difficult to define precisely (Bolker 2000), modularity implies greater interaction and integration within modules than among them, which virtually demands individuality concepts that accommodate degrees of interactions and integration. In modular organisms like plants, animal colonies, many fungi, and even some bacteria, modules with both structural and functional distinctness are generated iteratively and can become detached and live separately (Vuorisalo and Tuomi 1986; Andrews 1998; Hughes 2005; Barthélémy and Caraglio 2007; Clarke 2012; Booth 2014b). In modular genetic and protein networks, there exist clusters of enhanced interaction and function (Pereira-Leal et al. 2006; Wagner et al. 2007). It has even been argued that viruses can be modules, as correlated clusters of autocatalytic activity in networks of genetic and metabolic machinery (Krakauer and Zanotto 2009). Embryological modules (multicellular condensations) that mediate interactions between genotype and phenotype have been considered primordial individuals from which a part or an organ develops (Hall 2003a; de Kroon et al. 2005; Arnellos et al. 2014). Moreover, because interactions within modules rely on different mechanisms, individuality concepts involving modularity must accommodate developmental, genetic, structural, functional, or evolutionary contexts of modularity (Klingenberg 2008) for specific problems and problem spaces. Biologists and philosophers must ask, how should the boundaries or interactions or integration of modules be defined, and how should agency or control be apportioned? These things are nearly always relative and may work differently in different kinds of modular systems (Vuorisalo and Mutikainen 1999; Callebaut and Rasskin-Gutman 2005). How do modules interact with one another, or act as units of selection (Winther 2005; Wagner et al. 2007)? Modules thus pose both solutions and problems for biological individuality concepts involving part-part and part-whole relations.

Analogous problems arise from symbiotic interactions among different groups of organisms. For instance, lichens have generated problems of constitutional individuality for well over a century (de Bary 1887; Sanders 2001). Lichens are made up of at least one kind of fungus (the mycobiont) and one or more unicellular algae or cyanobacteria (the photobionts). In 1869, Schwendener described lichens as colonies with a sort of controlled parasitism, the fungal master living off the work of its green algal slaves (Schwenender in

Ahmadjian 1993, 4). The physiological dependence of partners is also shown by the highly variable morphology of the lichen thallus, whose morphogenesis is shaped by the interaction of the partners. The majority of these symbioses are thought to be obligate, the partners having simultaneous reproduction and not known to exist separately in nature, although a number of facultative lichen symbioses are known (Dal Grande et al. 2012). Some combinations of symbionts change periodically, with different photobionts present in different stages of the life of the fungal partner. Taken together, lichens thus display a phenomenal variability of structures and multifunctionality that bring about complicated choices for scientists in prioritizing developmental individuality, functional individuality, and genetic individuality of separate partners versus symbiotic wholes (Jahns and Ott 1997).

Holobionts are common (perhaps even ubiquitous) in the eukaryotic world (Zook 2015). They are typically considered in terms of eukaryotic hosts and their prokaryotic and eukaryotic symbionts, but understanding them as structural or functional individuals raises problems for conventional notions (Gilbert et al. 2012; McFall-Ngai et al. 2013; Bordenstein and Theis 2015; Gilbert and Epel 2015; Gilbert, this volume). Holobionts are composites, host and many symbionts that are each autonomous biological individuals in many respects—anatomically, physiologically, and genetically—yet are integral to sustaining and replicating the physiology, development, and evolution of the greater whole (Pradeu 2011; Booth 2014a; O'Malley 2015). Consider studies on the oligochaete worm *Olavius* and similar forms, in which neither mouth nor gut nor anus is present (Woyke et al. 2006; Kleiner et al. 2012). Different sulfate-reducing and sulfide-oxidizing bacterial endosymbionts in their oligochaete hosts cycle sulfur compounds and draw organic carbon from the external environment, also providing metabolic energy to the host, which may in turn move in and out of sulfur-containing habitats and provide compounds for the sustenance of the symbionts. Neither hosts nor sulfur-cycling bacteria seem to be viable independently, and in some systems, symbiont transmission takes place via different modes (vertical or environmental). Among animal-hosted holobionts, the community composition of gut microbiotas undergoes more or less continuous change through the host's lifetime, including environmental incursions and exchanges (Pradeu 2012; O'Malley 2014; Voreades et al. 2014; Gilbert and Epel 2015).

A superorganism, often understood as a biological object consisting of many multicellular organisms, poses yet another variant of a compositional problem. Eusocial insect colonies are a classic example of superorganisms (Hölldobler and Wilson 2009). Within such colonies, some castes of morphologically individuated workers leave no progeny, yet the division of labor among castes enables the functional unity and reproductive continuity of the colony as a whole, often considered an individual as well. However, a long-running dispute over group selection, cohesion, causal sufficiency, and fitness (A. Hamilton et al. 2009; Bouchard 2013; Haber 2013) still leaves concepts and criteria of superorganism individuality differing in explanatory roles. If the holobiont or superorganism is not always precisely a spatially and temporally discrete individual, it may nevertheless be a functional, unified-compositional, or evolutionary one. How should we select among or adjust our concepts for multispecies individuals, variably continuous and discontinuous in a physical sense, sometimes transient in a temporal sense?

Scientists studying microbes also confront constitution problems. Even in clonal cultures, microorganisms are heterogeneous, with extensive phenotypic variation and rapid gains and losses of genes, resulting in questionable assignment of absolute genetic identity and mechanistic function at a single-cell level. Recently, microbiologists have documented ecologically determined gene-transfer networks, suggesting that many clonal cultures are better understood as pangenomes (Mira et al. 2010; Polz et al. 2013). Yet at the same time, social communication and functional efficacy at levels of groups of interacting cells in bacterial populations are recognized and individuated (Avery 2006; West et al. 2006; Davidson and Surette 2008; Boon et al. 2014; Vliet and Ackermann 2015; Reynolds, this volume). Cell individuality is structured by the genome, but also by transcription and translation, enabling different phenotypes to be passed on through cell generations as a kind of "molecular memory" (Zengler 2009, 721). Multispecies microbial biofilms, lacking discrete parent-offspring lineages and reproductive bottlenecks, also challenge certain notions of evolutionary individuality. Biofilms show degrees of evolvability that may be accommodated by revised individuality concepts (Ereshefsky and Pedroso 2013; Sterner, this volume; but see Clarke 2016).

Individuality criteria relying on a predictable, stable, and heritable genetic composition at levels of the cell, organism, population, and genealogical lineage increasingly have to account for widespread occurrence of horizontal gene transfer. Especially prevalent in prokaryotes (Doolittle 2013; Bapteste 2014; O'Malley 2014), with rapidly advancing genomic data and bioinformatic analyses it is increasingly becoming recognized among eukaryotes (Bock 2010; Fitzpatrick 2012; Boto 2014). Transfer of genes, gene clusters, or entire chromosomes may occur between members of the same lineage, and sometimes even between different kingdoms and domains of life. Phylogenetic studies indicate that transfers detected in some eukaryotic species are recent as well as ancient (Keeling and Palmer 2008). The extent to which such transfers, including those that have occurred in deep time, will revise the work of biological individuality concepts and the scope of definitional criteria is unclear at present.

Constitution problems of genetic chimerism and mosaicism within organisms—each reflecting the presence of two or more genetically distinct cell populations in the body—are as important as those involving symbioses or microbial interactions. Chimerism and mosaicism are surprisingly common. Mosaicism is usually caused by somatic mutation, whereas chimerism most often originates from allogenic fusion of closely related organisms or maternal-fetal cellular exchange (Gill et al. 1995; Santelices 2004; Gamill and Nelson 2010). Mosaicism arises naturally in the genomes of multicellular eukaryotes as point mutations accumulate over time and organisms become a mosaic of slightly different and sporadically very different cells, particularly in long-lived modular organisms (Minelli 2014; Schweinsberg et al. 2015). Natural chimerism via fusion exists in rhodophytes, chlorophytes, plants, cellular slime molds, fungi, poriferans, cnidarians, bryozoans, ascidian chordates, and others. Heterokaryosis, the presence of genetically diverse nuclei in cell cytoplasm, is a form of chimerism especially widespread among fungi. It occurs regularly in filamentous hyphae, part of the fungal body. Growing through the soil, hyphae of genetically different bodies fuse. Nuclei and organelles are shuffled about in the common cytoplasm. In pioneering genomic studies, Boon and colleagues (Boon et al. 2010, 2015) show that single bodies can have extremely high genetic variation among nuclei, which is also reflected in transcriptomes. Single spores produced from different parts of the body are genetically diverse (Boon et al. 2010; L. Ma et al. 2016). Thus clonal offspring may have different fitnesses and not share the same evolutionary fate. While these results are preliminary, they offer an analogy to a pangenome in these organisms. Yet in a functional context, these genetically heterogeneous fungi are nonetheless integrated metabolically, reproductively, and developmentally in complementary senses of functional individuality.

Significant amounts of intraorganismal genetic heterogeneity in instances like these challenge evolutionary explanations of individuality that rest on received views about such issues as genetic uniformity, units of selection, cell lineage cooperation and conflict, policing mechanisms, and appropriate partitioning of fitness (Dupré 2010; Rinkevich 2011). Booth (2014b, 613) notes that heterokaryotic fungi "show that organisms need not always be composed of genetically identical (or similar) parts and that the evolution of such organisms is not rare or unusual." He argues that these organisms challenge views that genetically distinct parts of organisms tend not to cooperate, and tend to bring about within-organism conflict and thus degradation of the whole. In a multilevel selection context, Godfrey-Smith (2009), Clarke (2011), and Booth (2014b) offer thoughtful but different interpretations of these controversial problems of chimerism and mosaicism, their relative importance, and their relationship to an array of definitional criteria of individuality.

As these examples show, constitution, with its dominant focus on (non-hierarchical) interactions among parts occurring within commonly recognized biological individuals and beyond them, encompasses many challenges to our assumptions about biological individuals. Given how difficult it is for us to do without biological individuality concepts altogether, this suggests again the need for an epistemologically pluralist approach emphasizing particular problem contexts.

To sum up, looking at biological individuality and how it is used through the lens of these four basic kinds of problems—individuation, hierarchy, temporality, and constitution—shows the diversity of opportunities and challenges that biologists face when developing their problem spaces. A key point derived from consideration of all four kinds is that in many of the above cases and their problem spaces, entities are individuated and interact in ways that violate or simply don't apply to some biological individuality concepts, while adhering loosely to others. Different kinds of specific problems challenge by turn conventional criteria of indivisibility, boundedness, autonomy, unity, genetic homogeneity, units of selection, internal versus external, and continuity. This variety of puzzles and paradoxes strengthens our belief that the search for a single encompassing concept of biological individuality, one necessary and sufficient for all cases, is imprudent at best. Yet even pluralistic individuality concepts may not be enough. What sort of a theoretical world might we live in if our primary consideration was not "the" individual but in its place, interrelationships and processes?

Individuality as the Semblance of Systems, Networks, and Processes

Despite the profound utility of individuality concepts for biologists—arguably, their necessity (Kovaka 2015)—the inability of any given individuality concept to work universally has sometimes raised the question of whether so much attention should be focused on individuals as biological objects. The kinds of problems elaborated above, especially constitution, have invited alternative approaches that subordinate the physical thing-ness of biological individuality to the operations that sustain individuality—or that even dissolve individuality into the functions or mechanisms of life. Approaches that view life in terms of systems, networks, and processes all generally forego a more orthodox, reductionist individuation of biological objects relying principally on more atomized definitional criteria. Instead the focus is on system- or network-wide behavior, on functional interactions or causal forces, on *relative* degree or appearance of individuality, or on dynamic behavior over time. To varying degrees, the goal of such approaches is to present a biology of relationality and interaction. Organicism, "systems biology" and its close companion biological network analysis, process philosophy, and scaffolding phenomena together represent a spectrum of approaches in which individuals and individuality may remain significant or completely melt away into the flow of relationships (Gatherer 2010; Caporael et al. 2014; Seibt 2013; Siegal 2014).

Although organicism may be traced far back in time (R. Wheeler 1936), its modern incarnation emerged in the early twentieth century as a prominent middle way between unseen guiding forces of vitalism and the mechanists' reductionist idealization of machine-like organisms (Trewavas 2006; Drack 2009; Drack and Wolkenhauer 2011; Nicholson and Gawne 2015). Leading biologists and philosophers like J. S. Haldane, J. H. Woodger, J. Needham, J. von Uexküll, L. von Bertalanffy, P. A. Weiss, C. M. Child, and W. E. Ritter rejected both vitalism and reductionist mechanism. These organicists varied considerably in their support of different tenets of a systems theoretic understanding relevant to individuality and organism. Their ideas nonetheless are broadly consistent with and partly contribute to Bertalanffy's views on systems theory, as summarized much later: living beings are "open systems" that should be considered "continuous processes"; organisms are "homeostatic . . . hierarchically organized systems" (Wuketits 1989, 17). As Nicholson and Gawne have argued (2015, 345), these views are "thematically and methodologically continuous" with contemporary views of the organism; we see them in evolutionary developmental biology, systems biology, and theoretical biology (Nicholson 2014). Today's mainstream organicism "allow[s] for reduction as an explanatory strategy that is often but not necessarily always helpful" (El-Hani and Emmeche 2000, 238). Importantly for this discussion, some organicists and systems theorists recognized degrees of individuality. In many of these organicist and systems views, living systems are considered dynamic hierarchies, with systems at different levels (often as individuals) superposed and interacting with one another and among levels that might range from cells to organisms to ecological communities and beyond.

Modern systems biology also focuses on hierarchies of organization that exhibit emergent properties, functions, robustness, or evolvability of modules or of the whole. It seeks out biological functions in terms of network organization and component interaction—properties or traits that cannot be reduced to singly identifiable elements like genes or proteins. But in its short history over the past few decades, the domain of its practical work seems considerably narrower than organicism or systems thinking in general (O'Malley and Dupré 2005; O'Malley 2012; Brigandt 2013; Siegal 2014). There is not yet a concise, widely accepted definition of systems biology, but its focus has largely been on high throughput quantitative measurements at a molecular level (genomics, proteomics), mathematical modeling, and iterative sequences of prediction, data analysis, and model tweaking. Network analysis, together with systems biology, is making significant theoretical and empirical inroads across many areas, but in ways that have not yet received significant philosophical attention (O'Malley et al. 2014). These areas span ecology (Proulx et al. 2005; Bascompte and Jordano 2007); microbial adaptations, evolution, and phylogeny (Davidson and Surette 2008; Doolittle 2013; Bapteste 2014); cellular function, organization, and disease (Barabási and Oltvai 2004; Feist et al. 2009; Barabási et al. 2011); biochemistry, molecular genetics, and proteomics (Han et al. 2004; Huang et al. 2004; Chen et al. 2009; W. Ma et al. 2009); and evolutionary developmental biology (Davidson et al. 2002; Wagner et al. 2007; Wagner 2014).

Systems biology and network analyses frequently overlap in trying to understand how complex functions are achieved, and sometimes in seeking evolutionary explanations for current organization and functionality. In many of these, the notion of "individual" melts away into modules or clusters or hubs that are foci of relational interactions to which agency or control is ascribed. In systems biology particularly, causation via control mechanisms is frequently shared through constraining or facilitating relationships among these foci. Just as often, modules or hubs are defined at a specific level of a biological hierarchy in one or another sense of a particular function or operation, whether structural or regulatory, that contributes to an outcome of the greater whole. Boundaries of "individuals" as such are fuzzy rather than sharply defined by specific definitional criteria, and autonomy is gradational. It is of course possible to view modules or hubs or bacterial communities as individuals in a sense, by privileging their functional interactions or integration as a phenomenon uniting a group (e.g., Doolittle 2013; Arnellos et al. 2014; see also Huneman 2014a and b).

Another approach, scaffolding and material overlap, provides a different way to identify and understand individuals whose reproduction or evolutionary persistence depends on resources (scaffolds) "external" to material parts, some of which may pass on to new generations (cf. Greisemer 2000, 2002, 2014; Wimsatt and Griesemer 2007). Scaffolds are supporting processes or structures that can also facilitate the acquisition of new capacities, or lead to new levels of emergent individuality. They may be temporary in fulfilling their essential roles. Scaffolded systems could include, for example, genes or viruses reliant on the machinery of cells, obligate parasites or symbionts dependent on hosts, or multicellular biofilms and their extracellular matrix (Griesemer 2014; Godfrey-Smith 2015; Sterner, this volume). Griesemer et al. (2014, 363) point out that scaffolding perspectives are a response to reductionist claims, in part to population genetic explanations that black-box or ignore "the constraints and affordances of parts, processes, and activities in development." And thus scaffolding belongs among those approaches that acknowledge individuality but center on a biology of interactions and relationality.

Up until now in our chapter, we have talked mostly about biological individuals as things with relevant properties, within the dominant ontology in biology that philosophers connect to Aristotelian substance metaphysics. The minority position of process philosophy opposes substance metaphysics and thus sounds decidedly unfamiliar, such "that processes are more fundamental, or at any rate not less fundamental than things" (Rescher 2007, 144), and that all being is dynamic in the flow of time. Like systems thinking, the process view has a long history (Rescher 2007; Gare 2011; Seibt 2013; Delafield-Butt 2008): at the end of the eighteenth century Schelling, for example, presented nature as unending productive activity, and things as merely momentary eddies in the stream of this activity (Schelling 2004, 18). Contemporary process thinking in biology typically acknowledges Alfred North Whitehead and his early twentieth-century philosophy of organism (Delafield-Butt 2008). Modern process-oriented philosophers of biology such as John Dupré tend to see "life" as a hierarchy of processes—metabolism, development, evolution— rather than of substances, individuals, and organisms (Dupré and O'Malley 2007, 2009; Bapteste and Dupré 2013; Dupré 2014). Organisms, however defined in terms of individuality, are living developmental processes. In addition, the material constituents of organisms are constantly changing through a life cycle; cells die and are replaced, members of a microbiome come and go, and so an argument can be made that it is the relationships among the processes rather than among the parts themselves that more fundamentally constitute the organism.

What does this mean for biological individuality? In terms of the work of individuality, this approach may have advantages in understanding *what* is occurring and *how* it is occurring as living things inevitably undergo continuous change. Our earlier discussions of temporality and constitution suggested ways in which "individual" and "organism" have fuzzy boundaries. Perhaps in a process view, these can more readily be comprehended as cross-sections through a spatiotemporal flow of events, brief slices of a life cycle, interacting life cycles, internally changing or intersecting genomes, all responding to internal as well as external forces (de la Rosa 2010; Dupré 2012). For instance, Guay and Pradeu (2016) note that "genidentity," first developed in the early to mid-twentieth century, can be used to understand the identity of an individual that passes through a continuous connection of states through time; the individual is not presupposed, but rather is a byproduct of a specific process being followed. Rieppel (2009) and Lockwood (2012) have independently argued that species are processes; importantly, however, this need not exclude complementary perspectives of species as individuals or as homeostatic property cluster natural kinds (Rieppel 2007). Processes are sometimes also taken to subsume the intertwined roles of structure and function (Dupré 2014; see also Brigandt, this volume).

Our discussion in this section again suggests a need for explanatory pluralism. We offer neither critique nor endorsement of any particular one among systems, network, process, or scaffolded frames of understanding the nature of individuals or organisms. Indeed, we think that these approaches are mutually illuminating. Thus, we have tried to complete the circle by engaging with an expanded set of approaches in which biological individuality concepts and their problem spaces might be considered.

Problem Spaces in a Research Agenda

Our discussion suggests that each of the many proposed definitions and criteria of biological individuality (Table 1.1) that provides explanatory adequacy for one specific problem is likely to be countered by some sort of biological object(s) that either is a borderline case or doesn't fit at all. Moreover, the definition and criteria of individuality best suited to one specific problem are unnecessary or just inadequate for any number of other specific problems, all of which have biological importance. The long history (Fig. 1.1) of disagreement over the "right" or "most fundamental" concept of biological individuality hasn't stopped biologists from using such concepts in doing useful work, and shouldn't, provided that their assumptions and definitions are acceptably sufficient and explicit (Kovaka 2015; Love forthcoming). Where does this leave clarification of "the" concept of biological individuality?

We advocate a rethinking of biological individuality, its variety of definitions and relevant definitional criteria. Rather than continuing the sometimes acrimonious disputes seeking "the" one true meaning of the concept, biological individuality has a more positive, forward-looking role in the work of biologists—as an established problem agenda (Love 2008, 2014, forthcoming). We have shown that the conceptual repertoire of biological individuality is exceptionally broad, and that its work in specific problems and problem spaces is fundamentally contextual. It represents a stable but still growing domain of material and abstract representations of biological objects, their interactions and processes, that has been in place for over 170 years. The work of successfully providing structure and buttressing generality, predictability, and explanatory power in biologists' specific problems is due in part to useful variants of definition and criteria of individuality; it doesn't make sense to see different variants as competitors if they are answering different questions (cf. Lloyd 2012). We have shown how these specific questions can be grouped into stable kinds: biological individuality covers them all. Borrowing from Alan Love (2014, 42), we may summarize this rethinking of biological individuality thus, "there are broad clusters of questions that reflect generally delineated phenomena . . . whose explanatory characterization sets the agenda of research." Taking biological individuality seriously as a problem agenda, new specific problems will continue to arise in new contexts. Some individuality concepts will change and thrive doing their work for biologists, while others will persist but fade into background. It could hardly be otherwise.

Acknowledgments

This chapter owes much to the participants in two *E Pluribus Unum* workshops, including contributors to this volume and its anonymous reviewers, who offered us their insights, criticisms, and suggestions. I. Brigandt, J. Griesemer, J. Havstad, S. Leavitt, A. Love, T. Lumbsch, O. Rieppel, and B. Sterner helped clarify specific issues. Of course, the views that remain are our own. We acknowledge vital support from the Chemical Heritage Foundation; the Negaunee Foundation; the Anonymous Fund, the Robert F. and Jean E. Holtz Center for Science and Technology Studies, the Department of the History of Science, and the Graduate School, all at the University of Wisconsin-Madison; and a John Templeton Foundation grant on Principles of Complexity awarded to David Krakauer and Jessica Flack. We are grateful to all.

References for Chapter One

Ackermann, M. 2013. "Microbial Individuality in the Natural Environment." *ISME Journal* 7 (3): 465–67.
Agar, W. E. 1948. "The Wholeness of the Living Organism." *Philosophy of Science* 15 (3): 179–91.
Ahmadjian, Vernon. 1993. *The Lichen Symbiosis.* New York: Wiley.
Anderson, Warwick, and Ian R. Mackay. 2014. *Intolerant Bodies: A Short History of Autoimmunity.* Baltimore: Johns Hopkins University Press.

Andrews, J. H. 1998. "Bacteria as Modular Organisms." *Annual Reviews in Microbiology* 52 (1): 105–26.

Appel, Toby A. 1987. *The Cuvier-Geoffroy Debate: French Biology in the Decades before Darwin.* New York: Oxford University Press.

Arber, Agnes. 1930. "Root and Shoot in the Angiosperms: A Study of Morphological Categories." *New Phytologist* 29 (5): 297–315.

———. 1941. "The Interpretation of Leaf and Root in the Angiosperms." *Biological Reviews* 16 (2): 81–105.

Arnellos, Argyris, Alvaro Moreno, and Kepa Ruiz-Mirazo. 2014. "Organizational Requirements for Multicellular Autonomy: Insights from a Comparative Case Study." *Biology & Philosophy* 29 (6): 851–84.

Avery, Simon V. 2006. "Microbial Cell Individuality and the Underlying Sources of Heterogeneity." *Nature Reviews Microbiology* 4 (8): 577–87.

Bailey, L. H. 1906. "The Plant Individual in the Light of Evolution: The Philosophy of Bud-Variation, and Its Bearing upon Weismannism." In *The Survival of the Unlike: A Collection of Evolution Essays Suggested by the Study of Domestic Plants*, 5th ed., 81–106. New York: Macmillan.

Bapteste, Eric. 2014. "The Origins of Microbial Adaptations: How Introgressive Descent, Egalitarian Evolutionary Transitions and Expanded Kin Selection Shape the Network of Life." *Frontiers in Microbiology* 5: 1–4.

Bapteste, Eric, and John Dupré. 2013. "Towards a Processual Microbial Ontology." *Biology & Philosophy* 28 (2): 379–404.

Barabási, Albert-László, and Zoltán N. Oltvai. 2004. "Network Biology: Understanding the Cell's Functional Organization." *Nature Reviews Genetics* 5 (2): 101–13.

Barabási, Albert-László, Natali Gulbahce, and Joseph Loscalzo. 2011. "Network Medicine: A Network-Based Approach to Human Disease." *Nature Reviews Genetics* 12 (1): 56–68.

Barthélémy, Daniel, and Yves Caraglio. 2007. "Plant Architecture: A Dynamic, Multilevel and Comprehensive Approach to Plant Form, Structure and Ontogeny." *Annals of Botany* 99 (3): 375–407.

Bascompte, Jordi, and Pedro Jordano. 2007. "Plant-Animal Mutualistic Networks: The Architecture of Biodiversity." *Annual Review of Ecology, Evolution, and Systematics* 38: 567–93.

Baum, David A. 1998. "Individuality and the Existence of Species through Time." *Systematic Biology* 47 (4): 641–53.

Bechtel, William. 1984. "The Evolution of Our Understanding of the Cell: A Study in the Dynamics of Scientific Progress." *Studies in History and Philosophy of Science Part A* 15 (4): 309–56.

Bell, Graham. 1982. *The Masterpiece of Nature: The Evolution and Genetics of Sexuality.* Berkeley: University of California Press.

———. 1994. "The Comparative Biology of the Alternation of Generations." *Lectures on Mathematics in the Life Sciences* 25: 1–26.

Bergson, Henri. 1911. *Creative Evolution.* Translated by Arthur Mitchell. Lanham, MD: University Press of America.

Bock, Ralph. 2010. "The Give-and-Take of DNA: Horizontal Gene Transfer in Plants." *Trends in Plant Science* 15 (1): 11–22.

Bolker, Jessica A. 2000. "Modularity in Development and Why It Matters to Evo-Devo." *American Zoologist* 40 (5): 770–76.

Bonner, John Tyler. 1974. *On Development: The Biology of Form.* Cambridge, MA: Harvard University Press.

Boon, Eva, Sébastien Halary, Eric Bapteste, and Mohamed Hijri. 2015. "Studying Genome Heterogeneity within the Arbuscular Mycorrhizal Fungal Cytoplasm." *Genome Biology and Evolution* 7 (2): 505–21.

Boon, E., E. Zimmerman, B. F. Lang, and M. Hijri. 2010. "Intra-isolate Genome Variation in Arbuscular Mycorrhizal Fungi Persists in the Transcriptome." *Journal of Evolutionary Biology* 23 (7): 1519–27.

Boon, Eva, Conor J. Meehan, Chris Whidden, Dennis H.-J. Wong, Morgan G. I. Langille, and Robert G. Beiko. 2014. "Interactions in the Microbiome: Communities of Organisms and Communities of Genes." *FEMS Microbiology Reviews* 38 (1): 90–118.

Booth, Austin. 2014a. "Symbiosis, Selection, and Individuality." *Biology & Philosophy* 29 (5): 657–73.

———. 2014b. "Populations and Individuals in Heterokaryotic Fungi: A Multilevel Perspective." *Philosophy of Science* 81 (4): 612–32.

Bordenstein, Seth R., and Kevin R. Theis. 2015. "Host Biology in Light of the Microbiome: Ten Principles of Holobionts and Hologenomes." *PLOS Biology* 13 (8): e1002226.

Boto, Luis. 2014. "Horizontal Gene Transfer in the Acquisition of Novel Traits by Metazoans." *Proceedings of the Royal Society B* 281 (1777): 20132450.

Bouchard , Frédéric. 2008. "Causal Processes, Fitness, and the Differential Persistence of Lineages." *Philosophy of Science* 75: 560–70.

———. 2013. "What Is a Symbiotic Superindividual and How Do You Measure Its Fitness?" In *From Groups to Individuals: Evolution and Emerging Individuality*, edited by Frédéric Bouchard and Philippe Huneman, 243–64. Cambridge, MA: MIT Press.

Bouchard, Frédéric, and Philippe Huneman, eds. 2013a. *From Groups to Individuals: Evolution and Emerging Individuality.* Cambridge, MA: MIT Press.

———. 2013b. "Introduction." In *From Groups to Individuals: Evolution and Emerging Individuality*, edited by Frédéric Bouchard and Philippe Huneman, 1–14. Cambridge, MA: MIT Press.

Boulter, Stephen. 2013. *Metaphysics from a Biological Point of View.* New York: Palgrave Macmillan.

Brandon, Robert N. 1996. *Concepts and Methods in Evolutionary Biology.* Cambridge: Cambridge University Press.

Braun, Alexander. 1853. "On the Phenomenon of Rejuvenescence in Nature." In *Botanical and Physiological Memoirs*, translated by A. Henfrey, 1–342. London: Ray Society.

———. 1855–56. "The Vegetable Individual, in Its Relation to Species." *American Journal of Science and Arts* 19: 297–317.

Brigandt, Ingo. 2003. "Homology in Comparative, Molecular, and Evolutionary Developmental Biology: The Radiation of a Concept." *Journal of Experimental Zoology Part B: Molecular and Developmental Evolution* 299B (1): 9–17.

———. 2010a. "The Epistemic Goal of a Concept: Accounting for the Rationality of Semantic Change and Variation." *Synthese* 177 (1): 19–40.

———. 2010b. "Beyond Reduction and Pluralism: Toward an Epistemology of Explanatory Integration in Biology." *Erkenntnis* 73 (3): 295–311.

———. 2012. "The Dynamics of Scientific Concepts: The Relevance of Epistemic Aims and Values." In *Scientific Concepts and Investigative Practice*, edited by Uljana Feest and Friedrich Steinle, 3: 75–103. Berlin Studies in Knowledge Research. Berlin: De Gruyter.

———. 2013. "Systems Biology and the Integration of Mechanistic Explanation and Mathematical Explanation." *Studies in History and Philosophy of Science Part C: Studies in History and Philosophy of Biological and Biomedical Sciences* 44 (4, Part A): 477–92.

Brigandt, Ingo, and Alan C. Love. 2012. "Conceptualizing Evolutionary Novelty: Moving beyond Definitional Debates." *Journal of Experimental Zoology Part B: Molecular and Developmental Evolution* 318 (6): 417–27.

Brusini, Jérémie, Cécile Robin, and Alain Franc. 2013. "To Fuse or Not to Fuse? An Evolutionary View of Self-Recognition Systems." *Journal of Phylogenetics & Evolutionary Biology* 1 (1): 1–8.

Burnet, Macfarlane. 1959. "Auto-Immune Disease: I. Modern Immunological Concepts." *British Medical Journal* 2 (5153): 645–50.

———. 1969. *Self and Not-Self: Cellular Immunology*, Book One. Melbourne: Melbourne University Press.

Burnett, Waldo I. 1854. "Researches on the Development of Viviparous Aphides." *American Journal of Science and Arts* 17: 62–78.

Buss, Leo W. 1987. *The Evolution of Individuality*. Princeton, NJ: Princeton University Press.

Calcott, Brett. 2011. "Alternative Patterns of Explanation for Major Transitions." In *The Major Transitions in Evolution Revisited*, edited by Brett Calcott and Kim Sterelny, 35–52. Cambridge, MA: MIT Press.

Calcott, Brett, and Kim Sterelny. 2011. *The Major Transitions in Evolution Revisited*. Cambridge, MA: MIT Press.

Callebaut, Werner, and Diego Rasskin-Gutman, eds. 2005. *Modularity: Understanding the Development and Evolution of Natural Complex Systems*. Cambridge, MA: MIT Press.

Caporael, Linnda R., James R. Griesemer, and William C. Wimsatt, eds. 2014. *Developing Scaffolds in Evolution, Culture, and Cognition*. Cambridge, MA: MIT Press.

Carpenter, William B. 1848. "On the Development and Metamorphoses of Zoophytes." *British and Foreign Medico-Chirurgical Review* 1: 183–214.

Causey, David. 1935. "Individuality in the Animal Kingdom." *Biologist* 17 (1): 9–16.

Chen, Luonan, Rui-Sheng Wang, and Xiang-Sun Zhang. 2009. *Biomolecular Networks: Methods and Applications in Systems Biology*. Vol. 10. Wiley Series in Bioinformatics. Hoboken, NJ: Wiley.

Cheung, Tobias. 2010. "What Is an 'Organism'? On the Occurrence of a New Term and Its Conceptual Transformations 1680–1850." *History and Philosophy of the Life Sciences* 32: 155–94.

Child, Charles M. 1912. "The Process of Reproduction in Organisms." *Biological Bulletin* 23 (1): 1–39.

———. 1915. *Individuality in Organisms*. Chicago: University of Chicago Press.

Clarke, Ellen. 2010. "The Problem of Biological Individuality." *Biological Theory* 5 (4): 312–25.

———. 2011. "Plant Individuality and Multilevel Selection Theory." In *Major Transitions in Evolution Revisited*, edited by Brett Calcott and Kim Sterelny, 227–50. Cambridge, MA: MIT Press.

———. 2012. "Plant Individuality: A Solution to the Demographer's Dilemma." *Biology & Philosophy* 27 (3): 321–61.

———. 2013. "The Multiple Realizability of Biological Individuals." *Journal of Philosophy* 8: 413–35.

———. 2014. "Origins of Evolutionary Transitions." *Journal of Biosciences* 39 (2): 303–17.

———. 2016. "Levels of Selection in Biofilms: Multispecies Biofilms Are Not Evolutionary Individuals." *Biology & Philosophy* 31 (2): 191–212.

Claus, Carl. 1874. *Die Typenlehre und E. Haeckel's Gastraea-Theorie.* Vienna: Manz.

Cobbold, Thomas Spencer. 1869. "On the Question of Organic Individuality, Entozoologically Considered." In *Entozoa: Being a Supplement to the Introduction to the Study of Helminthology*, 81–89. London: Groombridge and Sons.

Conklin, Edwin G. 1916. "The Basis of Individuality in Organisms from the Standpoint of Cytology and Embryology." *Science* 43 (1111): 523–527.

Dal Grande, F., I. Widmer, H. H. Wagner, and C. Scheidegger. 2012. "Vertical and Horizontal Photobiont Transmission within Populations of a Lichen Symbiosis." *Molecular Ecology* 21 (13): 3159–72.

Darwin, Charles. 1839. *Journal of Researches into the Natural History and Geology of the Countries Visited During the Voyage of H.M.S. Beagle Round the World, Under the Command of Capt. Fitz Roy, R.N.* London: John Murray.

———. 1872. *On the Origin of Species by Means of Natural Selection, or the Preservation of Favoured Races in the Struggle for Life.* 6th ed. London: John Murray.

Darwin, Erasmus. 1800. *Phytologia, Or, The Philosophy of Agriculture and Gardening with the Theory of Draining Morasses.* London: J. Johnson.

Davidson, Carla J., and Michael G. Surette. 2008. "Individuality in Bacteria." *Annual Review of Genetics* 42: 253–68.

Davidson, Eric H., Jonathan P. Rast, Paola Oliveri, Andrew Ransick, Cristina Calestani, Chiou-Hwa Yuh, Takuya Minokawa, et al. 2002. "A Genomic Regulatory Network for Development." *Science* 295 (5560): 1669–78.

Dawkins, Richard. 1982. *The Extended Phenotype.* Oxford: Oxford University Press.

Delafield-Butt, Jonathan. 2008. "Biology." In *Handbook of Whiteheadian Process Thought*, edited by Michel Weber and Will Desmond, 2: 157–70. Frankfurt: De Gruyter.

de Bary, A. 1887. *Comparative Morphology and Biology of the Fungi, Mycetozoa and Bacteria.* Translated by Henry E. F. Garnsey. 2nd ed. Oxford: Clarendon Press.

de Kroon, Hans, Heidrun Huber, Josef F. Stuefer, and Jan M. Van Groenendael. 2005. "A Modular Concept of Phenotypic Plasticity in Plants." *New Phytologist* 166 (1): 73–82.

de la Rosa, Laura Nuño. 2010. "Becoming Organisms: The Organisation of Development and the Development of Organisation." *History and Philosophy of the Life Sciences* 32: 289–316.

Delisle, Richard G. 2008. "Expanding the Framework of the Holism/Reductionism Debate in Neo-Darwinism: The Case of Theodosius Dobzhansky and Bernhard Rensch." *History and Philosophy of the Life Sciences* 30 (2): 207–26.

de Sousa R. 2005. "Biological Individuality." *Croatian Journal of Philosophy* 14: 195–218.

Doolittle, W. Ford. 2013. "Microbial Neopleomorphism." *Biology & Philosophy* 28 (2): 351–78.

Drack, Manfred. 2009. "Ludwig von Bertalanffy's Early System Approach." *Systems Research and Behavioral Science* 26 (5): 563–72.

Drack, Manfred, and Olaf Wolkenhauer. 2011. "System Approaches of Weiss and Bertalanffy and Their Relevance for Systems Biology Today." *Seminars in Cancer Biology* 21: 150–55.

Dupré, John. 2010. "The Polygenomic Organism." *Sociological Review* 58 (s1): 19–31.

———. 2012. *Processes of Life: Essays in the Philosophy of Biology.* Oxford: Oxford University Press.

———. 2014. "A Process Ontology for Biology." *Philosophers' Magazine* (67): 81–88.

Dupré, John, and Maureen A. O'Malley. 2007. "Metagenomics and Biological Ontology." *Studies in History and Philosophy of Science Part C: Studies in History and Philosophy of Biological and Biomedical Sciences* 38 (4): 834–46.

———. 2009. "Varieties of Living Things: Life at the Intersection of Lineage and Metabolism." *Philosophy and Theory in Biology* 1: e003.

Eberl, G. 2010. "A New Vision of Immunity: Homeostasis of the Superorganism." *Mucosal Immunology* 3 (5): 450–60.

Ehrlich, P, and J. Morgenroth. 1900. "Ueber Haemolysine: Dritte Mittheilung." *Berliner Klinische Wochenschrift* 37: 453–58.

El-Hani, Charbel Niño, and Claus Emmeche. 2000. "On Some Theoretical Grounds for an Organism-Centered Biology: Property Emergence, Supervenience, and Downward Causation." *Theory in Biosciences* 119 (3–4): 234–75.

Ellegård, Alvar. 1957. "The Darwinian Theory and Nineteenth-Century Philosophies of Science." *Journal of the History of Ideas* 18 (3): 362–93.

Elwick, James. 2003. "Herbert Spencer and the Disunity of the Social Organism." *History of Science* 41: 35–72.

———. 2007. *Styles of Reasoning in the British Life Sciences: Shared Assumptions, 1820–1858.* London: Pickering and Chatto.

Emerson, Alfred E. 1939. "Social Coordination and the Superorganism." *American Midland Naturalist* 21 (1): 182–209.

Ereshefsky, Marc, and Makmiller Pedroso. 2013. "Biological Individuality: The Case of Biofilms." *Biology & Philosophy* 28 (2): 331–49.

Fagerström, Torbjörn. 1992. "The Meristem–Meristem Cycle as a Basis for Defining Fitness in Clonal Plants." *Oikos* 63: 449–53.

Feest, Uljana, and Friedrich Steinle. 2012a. *Scientific Concepts and Investigative Practice.* Berlin: de Gruyter.

———. 2012b. "Scientific Concepts and Investigative Practice: Introduction." In *Scientific Concepts and Investigative Practice*, edited by Uljana Feest and Friedrich Steinle, 1–22. Berlin Studies in Knowledge Research 3. Berlin: de Gruyter.

Feibleman, James Kern. 1954. "Theory of Integrated Levels." *British Journal for the Philosophy of Science* 5 (17): 59–66.

Feist, Adam M., Markus J. Herrgård, Ines Thiele, Jennie L. Reed, and Bernhard Ø. Palsson. 2009. "Reconstruction of Biochemical Networks in Microorganisms." *Nature Reviews Microbiology* 7 (2): 129–43.

Fisher, R. A. 1934. "Indeterminism and Natural Selection." *Philosophy of Science* 1 (1): 99–117.

Fitzpatrick, David A. 2012. "Horizontal Gene Transfer in Fungi." *FEMS Microbiology Letters* 329 (1): 1–8.

Folse, Henri, III, and Joan Roughgarden. 2010. "What Is an Individual Organism? A Multilevel Selection Perspective." *Quarterly Review of Biology* 85 (4): 447–72.

Gallesio, Giorgio. 1814. *Theorie der vegetabilischen Reproduktion, oder: Untersuchungen über die Natur und die Ursachen der Abarten und Mißgebilde.* Vienna: Felix Stöckholzer v. Hirschfeld.

Gamill, Hilary S., and J. Lee Nelson. 2010. "Naturally Acquired Microchimerism." *International Journal of Developmental Biology* 54 (2–3): 531–43.

Gardner, Andy. 2009. "Adaptation as Organism Design." *Biology Letters* 5 (6): 861–64.

Gardner, Andy, and Alan Grafen. 2009. "Capturing the Superorganism: A Formal Theory of Group Adaptation." *Journal of Evolutionary Biology* 22 (4): 659–71.

Gare, Arran. 2011. "From Kant to Schelling and Process Metaphysics: On The Way to Ecological Civilization." *Cosmos and History: The Journal of Natural and Social Philosophy* 7 (2): 26–69.

Gass, Gillian, and Brian K. Hall. 2007. "Collectivity in Context: Modularity, Cell Sociology, and the Neural Crest." *Biological Theory* 2 (4): 349–59.

Gatherer, Derek. 2010. "So What Do We Really Mean When We Say That Systems Biology Is Holistic?" *BMC Systems Biology* 4 (1): 22.

Gaudichaud, Charles. 1841. *Recherches Générales sur l'Organographie, la Physiologie, et l'Organogénie des Végétaux.* Paris: Imprimerie Royale Fortin Masson.

Geddes, Patrick, and P. Chalmers Mitchell. 1911. "Morphology." *Encyclopaedia Britannica.* 11th ed. Cambridge: Cambridge University Press.

Ghiselin, Michael T. 1974. "A Radical Solution to the Species Problem." *Systematic Zoology* 23 (4): 536–44.

Gibson, Abraham H. 2013. "Edward O. Wilson and the Organicist Tradition." *Journal of the History of Biology* 46 (4): 599–630.

Gibson, Abraham H., Christina L. Kwapich, and Martha Lang. 2012. "The Roots of Multilevel Selection: Concepts of Biological Individuality in the Early Twentieth Century." *History and Philosophy of the Life Sciences* 35 (4): 505–32.

Gilbert, Scott F. 2014. "Symbiosis as the Way of Eukaryotic Life: The Dependent Co-origination of the Body." *Journal of Biosciences* 39 (2): 201–9.

Gilbert, Scott F., and David Epel. 2015. *Ecological Developmental Biology.* Sunderland, MA: Sinauer.

Gilbert, Scott F., and Sahotra Sarkar. 2000. "Embracing Complexity: Organicism for the 21st Century." *Developmental Dynamics* 219 (1): 1–9.

Gilbert, Scott F., Jan Sapp, and Alfred I. Tauber. 2012. "A Symbiotic View of Life: We Have Never Been Individuals." *Quarterly Review of Biology* 87 (4): 1–17.

Gill, Douglas E., Lin Chao, Susan L. Perkins, and Jason B. Wolf. 1995. "Genetic Mosaicism in Plants and Clonal Animals." *Annual Review of Ecology and Systematics* 26: 423–44.

Gillett, Carl. 2010. "Moving beyond the Subset Model of Realization: The Problem of Qualitative Distinctness in the Metaphysics of Science." *Synthese* 177 (2): 165–92.

Godfrey-Smith, Peter. 2009. *Darwinian Populations and Natural Selection.* Oxford: Oxford University Press.

———. 2011. "Darwinian Populations and Transitions in Individuality." In *The Major Transitions in Evolution Revisited,* edited by Brett Calcott and Kim Sterelny, 65–81. Cambridge, MA: MIT Press.

———. 2013. "Darwinian Individuals." In *From Groups to Individuals: Evolution and Emerging Individuality,* edited by Frédéric Bouchard and Philippe Huneman, 17–36. Cambridge, MA: MIT Press.

———. 2015. "Reproduction, Symbiosis, and the Eukaryotic Cell." *Proceedings of the National Academy of Sciences USA* 112 (33): 10120–25.

Goodnight, Charles J. 2013. "Defining the Individual." In *From Groups to Individuals: Evolution and Emerging Individuality,* edited by Frédéric Bouchard and Philippe Huneman, 37–53. Cambridge, MA: MIT Press.

Gould, Stephen Jay. 1983. "The Hardening of the Modern Synthesis." In *Dimensions of Darwinism,* edited by Marjorie Grene, 71–93. Cambridge: Cambridge University Press.

Gould, Stephen J., and Elisabeth A. Lloyd. 1999. "Individuality and Adaptation across Levels of Selection: How Shall We Name and Generalize the Unit of Darwinism?" *Proceedings of the National Academy of Sciences USA* 96 (21): 11904–9.

Grafen, Alan. 2006. "Optimization of Inclusive Fitness." *Journal of Theoretical Biology* 238 (3): 541–63.

Griesemer, James. 2000. "Development, Culture, and the Units of Inheritance." *Philosophy of Science* 67:S348–68.

———. 2001. "The Units of Evolutionary Transition." *Selection* 1 (1): 67–80.

———. 2002. "Limits of Reproduction: A Reductionistic Research Strategy in Evolutionary Biology." In *Promises and Limits of Reductionism in the Biomedical Sciences*, edited by Marc H. V. Van Regenmortel and David L. Hull, 211–32. Chichester: John Wiley.

———. 2014. "Reproduction and the Scaffolded Development of Hybrids." In *Developing Scaffolds in Evolution, Culture, and Cognition*, edited by Linnda R. Caporael, James R. Griesemer, and William C. Wimsatt, 23–55. Cambridge, MA: MIT Press.

Griesemer, James R., Linnda R. Caporael, and William C Wimsatt. 2014. "Developing Scaffolds: An Epilogue." In *Developing Scaffolds in Evolution, Culture, and Cognition*, edited by Linnda R. Caporael, James R. Griesemer, and William C. Wimsatt, 363–88. Cambridge, MA: MIT Press.

Grosberg, Richard K., and Richard R. Strathmann. 1998. "One Cell, Two Cell, Red Cell, Blue Cell: The Persistence of a Unicellular Stage in Multicellular Life Histories." *Trends in Ecology & Evolution* 13 (3): 112–16.

Guay, Alexandre, and Thomas Pradeu. 2016. "To Be Continued: The Genidentity of Physical and Biological Processes." In *Individuality Across the Sciences*, edited by Alexandre Guay and Thomas Pradeu, 317–47. Oxford: Oxford University Press.

Haber, Matt. 2013. "Colonies Are Individuals: Revisiting the Superorganism Revival." In *From Groups to Individuals: Evolution and Emerging Individuality*, edited by Frédéric Bouchard and Philippe Huneman, 195–216. Cambridge, MA: MIT Press.

Haeckel, Ernst. 1866. *Die Generelle Morphologie der Organismen: Allgemeine Grundzüge der organischen Formen-Wissenschaft, mechanisch begründet durch die von Charles Darwin reformierte Deszendenz-Theorie.* 2 vols. Berlin: Georg Reimer.

Hall, Brian K. 2003a. "Unlocking the Black Box between Genotype and Phenotype: Cell Condensations as Morphogenetic (Modular) Units." *Biology & Philosophy* 18 (2): 219–47.

———. 2003b. "Evo-Devo: Evolutionary Developmental Mechanisms." *International Journal of Developmental Biology* 47 (7/8): 491–96.

Hall, Brian K., and Wendy M. Olson, eds. 2007. *Keywords and Concepts in Evolutionary Developmental Biology.* New Delhi: Discovery.

Hamilton, Andrew, and Jennifer Fewell. 2013. "Groups, Individuals, and the Emergence of Sociality: The Case of Division of Labor." In *From Groups to Individuals: Evolution and Emerging Individuality*, edited by Frédéric Bouchard and Philippe Huneman, 175–94. Cambridge, MA: MIT Press.

Hamilton, Andrew, Nathan Smith, and Matthew Haber. 2009. "Social Insects and the Individuality Thesis: Cohesion and the Colony as a Selectable Individual." In *Organization of Insect Societies: From Genome to Sociocomplexity*, edited by Juergen Gadau and Jennifer Fewell. Cambridge, MA: Harvard University Press.

Hamilton, N. R. S., B. Schmid, and J. L. Harper. 1987. "Life-History Concepts and the Population Biology of Clonal Organisms." *Proceedings of the Royal Society B* 232 (1266): 35–57.

Han, Jing-Dong J., Nicolas Bertin, Tong Hao, Debra S. Goldberg, Gabriel F. Berriz, Lan V. Zhang, Denis Dupuy, et al. 2004. "Evidence for Dynamically Organized Modularity in the Yeast Protein–Protein Interaction Network." *Nature* 430 (6995): 88–93.

Harper, John L. 1977. *Population Biology of Plants.* London: Academic Press.

Harwood, Jonathan. 1994. "Metaphysical Foundations of the Evolutionary Synthesis: A Historiographical Note." *Journal of the History of Biology* 27 (1): 1–20.

Havstad, Joyce C., Leandro Assis, and Olivier Rieppel. 2015. "The Semaphorontic View of Homology." *Journal of Experimental Zoology Part B: Molecular and Developmental Evolution* 9999: 1–10.

Herron, Matthew D., Armin Rashidi, Deborah E. Shelton, and William W. Driscoll. 2013. "Cellular Differentiation and Individuality in the 'Minor' Multicellular Taxa." *Biological Reviews* 88 (4): 844–61.

Hofmeister, Wilhelm. 1862. *On the Germination, Development, and Fructification of the Higher Cryptogamia, and on the Fructification of the Coniferæ.* Translated by Frederick Currey. London: Ray Society.

Hölldobler, Bert, and Edward O. Wilson. 2009. *The Superorganism: The Beauty, Elegance, and Strangeness of Insect Societies.* New York: Norton.

Huang, Sui. 2004. "Back to the Biology in Systems Biology: What Can We Learn from Biomolecular Networks?" *Briefings in Functional Genomics & Proteomics* 2 (4): 279–97.

———. 2009. "Non-genetic Heterogeneity of Cells in Development: More than Just Noise." *Development* 136 (23): 3853–62.

Hughes, Roger N. 2005. "Lessons in Modularity: The Evolutionary Ecology of Colonial Invertebrates." *Scientia Marina* 69: 169–79.

Hull, David L. 1978. "A Matter of Individuality." *Philosophy of Science* 45 (3): 335–60.

———. 1980. "Individuality and Selection." *Annual Review of Ecology and Systematics* 11 (1): 311–32.

———. 1992. "Individual." In *Keywords in Evolutionary Biology*, edited by Evelyn Fox Keller and Elizabeth A. Lloyd, 180–87. Cambridge, MA: Harvard University Press.

Huneman, Philippe. 2014a. "Individuality as a Theoretical Scheme. I. Formal and Material Concepts of Individuality." *Biological Theory* 9 (4): 361–73.

———. 2014b. "Individuality as a Theoretical Scheme. II. About the Weak Individuality of Organisms and Ecosystems." *Biological Theory* 9 (4): 374–81.

Huxley, Julian S. 1912. *The Individual in the Animal Kingdom.* London: Cambridge University Press.

———. 1926. "The Biological Basis of Individuality." *Journal of Philosophical Studies* 1 (3): 305–19.

———. 1949. "Individuality." *Encyclopaedia Britannica.* 12th ed. Chicago: Encyclopaedia Britannica.

Huxley, Thomas Henry. 1852. "Upon Animal Individuality." *Proceedings of the Royal Institution of London* (1851–1854) 1: 184–89.

Jackson, Jeremy B. C., Leo W. Buss, and Robert E. Cook, eds. 1986. *Population Biology and Evolution of Clonal Organisms.* New Haven: Yale University Press.

Jahns, H. M., and S. Ott. 1997. "Life Strategies in Lichens—Some General Considerations." *Bibliotheca Lichenologica* 67: 49–68.

Janzen, Daniel H. 1977. "What Are Dandelions and Aphids?" *American Naturalist* 111 (979): 586–89.

Jeuken, M. 1952. "The Concept 'Individual' in Biology." *Acta Biotheoretica* 10 (1): 57–86.

Keeling, Patrick J., and Jeffrey D. Palmer. 2008. "Horizontal Gene Transfer in Eukaryotic Evolution." *Nature Reviews Genetics* 9 (8): 605–18.

Keller, Evelyn Fox, and Margaret S. Ewing. 1993. "The Kinds of 'Individuals' One Finds in Evolutionary Biology." In *Evolutionary Ethics*, edited by Matthew H. Nitecki and Doris V. Nitecki, 349–57. Albany: State University of New York Press.

Kellert, Stephen H., Helen E. Longino, and C. Kenneth Waters. 2006. "Introduction: The Pluralist Stance." In *Scientific Pluralism*, edited by Stephen H. Kellert, Helen E. Longino, and C. Kenneth Waters, vii–xxix. Minnesota Studies in the Philosophy of Science 19. Minneapolis: University of Minnesota Press.

Kleiner, Manuel, Jillian M. Petersen, and Nicole Dubilier. 2012. "Convergent and Divergent Evolution of Metabolism in Sulfur-Oxidizing Symbionts and the Role of Horizontal Gene Transfer." *Current Opinion in Microbiology* 15 (5): 621–31.

Kleisner, K. 2007. "The Formation of the Theory of Homology in Biological Sciences." *Acta Biotheoretica* 55 (4): 317–40.

Klingenberg, Christian P. 2008. "Morphological Integration and Developmental Modularity." *Annual Review of Ecology, Evolution, and Systematics* 39: 115–32.

Korn, Robert W. 1999. "Biological Organization—A New Look at an Old Problem." *BioScience* 49 (1): 51–57.

———. 2002. "Biological Hierarchies, Their Birth, Death and Evolution by Natural Selection." *Biology & Philosophy* 17 (2): 199–221.

Kovaka, Karen. 2015. "Biological Individuality and Scientific Practice." *Philosophy of Science* 82 (5): 1092–103.

Krakauer, David, and Paolo Zanotto. 2009. "Viral Individuality and Limitations of the Life Concept." In *Protocells: Bridging Nonliving and Living Matter*, edited by Steen Rasmussen, 513–36. Cambridge, MA: MIT Press.

Leigh, E. G., Jr. 2010. "The Group Selection Controversy." *Journal of Evolutionary Biology* 23 (1): 6–19.

Leuckart, Rudolf. 1851. *Ueber den Polymorphismus der Individuen oder die Erscheinungen der Arbeitstheilung in der Natur. Ein Beitrag zur Lehre vom Generationswechsel.* Giessen: Ricker.

Lewontin, Richard C. 1970. "The Units of Selection." *Annual Review of Ecology and Systematics* 1: 1–18.

Lloyd, Elizabeth. 2012. "Units and Levels of Selection." *Stanford Encyclopedia of Philosophy.* http://plato.stanford.edu/archives/win2012/entries/selection-units/.

Lockwood, Jeffrey A. 2012. "Species Are Processes: A Solution to the 'Species Problem' via an Extension of Ulanowicz's Ecological Metaphysics." *Axiomathes* 22 (2): 231–60.

Loeb, Leo. 1921. "Transplantation and Individuality." *Biological Bulletin* 40 (3): 143–80.

———. 1937. "The Biological Basis of Individuality." *Science* 86 (2218): 1–5.

Love, Alan C. 2008. "Explaining Evolutionary Innovations and Novelties: Criteria of Explanatory Adequacy and Epistemological Prerequisites." *Philosophy of Science* 75 (5): 874–86.

———. 2014. "The Erotetic Organization of Developmental Biology." In *Towards a Theory of Development*, edited by Alessandro Minelli and Thomas Pradeu, 33–55. Oxford: Oxford University Press.

———. 2015a. "Developmental Biology." *Stanford Encyclopedia of Philosophy.* http://plato .stanford.edu/archives/fall2015/entries/biology-developmental/.

———. 2015b. "ChINs, Swarms, and Variational Modalities: Concepts in the Service of an Evolutionary Research Program." *Biology & Philosophy* 30 (6): 873–88.

———. Forthcoming. "Individuation, Individuality, and Experimental Practice in Developmental Biology." In *Individuation across Experimental and Theoretical Sciences*, edited by O. Bueno, R.-L. Chen, and M. B. Fagan, XXX–XXX. Oxford: Oxford University Press.

Ma, Linda, Boya Song, Thomas Curran, Nhu Phong, Emilie Dressaire, and Marcus Roper. 2016.

"Defining Individual Size in the Model Filamentous Fungus Neurospora Crassa." *Proceedings of the Royal Society B* 283: 20152470.

Ma, Wenzhe, Ala Trusina, Hana El-Samad, Wendell A. Lim, and Chao Tang. 2009. "Defining Network Topologies That Can Achieve Biochemical Adaptation." *Cell* 138 (4): 760–73.

Maienschein, Jane. 2011. "'Organization' as Setting Boundaries of Individual Development." *Biological Theory* 6 (1): 73–79.

Margulis, Lynn. 1993. "Origins of Species: Acquired Genomes and Individuality." *Biosystems* 31 (2–3): 121–25.

Margulis, Lynn, and Ricardo Guerrero. 1991. "Two plus Three Equal One." In *Gaia 2—Emergence: The New Science of Becoming*, edited by William I. Thompson, 50–67. Hudson, NY: Lindisfarne.

Martins, Bruno M. C., and James C. W. Locke. 2015. "Microbial Individuality: How Single-Cell Heterogeneity Enables Population Level Strategies." *Current Opinion in Microbiology* 24: 104–12.

Maynard Smith, John, and Eörs Szathmáry. 1995. *The Major Transitions in Evolution.* Oxford: Oxford University Press.

McFall-Ngai, Margaret, Michael G. Hadfield, Thomas C. G. Bosch, Hannah V. Carey, Tomislav Domazet-Lošo, Angela E. Douglas, Nicole Dubilier, et al. 2013. "Animals in a Bacterial World, a New Imperative for the Life Sciences." *Proceedings of the National Academy of Sciences USA* 110 (9): 3229–36.

McShea, D. W., and D. Anderson. 2005. "The Remodularization of the Organism." In *Modularity: Understanding the Development and Evolution of Natural Complex Systems*, edited by W. Callebaut and D. Rasskin-Gutman, 185–205. Cambridge, MA: MIT Press.

McShea, Daniel W., and Edward P. Venit. 2001. "What Is a Part?" In *The Character Concept in Evolutionary Biology*, edited by Günter P. Wagner, 259–84. San Diego: Academic.

Medawar, P. B. 1957. "The Uniqueness of the Individual." In *The Uniqueness of the Individual*, 143–85. New York: Basic Books.

Metchnikoff, Élie. 1905. *Immunity in Infective Diseases.* Translated by Francis G. Binnie. Cambridge: Cambridge University Press.

Michod, Richard E. 1999. *Darwinian Dynamics: Evolutionary Transitions in Fitness and Individuality.* Princeton: Princeton University Press.

———. 2007. "Evolution of Individuality during the Transition from Unicellular to Multicellular Life." *Proceedings of the National Academy of Sciences USA* 104: 8613–18.

———. 2011. "Evolutionary Transitions in Individuality: Multicellularity and Sex." In *The Major Transitions in Evolution Revisited*, edited by Brett Calcott and Kim Sterelny, 169–97. Cambridge, MA: MIT Press.

Michod, Richard E., and Matthew D. Herron. 2006. "Cooperation and Conflict during Evolutionary Transitions in Individuality." *Journal of Evolutionary Biology* 19 (5): 1406–9.

Michod, Richard E., and Aurora M. Nedelcu. 2003. "On the Reorganization of Fitness during Evolutionary Transitions in Individuality." *Integrative and Comparative Biology* 43 (1): 64–73.

Michod, Richard E., and Denis Roze. 2001. "Cooperation and Conflict in the Evolution of Multicellularity." *Heredity* 86 (1): 1–7.

Milam, Erika L. 2010. *Looking for a Few Good Males: Female Choice in Evolutionary Biology.* Baltimore: Johns Hopkins University Press.

Milne-Edwards, Henri. 1827. "Organisation." In *Dictionnaire classique d'histoire naturelle*, edited by Isidore Audouin et al., 12: 332–44. Paris: Baudouin.

Minelli, Alessandro. 2011. "Animal Development, an Open-Ended Segment of Life." *Biological Theory* 6 (1): 4–15.

———. 2014. "Developmental Disparity." In *Towards a Theory of Development*, edited by Alessandro Minelli and Thomas Pradeu, 227–45. Oxford: Oxford University Press.

———. 2015. "Grand Challenges in Evolutionary Developmental Biology." *Frontiers in Ecology and Evolution* 2: 1–11.

Minelli, Alessandro, and Thomas Pradeu. 2014. "Theories of Development in Biology–Problems and Perspectives." In *Towards a Theory of Development*, edited by Alessandro Minelli and Thomas Pradeu, 1–14. Oxford: Oxford University Press.

Mira, Alex, Ana B. Martín-Cuadrado, Giuseppe D'Auria, and Francisco Rodríguez-Valera. 2010. "The Bacterial Pan-Genome: A New Paradigm in Microbiology." *International Microbiology* 13 (2): 45–57.

Montgomery, Edmund. 1880. "II.-The Unity of the Organic Individual." *Mind* 5 (19): 318–36.

Moreno, Alvaro, and Matteo Mossio. 2015a. "Constraints and Organisational Closure." In *Biological Autonomy: A Philosophical and Theoretical Enquiry*, edited by Alvaro Moreno and Matteo Mossio, 1–38. Dordrecht: Springer.

———. 2015b. "Organisms and Levels of Autonomy." In *Biological Autonomy: A Philosophical and Theoretical Enquiry*, edited by Alvaro Moreno and Matteo Mossio, 141–65. History, Philosophy and Theory of the Life Sciences 12. Dordrecht: Springer.

Mumford, Edward P. 1925. "Some Remarks on the Conception of Individuality in Biology." *Science Progress in the Twentieth Century (1919–1933)* 20 (77): 83–91.

Nachtomy, Ohad. 2007. "Leibniz on Nested Individuals." *British Journal for the History of Philosophy* 15 (4): 709–28.

Nersessian, Nancy J. 2006. "The Cognitive-Cultural Systems of the Research Laboratory." *Organization Studies* 27 (1): 125–45.

Newell, Allen, and Herbert A. Simon. 1972. *Human Problem Solving*. Englewood Cliffs, NJ: Prentice-Hall.

Nicholson, Daniel J. 2014. "The Return of the Organism as a Fundamental Explanatory Concept in Biology." *Philosophy Compass* 9 (5): 347–59.

Nicholson, Daniel J., and Richard Gawne. 2015. "Neither Logical Empiricism nor Vitalism, but Organicism: What the Philosophy of Biology Was." *History and Philosophy of the Life Sciences* 37 (4): 345–81.

Niklas, Karl J., and Ulrich Kutschera. 2014. "Amphimixis and the Individual in Evolving Populations: Does Weismann's Doctrine Apply to All, Most or a Few Organisms?" *Naturwissenschaften* 101 (5): 357–72.

Nyhart, Lynn K. 1995. *Biology Takes Form: Animal Morphology and the German Universities, 1800–1900*. Chicago: University of Chicago Press.

Nyhart, Lynn K., and Scott Lidgard. 2011. "Individuals at the Center of Biology: Rudolf Leuckart's *Polymorphismus der Individuen* and the Ongoing Narrative of Parts and Wholes. With an Annotated Translation." *Journal of the History of Biology* 44: 373–443.

Okasha, Samir. 2006. *Evolution and the Levels of Selection*. Oxford: Oxford University Press.

O'Malley, Maureen A. 2012. "Evolutionary Systems Biology: Historical and Philosophical Perspectives on an Emerging Synthesis." In *Evolutionary Systems Biology*, edited by Orkun S. Soyer, 1–28. Advances in Experimental Medicine and Biology 751. New York: Springer.

———. 2014. *Philosophy of Microbiology*. Cambridge: Cambridge University Press.

———. 2015. "Endosymbiosis and Its Implications for Evolutionary Theory." *Proceedings of the National Academy of Sciences USA* 112 (33): 1270–77.

O'Malley, Maureen A., and J. Dupré. 2005. "Fundamental Issues in Systems Biology." *Bioessays* 27 (12): 1270–76.

O'Malley, Maureen A., Ingo Brigandt, Alan C. Love, John W. Crawford, Jack A. Gilbert, Rob Knight, Sandra D. Mitchell, and Forest Rohwer. 2014. "Multilevel Research Strategies and Biological Systems." *Philosophy of Science* 81 (5): 811–28.

Owen, Richard. 1849. *On Parthenogenesis; Or, The Successive Production of Procreating Individuals from a Single Ovum: A Discourse Introductory to the Hunterian Lectures on Generation and Development, for the Year 1849, Delivered at the Royal College of Surgeons of England.* London: J. Van Voorst.

———. 1851. "Professor Owen on Metamorphosis and Metagenesis. Being Abstract of a Lecture Delivered by Him to the Royal Institute of Great Britain in February 1851." *Edinburgh New Philosophical Journal* 50: 268–78.

Pearce, Trevor. 2014. "The Origins and Development of the Idea of Organism-Environment Interaction." In *Entangled Life*, edited by Gillian Barker, Eric Desjardins, and Trevor Pearce, 13–32. History, Philosophy and Theory of the Life Sciences 4. Dordrecht: Springer.

Pepper, John W., and Matthew D. Herron. 2008. "Does Biology Need an Organism Concept?" *Biological Reviews* 83 (4): 621–27.

Pereira-Leal, Jose B., Emmanuel D. Levy, and Sarah A. Teichmann. 2006. "The Origins and Evolution of Functional Modules: Lessons from Protein Complexes." *Philosophical Transactions of the Royal Society B: Biological Sciences* 361 (1467): 507–17.

Polz, Martin F., Eric J. Alm, and William P. Hanage. 2013. "Horizontal Gene Transfer and the Evolution of Bacterial and Archaeal Population Structure." *Trends in Genetics* 29 (3): 170–75.

Potochnik, Angela, and Brian McGill. 2012. "The Limitations of Hierarchical Organization." *Philosophy of Science* 79 (1): 120–40.

Pradeu, Thomas. 2011. "A Mixed Self: The Role of Symbiosis in Development." *Biological Theory* 6 (1): 80–88.

———. 2012. *The Limits of the Self: Immunology and Biological Identity.* Oxford: Oxford University Press.

———. 2013. "Immunity and the Emergence of Individuality." In *From Groups to Individuals: Evolution and Emerging Individuality*, edited by Frédéric Bouchard and Philippe Huneman, 77–96. Cambridge, MA: MIT Press.

Proulx, Stephen R., Daniel E. L. Promislow, and Patrick C. Phillips. 2005. "Network Thinking in Ecology and Evolution." *Trends in Ecology & Evolution* 20 (6): 345–53.

Queller, David C., and Joan E. Strassmann. 2009. "Beyond Society: The Evolution of Organismality." *Philosophical Transactions of the Royal Society B: Biological Sciences* 364 (1533): 3143–55.

Rainey, Paul B., and Benjamin Kerr. 2011. "Darwinian Populations and Transitions in Individuality." In *The Major Transitions in Evolution Revisited*, edited by Brett Calcott and Kim Sterelny, 141–62. Cambridge, MA: MIT Press.

Rescher, Nicholas. 2007. "The Promise of Process Philosophy." In *Columbia Companion to Twentieth-Century Philosophies*, edited by Constantin V. Boundas, 143–55. New York: Columbia University Press.

Rieppel, Olivier. 1994. "Homology, Topology, and Typology: The History of Modern Debates." In *Homology: The Hierarchical Basis of Comparative Biology*, edited by Brian K. Hall, 64–101. San Diego: Academic Press.

———. 2007. "Species: Kinds of Individuals or Individuals of a Kind." *Cladistics* 23 (4): 373–84.

———. 2009. "Species as a Process." *Acta Biotheoretica* 57 (1–2): 33–49.

Rinkevich, Baruch. 2011. "Quo Vadis Chimerism?" *Chimerism* 2 (1): 1–5.

Royce, Josiah. 1901. "Individual." In *Dictionary of Philosophy and Psychology*, edited by James M. Baldwin, 534–37. London: Macmillan.

Sanders, William B. 2001. "Lichens: The Interface between Mycology and Plant Morphology." *BioScience* 51 (12): 1025–35.

Santelices, Bernabé. 1999. "How Many Kinds of Individual Are There?" *Trends in Ecology & Evolution* 14 (4): 152–55.

———. 2004. "Mosaicism and Chimerism as Components of Intraorganismal Genetic Heterogeneity." *Journal of Evolutionary Biology* 17 (6): 1187–88.

Sapp, Jan. 1994. *Evolution by Association: A History of Symbiosis*. New York: Oxford University Press.

Schelling, Friedrich Wilhelm Joseph. 2004. *First Outline of a System of the Philosophy of Nature* (1799), translated by Keith R. Peterson. Albany: State University of New York Press.

Schneider, Gregory W., and Russell Winslow. 2014. "Parts and Wholes: The Human Microbiome, Ecological Ontology, and the Challenges of Community." *Perspectives in Biology and Medicine* 57 (2): 208–23.

Schwarz, Astrid. 2011. "History of Concepts for Ecology." In *Ecology Revisited*, edited by Astrid Schwarz and Kurt Jax, 19–28. Dordrecht: Springer.

Schweinsberg, Maximilian, Linda C. Weiss, Sebastian Striewski, Ralph Tollrian, and Kathrin P. Lampert. 2015. "More than One Genotype: How Common Is Intracolonial Genetic Variability in Scleractinian Corals?" *Molecular Ecology* 24 (11): 2673–85.

Scrosati, Ricardo. 2002. "An Updated Definition of Genet Applicable to Clonal Seaweeds, Bryophytes, and Vascular Plants." *Basic and Applied Ecology* 3 (2): 97–99.

Secord, James A. 2004. "Knowledge in Transit." *Isis* 95 (4): 654–72.

Seibt, Johanna. 2013. "Process Philosophy." *Stanford Encyclopedia of Philosophy*. http://plato.stanford.edu /archives/fall2013/entries/process-philosophy/.

Siegal, Mark L. 2014. "Systems Biology." In *Oxford Bibliographies in Evolutionary Biology*, edited by Jonathan Losos. New York: Oxford University Press. http://www.oxfordbibliographies.com /view/document/obo-9780199941728/obo-9780199941728–0010.xml?rskey=y56vlU &result=58&print.

Simon, Herbert A. 1973. "The Structure of Ill Structured Problems." *Artificial Intelligence* 4: 181–201.

Smith, Myron L., Johann N. Bruhn, and James B. Anderson. 1992. "The Fungus *Armillaria bulbosa* Is among the Largest and Oldest Living Organisms." *Nature* 356 (6368): 428–31.

Sober, E. 1991. "Organisms, Individuals, and Units of Selection." In *Organism and the Origins of Self*, edited by A. Tauber, 275–96. Dordrecht: Kluwer.

Spencer, Herbert. 1864. *The Principles of Biology*. Vol. 1. London: Williams and Norgate.

Steenstrup, Johannes Japetus Smith. 1845. *On the Alternation of Generations; or the Propagation and Development of Animals through Alternate Generations: A Peculiar Form of Fostering the Young in the Lower Classes of Animals*. Translated by George Busk. London: Ray Society.

Sterner, Beckett. 2015. "Pathways to Pluralism about Biological Individuality." *Biology & Philosophy* 30 (5): 609–28.

Strassmann, Joan E., and David C. Queller. 2010. "The Social Organism: Congresses, Parties, and Committees." *Evolution* 64 (3): 605–16.

Surman, Jan, Katalin Stráner, and Peter Haslinger. 2014. "Nomadic Concepts in the History of

Biology." *Studies in History and Philosophy of Science Part C: Studies in History and Philosophy of Biological and Biomedical Sciences* 48, Part B: 127–29.

Sutton, Walter S. 1902. "On the Morphology of the Chromosome Group in *Brachystola magna.*" *Biological Bulletin* 4 (1): 24–39.

Tauber, Alfred I. 1997. *The Immune Self: Theory or Metaphor?* Cambridge: Cambridge University Press.

———. 2015a. "The Biological Notion of Self and Non-self." *Stanford Encyclopedia of Philosophy.* http://stanford.library.usyd.edu.au /entries / biology-self /.

———. 2015b. "Reconceiving Autoimmunity: An Overview." *Journal of Theoretical Biology* 375: 52–60.

Thomson, John A. 1920. "The Criteria of Livingness. In *The System of Animate Nature: The Gifford Lectures Delivered in the University of St. Andrews in the Years 1915 and 1916,* 1: 79–106. London: Williams & Norgate.

Todd, Charles. 1930. "Cellular Individuality in the Higher Animals, with Special Reference to the Individuality of the Red Blood Corpuscle." *Proceedings of the Royal Society of London B* 106 (741): 20–44.

Todd, N. K., and A. D. M. Rayner. 1980. "Fungal Individualism." *Science Progress* 66: 331–54.

Trewavas, A. 2006. "A Brief History of Systems Biology." *Plant Cell* 18 (10): 2420–30.

Tuomi, J., and T. Vuorisalo. 1989. "Hierarchical Selection in Modular Organisms." *Trends in Ecology & Evolution* 4 (7): 209–13.

Turner, Scott. 2013. "Super Organisms and Superindividuality: The Emergence of Individuality in a Social Insect Assemblage." In *From Groups to Individuals: Evolution and Emerging Individuality,* edited by F. Bouchard and P. Huneman, 219–42. Cambridge, MA: MIT Press.

Van Gestel, Jordi, Hera Vlamakis, and Roberto Kolter. 2015. "Division of Labor in Biofilms: The Ecology of Cell Differentiation." In *Microbial Biofilms,* 2nd ed., edited by Pranab K. Mukherjee, Mahmoud Ghannoum, Marvin Whiteley, and Matthew Parsek, 67–97. Washington, DC: American Society of Microbiology.

Van Valen, Leigh M. 1989. "Three Paradigms of Evolution." *Evolutionary Theory* 9 (2): 1–17.

Verworn, Max. 1899. *General Physiology; an Outline of the Science of Life.* Translated by Frederic S. Lee. 2nd ed. London: Macmillan.

Vliet, Simon van, and Martin Ackermann. 2015. "Bacterial Ventures into Multicellularity: Collectivism through Individuality." *PLOS Biology* 13 (6): e1002162.

Voreades, Noah, Anne Kozil, and Tiffany L. Weir. 2014. "Diet and the Development of the Human Intestinal Microbiome." *Frontiers in Microbiology* 5 (494): 1–9.

Vuorisalo, Timo O., and Pia K. Mutikainen. 1999. "Modularity and Plant Life Histories." In *Life History Evolution in Plants,* edited by Timo O. Vuorisalo and Pia K. Mutikainen, 1–25. Dordrecht: Kluwer.

Vuorisalo, T., and J. Tuomi. 1986. "Unitary and Modular Organisms—Criteria for Ecological Division." *Oikos* 47 (3): 382–85.

Wagner, Günter P. 1989a. "The Biological Homology Concept." *Annual Review of Ecology and Systematics* 20 (1): 51–69.

———. 1989b. "The Origin of Morphological Characters and the Biological Basis of Homology." *Evolution* 43 (6): 1157–71.

———. 2014. *Homology, Genes, and Evolutionary Innovation.* Princeton: Princeton University Press.

Wagner, Günter P., Mihaela Pavlicev, and James M. Cheverud. 2007. "The Road to Modularity." *Nature Reviews Genetics* 8 (12): 921–31.

Waters, C. Kenneth. 2014. "Shifting Attention from Theory to Practice in Philosophy of Biology." In *New Directions in the Philosophy of Science*, edited by Maria Carla Galavotti, Dennis Dieks, Wenceslao J. Gonzalez, Stephan Hartmann, Thomas Uebel, and Marcel Weber, 121–39. The Philosophy of Science in a European Perspective 5. Cham: Springer.

Weismann, August. 1889. "The Continuity of the Germ-Plasm as the Foundation of a Theory of Heredity, 1885." In *Essays upon Heredity and Kindred Biological Problems*, edited by Edward B. Poulton, Selmar Schönland, and Arthur E. Shipley, translated by Selmar Schönland, 1: 163–254. Oxford: Clarendon Press.

———. 1893. *The Germ-Plasm: A Theory of Heredity*. New York: Charles Scribner's Sons.

West, Stuart A., and E. Toby Kiers. 2009. "Evolution: What Is an Organism?" *Current Biology* 19 (23): R1080–82.

West, Stuart A., Roberta M. Fisher, Andy Gardner, and E. Toby Kiers. 2015. "Major Evolutionary Transitions in Individuality." *Proceedings of the National Academy of Sciences USA* 112 (33): 10112–19.

West, Stuart A., Ashleigh S. Griffin, Andy Gardner, and Stephen P. Diggle. 2006. "Social Evolution Theory for Microorganisms." *Nature Reviews Microbiology* 4 (8): 597–607.

Wheeler, Raymond H. 1936. "Organismic Logic in the History of Science." *Philosophy of Science* 3 (1): 26–61.

Wheeler, William M. 1911. "The Ant-Colony as an Organism." *Journal of Morphology* 22 (2): 307–25.

Williams, George C. 1992. *Natural Selection: Domains, Levels, and Challenges*. New York: Oxford University Press.

Williams, Raymond. 1985. *Keywords: A Vocabulary of Culture and Society*. New York: Oxford University Press.

Wilson, David Sloan, and Elliott Sober. 1989. "Reviving the Superorganism." *Journal of Theoretical Biology* 136 (3): 337–56.

Wilson, Jack. 1999. *Biological Individuality: The Identity and Persistence of Living Entities*. Cambridge: Cambridge University Press.

Wilson, Robert A. 2005. *Genes and the Agents of Life: The Individual in the Fragile Sciences*. Cambridge: Cambridge University Press.

Wilson, Robert A., and Matthew Barker. 2013. "The Biological Notion of Individual." *Stanford Encyclopedia of Philosophy*. http://plato.stanford.edu/archives/spr2013/entries/biology-individual/.

Wimsatt, William C. 1994. "The Ontology of Complex Systems: Levels, Perspectives, and Causal Thickets." In *Biology and Society: Reflections on Methodology*, 207–74. Canadian Journal of Philosophy, supplementary vol. 20.

———. 2007. *Re-Engineering Philosophy for Limited Beings*. Cambridge, MA: Harvard University Press.

Wimsatt, William C., and James R. Griesemer. 2007. "Reproducing Entrenchments to Scaffold Culture: The Central Role of Development in Cultural Evolution." In *Integrating Evolution and Development: From Theory to Practice*, edited by Roger Sansom and Robert N. Brandon, 227–323. Cambridge, MA: MIT Press.

Winsor, Mary P. 1976. *Starfish, Jellyfish, and the Order of Life: Issues in Nineteenth-Century Science*. New Haven: Yale University Press.

Winther, Rasmus G. 2001. "Varieties of Modules: Kinds, Levels, Origins, and Behaviors." *Journal of Experimental Zoology* 291 (2): 116–29.

———. 2005. "Evolutionary Developmental Biology Meets Levels of Selection: Modular Integration or Competition, or Both?" In *Modularity: Understanding the Development and Evolution of Natural Complex Systems*, edited by W. Callebaut and D. Rasskin-Gutman, 61–97. Cambridge, MA: MIT Press.

———. 2006. "Parts and Theories in Compositional Biology." *Biology & Philosophy* 21 (4): 471–99.

———. 2011. "Part-Whole Science." *Synthese* 178 (3): 397–427.

Wolfe, Charles T. 2010. "Do Organisms Have an Ontological Status?" *History and Philosophy of the Life Sciences* 32 (2–3): 195–231.

———. 2014. "The Organism as Ontological Go-between: Hybridity, Boundaries and Degrees of Reality in Its Conceptual History." *Studies in History and Philosophy of Science Part C: Studies in History and Philosophy of Biological and Biomedical Sciences* 48, Part B: 151–61.

Woyke, Tanja, Hanno Teeling, Natalia N. Ivanova, Marcel Huntemann, Michael Richter, Frank Oliver Gloeckner, Dario Boffelli, et al. 2006. "Symbiosis Insights through Metagenomic Analysis of a Microbial Consortium." *Nature* 443 (7114): 950–55.

Wuketits, Franz M. 1989. "Organisms, Vital Forces, and Machines: Classical Controversies and the Contemporary Discussion 'Reductionism vs. Holism.'" In *Reductionism and Systems Theory in the Life Sciences: Some Problems and Perspectives*, edited by Paul Heuningen-Huene and Franz M. Wuketits, 3–28. Dordrecht: Kluwer.

Zengler, Karsten. 2009. "Central Role of the Cell in Microbial Ecology." *Microbiology and Molecular Biology Reviews* 73 (4): 712–29.

Zook, Douglas. 2015. "Symbiosis—Evolution's Co-Author." In *Reticulate Evolution: Symbiogenesis, Lateral Gene Transfer, Hybridization and Infectious Heredity*, edited by Nathalie Gontier, 3: 41–80. Interdisciplinary Evolutionary Research. Cham: Springer.

Zylstra, Uko. 1992. "Living Things as Hierarchically Organized Structures." *Synthese* 91 (1–2): 111–33.

2

Cells, Colonies, and Clones: Individuality in the Volvocine Algae

MATTHEW D. HERRON

I saw a very many great round particles, of the bigness of a great corn of sand drive and move in the water . . . This was to me a very pleasant sight, because the said particles, as often as I did look on them, did neither lye still, and that their motion did proceed from their turning round; and that the more, because I did fancy at first that they were small animals, and the smaller these particles were, the greener was their colour.

VAN LEEUWENHOEK (1700, 511)

Introduction

Life is organized hierarchically, with genes in genomes, organelles within eukaryotic cells, cells within multicellular organisms, and organisms within societies. At various points in evolutionary history, the higher-level units in this hierarchy (collectives) must have emerged from interactions among the lower-level units (components). These transitions, for example from prokaryote to eukaryote and from unicellular to multicellular organisms, have led to entirely new kinds of biological individuals with new, emergent properties. Collectively they are known as "major transitions" (Maynard Smith and Szathmáry 1997) or "evolutionary transitions in individuality" (hereafter = ETIs; Michod and Roze 1997). In each case, groups are formed by previously existing individuals, either by coming together or by staying together (Tarnita et al. 2013), scenarios that Queller (1997) referred to as egalitarian and fraternal transitions, respectively. Over evolutionary time, such groups begin to take on characteristics of individuals, perhaps eventually emerging as individuals in their own right. If we conceive of ETIs as a process in which groups evolve new levels, or at least increased degrees, of individuality, it is worth thinking about what we mean by individuality.

One particularly common type of ETI is the transition from single-celled organisms to multicellular organisms, which in some cases have evolved cellular differentiation (i.e., multiple functionally differentiated cell types). Cellular differentiation is a prerequisite for large, complex body plans, as scaling laws dictate that large organisms will face problems not experienced by their

smaller relatives (Bonner 2004). Large motile organisms need specialized cells
to move (Kirk 1998), large sessile organisms need specialized structures to
stay put (Butterfield 2000; Rozhnov 2001), and any large organisms with substantial three-dimensional structure face problems of nutrient transport due
to a reduced surface area to volume ratio (Bonner 2004). In fact, the number
of cell types in a multicellular organism has frequently been used as a metric of complexity (e.g., McShea 1996; Bell and Mooers 1997; Bonner 2003).
Multicellularity has evolved independently in at least 25 separate lineages, including the Eubacteria, Archaea, and several lineages spanning the deepest
divergences within the eukaryotes, but only a handful of these lineages have
evolved cellular differentiation (Bonner 1998; Grosberg and Strathmann 1998;
Herron et al. 2013).

Although the origins of differentiated multicellularity have been few in
number, the consequences have been immense. Each of the major macroscopic groups had its origin in such a transition, and each has subsequently
diversified into thousands to millions of species. Aside from their own diversification, the multicellular red, green, and brown algae, land plants, animals,
and fungi have profoundly impacted the evolution of other lineages as well,
drastically altering their environments with often dramatic effects on their
evolutionary trajectories.

Here, I use the volvocine green algae as a case study to consider the emergence of a new level of individuality during the transition to multicellularity.
Although a distinction is sometimes made between organisms and individuals
(e.g., Queller and Strassmann 2009), I will treat these terms as synonyms in the
context of individual multicellular organisms. After briefly reviewing efforts to
define individuality, I introduce the volvocine algae as a model system, review
the evolutionary history of developmental changes in this group, and consider how these changes relate to the emergence of a new kind of individual.

What Is an Individual?

The deceptively simple question of what kinds of biological units should be
considered individuals in fact has a long and complex history. A number of
criteria have been proposed or implemented, and no universal agreement
exists. Various authors have presented structural, genetic, ecological, physiological, reproductive, functional, and behavioral criteria as decisive (for a
more comprehensive review, see Clarke 2010). Among the many criteria that
have been proposed for individuality, most can be classified into a few major
categories.

Santelices (1999) proposed a classification with three categories: genetic uniqueness (individuals are genetically distinct from others of the same species), genetic homogeneity (all of the parts of an individual are genetically identical or nearly so), and physiological autonomy. The presence or absence of these attributes defines eight possible combinations, most of which are represented on Earth. For example, "unitary" individuals, typified by vertebrates, possess all three attributes. Clonal reproducers, such as some plants and multicellular protists, are genetically homogeneous and physiologically autonomous but lack genetic uniqueness. The genetic criteria contribute to the conception of an individual as a unit of selection, while physiological autonomy principally relates to the ecological interactions of an individual with its environment and with other individuals.

Recent efforts to understand individuality have tended to focus on evolutionary factors, treating individuals as units of selection or units of adaptation. Michod and Nedelcu (2003) treat individuals as units of selection as determined by heritable variation in fitness. A division of labor between reproductive (germ) and vegetative (somatic) functions limits within-organism genetic variation, thereby preventing within-organism conflicts that might otherwise disrupt organismal function. By exporting fitness from the components to the collective, such a germ-soma distinction maintains the unity of the individual (Michod 1997; Michod et al. 2006).

In the view of Queller and Strassmann, individuals are characterized by high levels of cooperation and low levels of conflict among their component parts (Queller and Strassmann 2009; Strassmann and Queller 2010). High cooperation and low conflict indicate that organisms are highly functionally integrated and allow them to function as "bundles of adaptation" (Strassmann and Queller 2010, 605). The two factors are treated as independent and continuous, setting up a two-dimensional space with organisms occupying one quadrant. Germ-soma specialization and single-cell bottlenecks (Hamilton 1964), which for some other authors are decisive, are for Strassmann and Queller merely mechanisms that contribute to the crucial criteria of high within-organism cooperation and low within-organism conflict.

Folse and Roughgarden (2010) argue that the crucial criteria for individuality are alignment of fitness interests of the lower-level units so that little or no within-organism conflict occurs, interdependence of the parts due to germ-soma differentiation, and functional integration as evidence of adaptation. As in the view of Queller and Strassmann (2009; Strassmann and Queller 2010), genetic homogeneity and unicellular bottlenecks are ways of preventing conflict, but not necessarily the only ways. Adaptive functional integration at the organism level indicates that the organism is a unit of fitness. Folse and Roughgarden envision these traits arising in order during

ETIs: first high genetic relatedness among components, then a germ-soma division of labor, and finally adaptation at the new, emergent level.

In the framework of Godfrey-Smith (2009) and Clarke (2012), individuals are defined by their membership in Darwinian populations, those that are capable of adaptive evolution. At a minimum, such populations must possess Lewontin's (1970) criteria for evolution by natural selection: heritable variation in phenotypes that affect fitness. Other attributes are sometimes associated with the capacity for adaptive change, but it is the capacity itself that is central. Godfrey-Smith recognizes a continuum of Darwinian and Darwinian-like processes, from marginal cases that meet only the minimal criteria to "paradigm" cases that are capable of producing complex adaptations. Populations vary more or less continuously along several axes, including the amount of phenotypic variation present, the reliability of inheritance, the strength of intra-specific ecological interactions, the extent to which fitness depends on intrinsic features, and the smoothness of the adaptive landscape (i.e., the degree to which small changes in phenotype cause small changes in fitness). Populations that possess all of these features in high degree have the potential for sustained and complex adaptive change, while those with low degrees of one or more criteria are at best capable of less interesting evolutionary outcomes. Other criteria may also be important for particular kinds of organisms. For collective reproducers, those that are composed of parts that are themselves capable of reproduction, high degrees of individuality are characterized by development that includes a bottleneck, a germ-soma division of labor, and functional integration. High degrees of these criteria indicate that Darwinian processes will be more powerful, and thus adaptations will tend to occur, at the level of the collective rather than of its components.

The criteria of these various systems are largely overlapping, as for example genetic homogeneity (Santelices 1999) aligns fitness interests (Folse and Roughgarden 2010), which is likely to reduce within-individual conflict (Queller and Strassmann 2009; Strassmann and Queller 2010). Functional integration among components plays a role in all of them, either as evidence of adaptation at the level of the collective (Queller and Strassmann 2009; Folse and Roughgarden 2010; Strassmann and Queller 2010) or by contributing to the evolvability of the collective (Michod and Nedelcu 2003; Godfrey-Smith 2009). Finally, all of these concepts share a diachronic view of individuality (Okasha 2005), viewing individuality as a derived trait (Buss 1987) in the context of major transitions (Maynard Smith and Szathmáry 1997; or ETIs, Michod and Roze 1997).

With their focus on change over time, these recent syntheses lend themselves well to the diachronic ETI tradition, treating the levels of the biological hierarchy as outcomes rather than starting conditions of the evolutionary process (Buss 1987). In addition, the criteria they propose were derived without reference to any particular taxonomic group, a crucial advantage if we want to draw non-trivial conclusions about the evolution of individuality (Herron et al. 2013). For these reasons, the discussion that follows will privilege these recent, evolutionary views, while occasionally referencing more traditional criteria.

The Volvocine Algae

The independent origins of differentiated multicellularity are replicate experiments with the potential to reveal general principles involved in this transition. Unfortunately, for most lineages the evidence of the intermediate stages has been obscured by extinction and by an inadequate fossil record. In the volvocine green algae (Fig. 2.1), though, abundant evidence remains

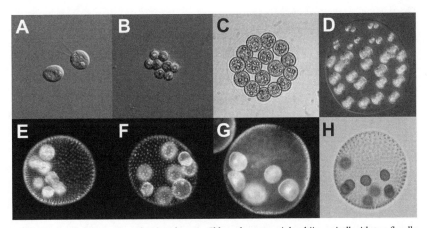

FIGURE 2.1. Representative volvocine algae. A: *Chlamydomonas reinhardtii*, a unicell with two flagella at the anterior. *C. reinhardtii* diverged from the multicellular volvocine algae around 250 million years ago. B: *Gonium pectorale*, a flat or slightly curved plate of 8 to 32 cells (8 in this example), all oriented in the same direction (photos A and B by Deborah Shelton). C: *Eudorina elegans*, a spheroid with up to 32 undifferentiated cells (16 in this example). D: *Pleodorina starrii*, a partially differentiated spheroid with up to 64 cells (32 in this example). The small cells near the anterior pole (top) are terminally differentiated somatic cells specialized for motility; the larger cells perform both reproductive and motility functions. E: *Volvox carteri*, a spheroid with ca. 2000 small somatic cells arranged at the periphery and a handful of much larger reproductive cells (gonidia). F: *Volvox tertius*, a spheroid with ~1000 small somatic cells arranged at the periphery and a handful of much larger reproductive cells. The germ cells in this colony have begun to develop into daughter colonies, and one is in the process of inversion. G: *Volvox barberi*, a spheroid with ca. 30,000 small somatic cells arranged at the periphery and a handful of much larger reproductive cells. The germ cells in this colony have begun to develop into daughter colonies, and some are in the process of inversion. H. *Volvox aureus*, a spheroid with up to ~2000 small somatic cells arranged at the periphery and a handful of much larger reproductive cells.

in the form of extant species with various mixtures of derived and ancestral traits.

The volvocine algae (Chlorophyceae: Chlamydomonadales) are an informal grouping of facultatively sexual, haploid, photosynthetic algae found mostly in freshwater. Following the taxonomy of Nozaki and Itoh (1994), I will take this grouping to include the families Volvocaceae (*Eudorina, Pandorina, Platydorina, Pleodorina, Volvox, Volvulina, Yamagishiella*), Goniaceae (*Astrephomene, Gonium*) and Tetrabaenaceae (*Basichlamys, Tetrabaena*) along with the few most closely related unicells in the genera *Chlamydomonas* and *Vitreochlamys*.

The ~50 species of volvocine algae exhibit nearly every conceivable intermediate between unicellular and fully differentiated multicellular life, including single-celled forms and multicellular forms with and without cellular differentiation. This diversity makes the volvocine algae a uniquely useful model system for understanding the evolution of multicellularity and cellular differentiation. I will refer to the members of the families Tetrabaenaceae, Goniaceae, and Volvocaceae as colonial, and to their physically discrete units as colonies; these same units are sometimes referred to as spheroids or coenobia (Cohn 1875). I intend this usage not to imply any particular degree of individuality, but only as a term of convenience.

Partitioning Individuality

The volvocine algae illustrate two of the classic problems plaguing discussions of individuality. First, a typical volvocine life cycle involves many rounds of asexual reproduction punctuated by occasional rounds of sexual reproduction (Fig. 2.2), and so many genetically identical colonies may descend from a single zygote. This is an instance of the ramet vs. genet problem: the genetically unique and (largest) genetically homogeneous units are the descendants of a given zygote (i.e., a genet; Sarukhán and Harper 1973), which may include a large number of physiologically discrete and autonomous colonies (i.e., ramets; Stout 1929). This is a much-discussed problem in plants and colonial invertebrates (e.g., White 1979; Cook 1980; Harper 1980; Tuomi and Vuorisalo 1989; Clarke 2012; Gorelick 2012), but it is relevant for any life cycle that includes both sexual and asexual phases. If we consider genetic homogeneity to be central to individuality, we should bear in mind that the clone or genet, not the colony, is the largest genetically homogeneous unit.

The second problem is whether we should consider a colony of a given species to be a group of individuals (the cells) or an individual in its own right. This question has deep historical roots, as Ehrenberg, contrary to

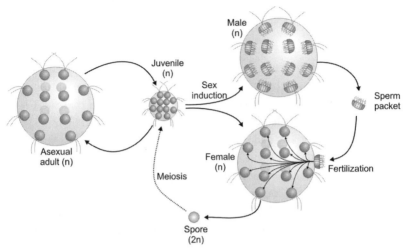

FIGURE 2.2. Example of a volvocine life cycle (based on that of *Eudorina*). Each cell in a haploid asexual spheroid undergoes a series of rapid cell divisions early in development, eventually producing a juvenile spheroid, which is eventually released from the mother spheroid. In species with cellular differentiation, only the reproductive cells divide. Juveniles escape from the parental spheroid possessing all of the cells they will have as adults; continued growth occurs by increases in cell size and in the volume of extracellular matrix rather than by cell division. The trigger to enter the sexual cycle varies among species, including low nutrient levels, heat shock, and chemical signals released by colonies. In isogamous species, cells differentiate into gametes of opposite mating types. In anisogamous species, cells differentiate into motile sperm packets or immotile eggs. In either case, fertilization results in a diploid zygote that eventually matures into a dormant, desiccation-resistant spore. Spores germinate through meiosis upon the return of optimal growth conditions. In some species, germination results in four viable offspring; in others only a single offspring and three polar bodies are produced.

van Leewenhoek (1700) and Linnaeus (1758), considered a *Volvox* spheroid a colony of hundreds or thousands of individuals (Ehrenberg 1832). Some form of this problem will be relevant for any ETI. For example, in eusocial insects the problem is whether to consider a colony a group of individual animals or a superorganism composed of numerous animals. Similarly, certain obligately endosymbiotic bacteria with extremely reduced genomes can be thought of as individuals in their own right or as organelles of the host cells (Andersson 2000).

Within the colonial volvocine algae, there are three kinds of biological units that can contend for at least a degree of individuality: the cells, the colonies, and the clones (or genets). Previous discussions have tended to ask (implicitly or explicitly), "Is a *Volvox* spheroid an individual?" or at least "How much of an individual is a *Volvox* spheroid?" I suggest that the relevant question is rather "How much individuality is present at each of the three (or more) levels?"

Cells and Colonies

The transition to differentiated multicellularity in the volvocine algae has been broken down into a series of developmental changes (Kirk 2005), each plausibly adaptive, as Darwin did for the vertebrate eye (Darwin 1872; see also Nilsson and Pelger 1994; Dawkins 1997). The history of these and other developmental changes has been reconstructed in a phylogenetic framework, showing that some traits had multiple, independent origins, as well as reversions from derived back to ancestral states (Herron and Michod 2008; Herron 2009; Herron et al. 2010). Timing of developmental changes was inferred from a fossil-calibrated molecular clock analysis (Herron et al. 2009).

Here I review the evolutionary history of developmental changes in the volvocine algae and consider how these changes relate to various concepts of individuality. In this section, I propose adaptive explanations, or at least adaptive values, for several of the derived traits of volvocine algae. Bearing in mind the pitfalls of inferring adaptive origins of traits (Gould and Lewontin 1979), these speculations should be considered only as plausible adaptive hypotheses.

Volvocine life cycles include both asexual and sexual reproduction, with many rounds of asexual reproduction often preceding a single round of sexual reproduction (Fig. 2.2). In the discussion that follows, I consider the asexual developmental changes identified by Kirk (2005) in roughly chronological order (Fig. 2.3). The retention of cytoplasmic bridges in adults (Herron et al. 2010) is appended to this series as step 13.

CONVERSION OF THE CELL WALL INTO EXTRACELLULAR MATRIX (STEP 5)

A requirement for the evolution of multicellularity in any origin of multicellularity is a mechanism for keeping cells attached to each other (Abedin and King 2010); in the volvocine algae, this is accomplished by the formation of an extracellular matrix (ECM). In unicellular volvocines, the protoplast is surrounded by a cell wall with two distinct layers. The outer cell wall has a highly organized, quasi-crystalline structure and provides structural support for the cell (Imam et al. 1985), while the inner layer has an amorphous structure. Following one or more rounds of mitotic cell division within the cell wall of the mother cell, the cell wall ruptures, releasing the daughter cells. In one of the smallest colonial volvocines, *Basichlamys*, four-celled colonies are formed when the daughter cells remain attached to the mother cell wall (Stein 1959; Iyengar and Desikachary 1981). In the other four-celled species,

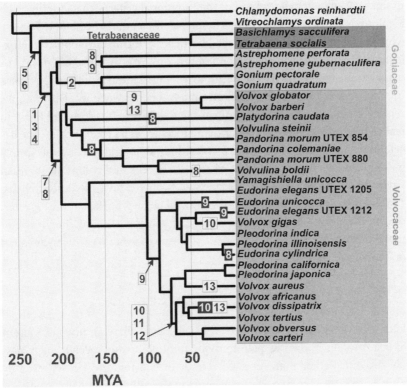

FIGURE 2.3. Evolutionary relationships and estimated divergence times among volvocine algae (adapted from Herron et al. 2009). Shaded boxes identify the 3 multicellular families; species not highlighted in this manner are unicellular. Character state changes are those supported by hypothesis tests in Herron and Michod (2008) and Herron et al. (2010). 1: incomplete cytokinesis; 2: partial inversion; 3: rotation of the basal bodies; 4: establishment of organismal polarity; 5: transformation of the cell wall into extracellular matrix; 6: genetic control of cell number; 7: complete inversion; 8: increased volume of extracellular matrix; 9: sterile somatic cells; 10: specialized germ cells; 11: asymmetric division; 12: bifurcated cell division program (steps 11 and 12 may have had 2 separate origins in the clade including *V. africanus* and *V. carteri*); 13: small gonidia, growth between divisions, and retention of cytoplasmic bridges in adult spheroids; 14: slow divisions and light-dependent divisions. Steps 1–12 are described in Kirk (2005), and I have retained Kirk's numbering for consistency. Step 13 is described in Herron et al. (2010).

Tetrabaena, daughter cells remain attached to the mother cell wall as in *Basichlamys,* but they are also connected to each other by connections between their own outer cell walls, as in *Gonium* (Stein 1959; Nozaki 1990). In the Volvocaceae, the outer layers of the daughter cell walls have fused to become a colonial boundary, and individual cells are surrounded only by a homolog of the inner layer, now referred to as an extracellular matrix (D. Kirk et al. 1986).

The simple step of daughter cells remaining attached to each other has important implications for their degree of individuality, and colonies formed in this way already meet several of the traditional criteria. Structurally, such a colony is a discrete, contiguous, and spatially bounded unit (Hull 1980), which passes through a single-celled bottleneck during development (Huxley 1912, p. 45) and is genetically homogeneous (Weismann 1904, p. 347) or nearly so. In addition, this initial shift to a colonial lifestyle aligned the fitness interests of the component cells (Folse and Roughgarden 2010). For example, relative to independently living offspring, the cells in a colony are likely to realize the same viability, living or dying together as a group. Finally, the development of an ECM may have increased physiological integration: simply by keeping cells together, the ECM forces the cells in a colony to share an environment in which each is affected by the others' physiological activities. *Basichlamys*, in which the ECM is the only thing keeping the colony together, may be the least physiologically integrated volvocine colony, but it is still more integrated than the independent offspring of a mother *Chlamydomonas* or *Vitreochlamys* cell.

GENETIC CONTROL OF CELL NUMBER (STEP 6)

Except for a few species of *Volvox*, volvocine algae share an unusual pattern of cell division known as palintomy or multiple fission (Sleigh 1989; Desnitski 1992). Rather than repeated rounds of two-fold growth followed by a single cell division (binary fission), palintomic division consists of many-fold growth followed by multiple rounds of cell division without substantial growth between rounds of division. Thus, a cell grows $\sim2^N$-fold then divides N times without intervening growth to produce 2^N daughter cells. Because each colonial species appears to have a genetically fixed maximum cell number, the number of rounds of cell division has been inferred to have undergone a shift from environmental control (in unicells) to genetic control (in colonies) (Kirk 2005). However, in both unicellular and colonial species, N is controlled by both genetic and environmental factors. Just as in the colonial species, each species of *Chlamydomonas* and *Vitreochlamys* has a characteristic maximum number of rounds of cell division, suggesting a genetic influence (e.g., Lien and Knutsen 1979; Iyengar and Desikachary 1981; Ettl 1983). It is possible that the same mechanism that controls maximum offspring number in the unicellular species controls cell number in the colonial species, as suggested by Koufopanou (1994).

Genetic control of cell number could play a role in maintaining genetic homogeneity among the cells in a colony (by limiting the number of cell divisions and thus the opportunity for mutation). Should a larger colony be considered less of an individual than a small one, all else equal? Strassmann and

Queller (2010) seem to suggest that it should, since large colonies are more likely to deviate from perfect genetic homogeneity and therefore to experience conflicts among cells. Genetic homogeneity is an important criterion in many treatments of individuality (e.g., Santelices 1999; Folse and Roughgarden 2010), and factors that increase genetic homogeneity are also likely to reduce intra-organismal conflict (thereby aligning the fitness interests of the components; Folse and Roughgarden 2010).

INCOMPLETE CYTOKINESIS (STEP 1) AND RETENTION OF CYTOPLASMIC BRIDGES IN ADULTS (STEP 13)

In the Goniaceae and Volvocaceae, cells in the developing embryos remain connected by cytoplasmic bridges due to incomplete cytokinesis (Hoops et al. 2006). In most species, these bridges are lost later in development, and in the Tetrabaenaceae they never form (Stein 1959). Cytoplasmic bridges play an important role in the developmental process of inversion (steps 2 and 7) (Viamontes 1977; Viamontes et al. 1979; Green et al. 1981), and in three independent lineages of *Volvox* they are retained in adult spheroids (step 13) (Hoops et al. 2006; Herron et al. 2010).

The development of cytoplasmic bridges may represent a substantial increase in physiological integration, although little is known about the nature or extent of the resulting interactions. The cytoplasmic bridges in the adults of some *Volvox* species could provide a means for direct cell-cell interaction or communication, but as far as I am aware there is no evidence for or against this.

ROTATION OF THE BASAL BODIES (STEP 3) AND ESTABLISHMENT OF ORGANISMAL POLARITY (STEP 4)

Most colonial volvocine algae have a discernable anterior-posterior polarity. In the unicellular *Chlamydomonas* and *Vitreochlamys*, the two flagella are oriented, and beat, in opposite directions. As a result, the cell is pulled through the water in a breaststroke, and the flagella-bearing end of the cell is the anterior (as defined by direction of movement). This arrangement would be suboptimal for a colony with more than a few cells, and in the Goniaceae and Volvocaceae, the flagella are rotated to beat in the same direction (Gerisch 1959; Greuel and Floyd 1985; Hoops 1993). As a result of this rotation, *Gonium* has a center-to-edge polarity defined by cells with the two different orientations, while the spheroidal colonies have an anterior-posterior polarity. Other manifestations of organismal polarity are apparent in at least some members of the Volvocaceae, and these have presumably evolved after the rotation of the basal bodies. In many species, cell size increases from anterior to posterior, and the eyespots are graded in both size and their relative position in the cell (Kirk 2005). In some species of *Eudorina*, the anterior-most cells facultatively differentiate as soma, depending on environmental conditions (Goldstein 1967).

Although rotation of the basal bodies is a change that occurs within the cells, the resulting center-edge or anterior-posterior polarity is an emergent trait that is defined, and makes adaptive sense, at the colony level. Without this reorganization, spheroidal colonies would be incapable of swimming. This is a particularly clear example of functional integration and thus evidence of a colony-level adaptation in the sense of Folse and Roughgarden (2010).

PARTIAL (STEP 2) AND COMPLETE (STEP 7) INVERSION

The embryos of *Gonium* and the Volvocaceae are all inside-out to some degree at the end of cell division, with the flagella on the concave side of a bowl shape or completely inside of a rough sphere. As with the ancestral flagellar orientation, this situation is likely suboptimal for motility, and so the developing colonies turn themselves partially (*Gonium*) or completely (Volvocaceae) inside-out, ending with the flagella on the convex or outer surface. Partial inversion in *Gonium* and complete inversion in the Volvocaceae may have evolved independently (Herron and Michod 2008), or partial inversion may have been an intermediate step preceding complete inversion (Kirk 2005).

Inversion in the Volvocaceae results from cells moving relative to the cytoplasmic bridges arising from incomplete cytokinesis (Marchant 1977; Viamontes 1977). This process requires a remarkable degree of functional integration among cells, as the cells in an inverting embryo change in shape and move relative to one another in a pattern that is both spatially and temporally coordinated (Höhn and Hallmann 2011).

INCREASED VOLUME OF ECM (STEP 8)

In contrast to the relatively compact colonies of *Gonium, Pandorina,* and *Platydorina*, in which the majority of the colony volume is made up of cells, in the remaining members of the Volvocaceae and in *Astrephomene*, the majority of the colony volume is made up of ECM. This gives the largest spheroids the appearance of a hollow ball, with the cells arranged on the periphery. In the various species of *Volvox*, ECM makes up >99% of the colony volume. The volume of ECM appears to be an evolutionarily labile trait, having undergone several expansions and contractions throughout the history of the group (Herron and Michod 2008).

The increased volume of ECM in large spheroids would be one way of increasing organismal size without the long growth periods or extremely large reproductive cells that would be required to achieve the same size by increasing cell number. ECM may serve as a storage medium for nutrients, which are shared by all the cells in a colony (Bell 1985). If either of these factors played a role, evolutionary changes in ECM volume could reasonably be considered colony-level adaptations in the sense of Folse and Roughgarden (2010).

PARTIAL (STEP 9) AND COMPLETE (STEP 10) CELLULAR DIFFERENTIATION

One of the most-studied aspects of volvocine biology, both from a theoretical and from a molecular-genetic perspective, is the development of sterile somatic cells. Obligate somatic cells (that is, cells that develop as soma regardless of environmental conditions) are present in *Astrephomene*, *Pleodorina*, and *Volvox* and are thought to be important for these large colonies to maintain motility throughout the life cycle (Koufopanou 1994; Solari et al. 2006). As a form of cellular differentiation, somatic cells are relevant to the evolution of complex multicellularity. As an extreme form of altruism, in that somatic cells sacrifice their reproductive capacity, their evolution is of broad interest to biologists and theoreticians interested in the evolution of cooperation. Somatic cells have arisen in at least three separate lineages within the volvocine algae, and they appear to have been lost in some lineages of *Eudorina* (Fig. 2.3).

In *V. africanus*, *V. carteri*, *V. gigas*, *V. obversus*, and *V. tertius*, the reproductive cells (gonidia) lack flagella and do not contribute to motility (Kirk 1998; Herron et al. 2010). In contrast, in *Astrephomene*, *Pleodorina*, and the remaining species of *Volvox*, the reproductive cells contribute to motility for at least part of the life cycle (Kirk 1998). In this sense, the former *Volvox* species are completely differentiated, with one set of cells (the soma) specialized for motility and not contributing to reproduction, and the other set (the gonidia) specialized for reproduction and contributing nothing to motility (Kirk 1998). This trait had two independent origins and was apparently lost in *V. dissipatrix* (Fig. 2.3).

Michod (2003) considers this type of germ-soma specialization to be crucial to individuality. In addition, the changes related to cellular differentiation (steps 9–12) have increased interdependence, as reproductive cells became dependent on somatic cells for motility and somatic cells became dependent on gonidia for reproduction (Michod et al. 2006; Folse and Roughgarden 2010). Finally, the changes related to cellular specialization have been interpreted as colony-level adaptations to maintain or increase motility (Koufopanou 1994; Solari et al. 2006).

ASYMMETRIC DIVISION / BIFURCATED CELL DIVISION PROGRAM (STEPS 11 AND 12)

In *V. carteri*, *V. obversus*, and *V. africanus*, cell fate is determined by a series of asymmetric divisions, after which the smaller division products continue dividing to produce somatic cells, while the larger products stop dividing and differentiate into gonidia (Starr 1969, 1970; M. Kirk et al. 1993; Ransick 1993; Herron et al. 2010). Cell fate is strictly determined by cell size: cells larger than ~8 μm in diameter differentiate as gonidia, smaller cells as soma (M. Kirk et al. 1993). It is unclear whether these traits arose independently in *V. africanus* or if they were lost in the ancestor of *V. tertius* and *V. dissipatrix* (Fig. 2.3).

Buss (1987) and Michod et al. (2003) have interpreted early germ-line segregation as an adaptation to maintain individuality in the face of this countervailing pressure as cell number increases. Michod has expanded on this idea and considers the bifurcated cell division program observed in *V. carteri*, *V. obversus*, and *V. africanus* as an adaptation to reduce or prevent intraorganismal conflict by reducing the number of cell divisions in the germ line and thus the genetic heterogeneity among cells (Michod and Nedelcu 2003).

Colonies and Clones

A number of authors concerned with individuality see individuals as the units of evolution. In the views of Godfrey-Smith (2009), Clarke (2012), and Michod and Nedelcu (2003), this is reflected in heritable variation in fitness among individuals. In the colonial volvocine algae, the way in which heritable variation in fitness is partitioned within a particular population depends on the developmental program, the mutation rate, and the demography of the population.

At one extreme, we can imagine a pond in which a population is founded by a single colony, which begins reproducing asexually. This is biologically plausible, for example if a pond is colonized through long-distance dispersal. Heritable variation will arise through new mutations and will reside primarily among colonies. Any genetic variation among cells will be fleeting: a mutation arising during development will give rise to a chimeric colony, but each of that colony's offspring will be genetically homogeneous (some with and some without the mutation), since each daughter colony derives from a single cell.

At the opposite extreme, we can imagine a pond in which the founding population is large and genetically diverse. This situation too is biologically plausible, for example when a summer bloom is initiated by descendants of sexually produced spores from the previous year's population. In this case, heritable variation in fitness will be found mainly among the clonal lineages descending from different spores. Depending on the mutation rate, additional genetic variation will eventually arise within these lineages due to new mutations.

Considering these two scenarios as extremes along a continuum, we see that the individuals-as-units-of-evolution view implies that the degree of individuality of a given unit is not entirely an inherent property of the unit itself. Rather, the degree of individuality is contingent on the particular ecological and demographic circumstances. Furthermore, as these circumstances change over ecological time, the proportion of heritable variation in fitness found at each level changes as well. This is not only an epistemological distinction. Heritability itself, and not just measurements of heritability, really does change from one generation to the next as allele frequencies change (Lynch and Walsh 1998). As it does, the partitioning of heritable variation in fitness among cells, colonies, and clones will change as well. Contrary to our intuitions, then, the units-of-evolution view suggests that the degree of individuality can change not only through long-term evolutionary change but also through short-term changes in population structure.

Conclusions

Most recent discussions of individuality recognize that the relevant criteria are continuous rather than categorical and, as a result, that intermediate degrees of individuality are possible (Santelices 1999; Pepper and Herron 2008; Godfrey-Smith 2009; Queller and Strassmann 2009; Folse and Roughgarden 2010; Strassmann and Queller 2010; Clarke 2012). This recognition is crucial to understanding the emergence of individuality at a new, higher level. The outcome of an ETI is a new individual composed of what, before the transition, were individuals in their own right. But this dichotomous view ignores the gray areas characterized by the intermediate steps in the transition. The volvocine algae give us a unique view of these gray areas through the existence of living species with intermediate degrees of individuality. The point of using the volvocine algae as a model system is that we can see intermediate stages of individuation (or "organismality"; Pepper and Herron 2008; Queller and Strassmann 2009).

There is no line we can draw, or at least none that everyone will agree on, "below" which colonies are groups of cells and "above" which they are individuals in their own right. However, evolutionary reconstructions show how the degree of individuality of volvocine colonies has changed over time in some lineages. By most criteria, the initial formation of simple clusters of cells was a crucial step, creating a new spatially contiguous unit with fitness interests largely aligned among its members. The extant Tetrabaenaceae have retained this condition for ca. 200 million years while their sister group, the Goniaceae + Volvocaceae, underwent further developmental changes (Herron et al. 2009). Several of these changes—especially organismal polarity, inversion, and cellular differentiation—increased the functional integration of volvocine colonies and likely represent adaptations at the colony level.

Groups that are undergoing an ETI should be expected to have intermediate levels of individuality (Pepper and Herron 2008). This may even be an indication that such a transition is in progress, although the alternative, that the group in question is evolutionarily stable at an intermediate level of individuality, should be considered as well (Herron et al. 2013). Several lineages of volvocine algae are living fossils that have undergone little change in 100 million years or more (e.g., *Astrephomene perforata*, *Gonium pectorale*, *Volvulina steinii*, *Yamagishiella unicocca*; see Fig. 2.3). Reversals from derived to ancestral states, for example losses of cellular differentiation, indicate that the path to a new level of individuality is not, or at least not always, straightforward. Along with the long-term existence of taxa with intermediate degrees of individuality (Herron et al. 2009), this suggests that transitions in individuality, once begun, do not lead inevitably to fully individuated wholes.

Individuality in the volvocine algae can, in principle, be partitioned among at least three levels of the biological hierarchy: cells, colonies, and clones. This is likely to be true for any clonally developing multicellular organisms that undergo facultative sexual reproduction. Furthermore, the ways in which heritable variation in fitness is partitioned among these three levels will vary among species and according to ecological and demographic circumstances.

The volvocine algae, of course, represent only one modestly large clade, and so we should be cautious about generalizing. However, a survey of Earth's biodiversity shows that this group is not a pathological case. Possibly nothing matches the ideal of a unitary individual: mutation inevitably disrupts genetic homogeneity (Otto and Hastings 1998; Pineda-Krch and Lehtilä 2004), and any group that alternates sexual and asexual reproduction will have some form of the ramet/genet problem (Santelices 1999). Finally, all extant living things are products of ETIs and include subunits that once reproduced independently. We should not be surprised when the ghosts of previous transitions lead these subunits to retain some degree of individuality.

Acknowledgments

I would like to thank the participants and organizers of the Cain Conference "*E Pluribus Unum*: Bringing Biological Parts and Wholes into Historical and Philosophical Perspective" and the workshop "What Is an Individual? Where Philosophy, History, and Biology Coincide," particularly Lynn Nyhart, Scott Lidgard, Peter Godfrey-Smith, and Rick Grosberg, for valuable input and discussion. This work was conducted at the Division of Biological Sciences, University of Montana. While writing this chapter, I was supported by a NASA Astrobiology Institute postdoctoral fellowship and by grants from the John Templeton Foundation (43285), the NASA Astrobiology Institute (Cooperative Agreement Notice 7), and the National Science Foundation (DEB-1457701), for which I am grateful.

References for Chapter Two

Abedin, M., and N. King. 2010. "Diverse Evolutionary Paths to Cell Adhesion." *Trends in Cell Biology* 20: 742–34.

Andersson, J. O. 2000. "Evolutionary Genomics: Is *Buchnera* a Bacterium or an Organelle?" *Current Biology* 10: R866–68.

Bell, G. 1985. "The Origin and Early Evolution of Germ Cells as Illustrated by the Volvocales." In *The Origin and Evolution of Sex*, edited by H. Halvorson and A. Monroy, 221–56. New York: Alan R. Liss.

Bell, G., and A. O. Mooers. 1997. "Size and Complexity among Multicellular Organisms." *Biological Journal of the Linnean Society* 60: 345–63.

Bonner, J. T. 1998. "The Origins of Multicellularity." *Integrative Biology* 1: 27–36.

———. 2003. "On the Origin of Differentiation." *Journal of Biosciences* 28: 523–28.

———. 2004. "Perspective: The Size-Complexity Rule." *Evolution* 58: 1883–90.

Buss, L. W. 1987. *The Evolution of Individuality*. Princeton, NJ: Princeton University Press.

Butterfield, N. J. 2000. "*Bangiomorpha pubescens* n. gen., n. sp.: Implications for the Evolution of Sex, Multicellularity, and the Mesoproterozoic/Neoproterozoic Radiation of Eukaryotes." *Paleobiology* 26: 386–404.

Clarke, E. 2010. "The Problem of Biological Individuality." *Biological Theory* 5: 312–25.

———. 2012. "Plant Individuality: A Solution to the Demographer's Dilemma." *Biology & Philosophy* 27: 321–61.

Cohn, F. 1875. *Die Entwickelungsgeschichte der Gattung Volvox: Festschrift dem Geheimen Medicinalrath Prof. Dr. Göppert zu seinem fünfzigjährigen Doctorjubiläum am 11 Januar 1875 gewidmet von der philosophischen Facultät der Königl. Universität zu Breslau*. Breslau: Grass, Barth u. Co.

Cook, R. E. 1980. "Reproduction by Duplication." *Natural History* 89: 88–93.

Darwin, C. R. 1872. *The Origin of Species by Means of Natural Selection, or the Preservation of Favoured Races in the Struggle for Life.* 6th ed. London: John Murray.

Dawkins, R. 1997. *Climbing Mount Improbable.* New York: W. W. Norton.

Desnitski, A. G. 1992. "Cellular Mechanisms of the Evolution of Ontogenesis in *Volvox.*" *Archiv für Protistenkunde* 141: 171–78.

Ehrenberg, C. G. 1832. "Über die Entwickelung und Lebensdauer der Infusionsthiere; nebst ferneren Beiträgen zu einer Vergleichung ihrer organischen Systeme." *Abhandlungen der Königlichen Akademie Wissenschaften zu Berlin, Physikalische Klasse* 1831: 1–154.

Ettl, H. 1983. "Chlorophyta I—Phytomonidia." In *Süsswasserflora von Mitteleuropa,* edited by H. Ettl, J. Gerloff, H. Heynig, and D. Mollenhauer, 1–807. Stuttgart: Gustav Fischer.

Folse, H. J., III, and J. Roughgarden. 2010. "What Is an Individual Organism? A Multilevel Selection Perspective." *Quarterly Review of Biology* 85: 447–72.

Gerisch, G. 1959. "Die Zelldifferenzierung bei *Pleodorina californica* Shaw und die Organisation der Phytomonadinenkolonien." *Archiv für Protistenkunde* 104: 292–358.

Godfrey-Smith, P. 2009. *Darwinian Populations and Natural Selection.* Oxford: Oxford University Press.

Goldstein, M. 1967. "Colony Differentiation in *Eudorina.*" *Canadian Journal of Botany* 45: 1591–96.

Gorelick, R. 2012. "Mitosis Circumscribes Individuals; Sex Creates New Individuals." *Biology & Philosophy* 27: 871–90.

Gould, S. J., and R. C. Lewontin. 1979. "The Spandrels of San Marco and the Panglossian Paradigm: A Critique of the Adaptationist Programme." *Proceedings of the Royal Society B* 205: 581–98.

Green, K. J., G. I. Viamontes, and D. L. Kirk. 1981. "Mechanism of Formation, Ultrastructure and Function of the Cytoplasmic Bridge System during Morphogenesis in *Volvox.*" *Journal of Cell Biology* 91: 756–69.

Greuel, B. T., and G. L. Floyd. 1985. "Development of the Flagellar Apparatus and Flagellar Orientation in the Colonial Green Alga *Gonium pectorale* (Volvocales)." *Journal of Phycology* 21: 358–71.

Grosberg, R. K., and R. R. Strathmann. 1998. "One Cell, Two Cell, Red Cell, Blue Cell: The Persistence of a Unicellular Stage in Multicellular Life Histories." *Trends in Ecology and Evolution* 13: 112–16.

Hamilton, W. D. 1964. "The Genetical Evolution of Social Behaviour. II." *Journal of Theoretical Biology* 7: 17–52.

Harper, J. L. 1980. "Plant Demography and Ecological Theory." *Oikos* 35: 244–53.

Herron, M. D. 2009. "Many from One: Lessons from the Volvocine Algae on the Evolution of Multicellularity." *Communicative & Integrative Biology* 2: 368–70.

Herron, M. D., and R. E. Michod. 2008. "Evolution of Complexity in the Volvocine Algae: Transitions in Individuality through Darwin's Eye." *Evolution* 62: 436–51.

Herron, M. D., A. G. Desnitskiy, and R. E. Michod. 2010. "Evolution of Developmental Programs in *Volvox* (Chlorophyta)." *Journal of Phycology* 46: 316–24.

Herron, M. D., J. D. Hackett, F. O. Aylward, and R. E. Michod. 2009. "Triassic Origin and Early Radiation of Multicellular Volvocine Algae." *Proceedings of the National Academy of Sciences USA* 106: 3254–58.

Herron, M. D., A. Rashidi, D. E. Shelton, and W. W. Driscoll. 2013. "Cellular Differentiation and Individuality in the 'Minor' Multicellular Taxa." *Biological Reviews of the Cambridge Philosophical Society* 88: 844–61.

Höhn, S., and A. Hallmann. 2011. "There Is More than One Way to Turn a Spherical Cellular Monolayer Inside Out: Type B Embryo Inversion in *Volvox globator*." *BMC Biology* 9: 89.

Hoops, H. J. 1993. "Flagellar, Cellular and Organismal Polarity in *Volvox carteri*." *Journal of Cell Science* 104: 105–17.

Hoops, H. J., I. Nishii, and D. L. Kirk. 2006. "Cytoplasmic Bridges in *Volvox* and Its Relatives." In *Cell-Cell Channels*, edited by F. Baluska, D. Volkmann, and P. W. Barlow, 65–84. Georgetown, TX: Landes Bioscience/Springer.

Hull, D. L. 1980. "Individuality and Selection." *Annual Review of Ecology and Systematics* 11: 311–32.

Huxley, J. S. 1912. *The Individual in the Animal Kingdom*. Cambridge: Cambridge University Press.

Imam, S. H., M. J. Buchanan, H. C. Shin, and W. J. Snell. 1985. "The *Chlamydomonas* Cell Wall: Characterization of the Wall Framework." *Journal of Cell Biology* 101: 1599–607.

Iyengar, M. O. P., and T. V. Desikachary. 1981. *Volvocales*. New Delhi: Indian Council of Agricultural Research.

Kirk, D. L. 1998. *Volvox: Molecular-Genetic Origins of Multicellularity*. Cambridge: Cambridge University Press.

———. 2005. "A Twelve-Step Program for Evolving Multicellularity and a Division of Labor." *BioEssays* 27: 299–310.

Kirk, D. L., R. Birchem, and N. King. 1986. "The Extracellular Matrix of *Volvox*: A Comparative Study and Proposed System of Nomenclature." *Journal of Cell Science* 80: 207–31.

Kirk, M. M., A. Ransick, S. E. McRae, and D. L. Kirk. 1993. "The Relationship between Cell Size and Cell Fate in *Volvox carteri*." *Journal of Cell Biology* 123: 191–208.

Koufopanou, V. 1994. "The Evolution of Soma in the Volvocales." *American Naturalist* 143: 907–31.

Lewontin, R. C. 1970. "The Units of Selection." *Annual Review of Ecology and Systematics* 1: 1–18.

Lien, T., and G. Knutsen. 1979. "Synchronous Growth of *Chlamydomonas reinhardtii* (Chlorophyceae): A Review of Optimal Conditions." *Journal of Phycology* 15: 191–200.

Linnaeus, C. 1758. *Systema naturae per regna tria naturae, secundum classes, ordines, genera, species, cum characteribus, differentiis, synonymis, locis*. 10th ed. Vol. 1. Stockholm: Laurent Salvius.

Lynch, M., and B. Walsh. 1998. *Genetics and Analysis of Quantitative Traits*. Sunderland, MA: Sinauer.

Marchant, H. J. 1977. "Colony Formation and Inversion in the Green Alga *Eudorina elegans*." *Protoplasma* 93: 325–39.

Maynard Smith, J., and E. Szathmáry. 1997. *The Major Transitions in Evolution*. Oxford: Oxford University Press.

McShea, D. W. 1996. "Perspective: Metazoan Complexity and Evolution: Is There a Trend?" *Evolution* 50: 477–92.

Michod, R. E. 1997. "Cooperation and Conflict in the Evolution of Individuality. I. Multilevel Selection of the Organism." *American Naturalist* 149: 607–45.

———. 2003. "Cooperation and Conflict Mediation during the Origin of Multicellularity." In *Genetic and Cultural Evolution of Cooperation*, edited by P. Hammerstein, 291–307. Cambridge, MA: MIT Press.

Michod, R. E., and A. M. Nedelcu. 2003. "On the Reorganization of Fitness during Evolutionary Transitions in Individuality." *Integrative and Comparative Biology* 43: 64–73.

Michod, R. E., and D. Roze. 1997. "Transitions in Individuality." *Proceedings of the Royal Society of London B* 264: 853–57.

Michod, R. E., A. M. Nedelcu, and D. Roze. 2003. "Cooperation and Conflict in the Evolution of individuality. IV. Conflict Mediation and Evolvability in *Volvox carteri*." *BioSystems* 69: 95–114.

Michod, R. E., Y. Viossat, C. A. Solari, M. Hurand, and A. M. Nedelcu. 2006. "Life-History Evolution and the Origin of Multicellularity." *Journal of Theoretical Biology* 239: 257–72.

Nilsson, D.-E., and S. Pelger. 1994. "A Pessimistic Estimate of the Time Required for an Eye to Evolve." *Proceedings of the Royal Society of London B* 256: 53–58.

Nozaki, H. 1990. "Ultrastructure of the Extracellular Matrix of *Gonium* (Volvocales, Chlorophyta)." *Phycologia* 29: 1–8.

Nozaki, H., and M. Itoh. 1994. "Phylogenetic Relationships within the Colonial Volvocales (Chlorophyta) Inferred from Cladistic Analysis Based on Morphological Data." *Journal of Phycology* 30: 353–65.

Okasha, S. 2005. "Multilevel Selection and the Major Transitions in Evolution." *Philosophy of Science* 72: 1013–25.

Otto, S., and I. Hastings. 1998. "Mutation and Selection within the Individual." *Genetica* 102–103: 507–52.

Pepper, J. W., and M. D. Herron. 2008. "Does Biology Need an Organism Concept?" *Biological Reviews of the Cambridge Philosophical Society* 83: 621–27.

Pineda-Krch, M., and K. Lehtilä. 2004. "Challenging the Genetically Homogeneous Individual." *Journal of Evolutionary Biology* 17: 1192–94.

Queller, D. C. 1997. "Cooperators since Life Began." *Quarterly Review of Biology* 72: 184–88.

Queller, D. C., and J. E. Strassmann. 2009. "Beyond Society: The Evolution of Organismality." *Philosophical Transactions: Biological Sciences* 364: 3143–55.

Ransick, A. 1993. "Specification of Reproductive Cells in *Volvox*." In *Evolutionary Conservation of Developmental Mechanisms: 50th Symposium of the Society for Developmental Biology, Marquette University, June 20–23, 1991*, edited by A. Spradling, 55–70. New York: Wiley-Liss.

Rozhnov, S. V. 2001. "Evolution of the Hardground Community." In *The Ecology of the Cambrian Radiation*, edited by A. Y. Zhuravlev and R. Riding, 238–53. New York: Columbia University Press.

Santelices, B. 1999. "How Many Kinds of Individual Are There?" *Trends in Ecology and Evolution* 14: 152–55.

Sarukhán, J., and J. L. Harper. 1973. "Studies on Plant Demography: *Ranunculus repens* L., *R. bulbosus* L. and *R. acris* L.: I. Population Flux and Survivorship." *Journal of Ecology* 61: 675.

Sleigh, M. A. 1989. *Protozoa and Other Protists*. New York: Edward Arnold.

Solari, C. A., J. O. Kessler, and R. E. Michod. 2006. "A Hydrodynamics Approach to the Evolution of Multicellularity: Flagellar Motility and Germ-Soma Differentiation in Volvocalean green Algae." *American Naturalist* 167: 537–54.

Starr, R. C. 1969. "Structure, Reproduction and Differentiation in *Volvox carteri* f. *nagariensis* Iyengar, strains HK9 & 10." *Archiv für Protistenkunde* 111: 204–22.

———. 1970. "Control of Differentiation in *Volvox*." *Developmental Biology* 29: S59–100.

Stein, J. R. 1959. "The Four-Celled Species of *Gonium*." *American Journal of Botany* 46: 366–71.

Stout, A. B. 1929. "The Clon in Plant Life." *Journal of the New York Botanical Garden* 30: 25–37.

Strassmann, J. E., and D. C. Queller. 2010. "The Social Organism: Congresses, Parties, and Committees." *Evolution* 64: 605–16.

Tarnita, C. E., C. H. Taubes, and M. A. Nowak. 2013. "Evolutionary Construction by Staying Together and Coming Together." *Journal of Theoretical Biology* 320: 10–22.

Tuomi, J., and T. Vuorisalo. 1989. "What Are the Units of Selection in Modular Organisms?" *Oikos* 54: 227–33.

Van Leeuwenhoek, A. 1700. "Part of a Letter from Mr Antony van Leeuwenhoek, concerning the Worms in Sheeps Livers, Gnats, and Animalcula in the Excrements of Frogs." *Philosophical Transactions of the Royal Society of London* 22: 509–18.

Viamontes, G. I. 1977. "Cell Shape Changes and the Mechanism of Inversion in *Volvox*." *Journal of Cell Biology* 75: 719–30.

Viamontes, G. I., L. J. Fochtmann, and D. L. Kirk. 1979. "Morphogenesis Variables in *Volvox*: Analysis of Critical Variables." *Cell* 17: 537–50.

Weismann, A. 1904. *The Evolution Theory*. London: Edward Arnold.

White, J. 1979. "The Plant as a Metapopulation." *Annual Review of Ecology and Systematics* 10: 109–45.

3

Individuality and the Control of Life Cycles

BECKETT STERNER

Introduction

The units of evolution have themselves evolved. In some cases, one unit of evolution becomes integrated into a larger one and loses much of its autonomy. For instance, single-celled organisms evolved into multicellular organisms, insect colonies evolved out of cooperation among individual insects, and eukaryotes evolved out of a symbiotic relationship that formed after one prokaryote engulfed another (Maynard Smith and Szathmáry 1995). Biologists call these events evolutionary transitions in individuality, and together each transition builds on another to form a compositional hierarchy of the units of evolution (Buss 1987; Maynard Smith and Szathmáry 1995; Calcott and Sterelny 2011). At the most general level, evolutionary transitions matter because the nature of biological individuality is essential to understanding the nature of life and the scope of the biological sciences. The particular challenge posed by the transitions is to explain how new kinds of biological individuality, composed out of modified versions of preexisting individuals, actually evolved and persisted over time.

Any method for approaching this problem faces a number of distinctive challenges, of which I list only three here. One is level neutrality: the method must be able to explain how and why evolutionary transitions happened and persisted at each level of the compositional hierarchy. It must, for instance, work as well for explaining the transition from genes to chromosomes, prokaryotes to eukaryotes, and single-celled to multicellular organisms. Another challenge is that the method must be able to determine whether something is an individual or not.[1] In the context of the evolutionary transitions, this is necessary to evaluating whether a transition has happened and to what degree. A third challenge is to address a range of different explanatory questions. We want to know how and why the transition occurred, along with why the new kind of individual persisted over time. Moreover, we will also want to explain what happened after the transition in terms of how the transition occurred. For example, some transitions lead to a diversification of subtypes within the new kind of individual, such as after the formation of eukaryotes, while other transitions result in a comparatively static form.

The current, dominant approach to explaining evolutionary transitions, multilevel selection (MLS) theory, answers aspects of these challenges using mathematical models from population genetics. For example, an MLS model can address whether an evolved cooperative trait among single cells in a larger group can become universal and remain stable given its regular loss through mutation as well as competition between cells (Michod 1997). Variations on such a model can also address other problems, such as the benefits and costs of a multicellular individual reproducing using propagules made of one or many cells. Moreover, MLS theory provides a principled way for determining whether individuality exists at some level in terms of whether we can ascribe fitness to groups at that level (Okasha 2006).

However, I will argue that MLS theory does not provide a complete, self-sufficient approach to theorizing about evolutionary transitions. As a formal, mathematical theory about evolution within a population, it presupposes but does not address the material structure of the population that realizes the model. An MLS model might tell us whether a cooperative trait could become fixed in a population, for example, but it won't be able to explain how the cooperation actually works to produce an adaptive effect on the group's fitness. It also won't be able to account for the sources of variety in the possible modes of cooperation available to a population. MLS theory can tell us when fitness has transferred from one level of units to another, but it can give misleading answers unless we have some other, principled guide for picking out units (Clarke 2012; Clarke 2014). Furthermore, it is commonly acknowledged among biologists that actually measuring the fitness of an individual— sometimes even obtaining and interpreting proxy measurements—is difficult in practice and prone to error (Hendry 2005; Orr 2009). Hence even if MLS theory were sufficient in principle, there would still be room for other approaches in practice that avoided the difficulties and limitations of depending on fitness alone.

I will also argue for a positive complement to MLS theory based on the material and causal structures that are responsible for the control of events within an individual's life cycle. I introduce the concept of a demarcator as a material entity or causal process responsible for a biological individual's nature as a complex whole that is composed out of a set of parts.[2] As defined here, a demarcator is a necessary participant in key events of an individual's life cycle, such as in reproduction, and it also serves as a focal point for the control of the life cycle overall. That is, the causal processes that influence where, when, and how different events in a life cycle occur do so by acting through or on the demarcator. I will show how this perspective allows us to individuate biological entities based on their possession of one or more demarcators and the extent to which these demarcators are focal points for control.

Besides not defining individuality in terms of fitness, one benefit of the demarcator approach is systematizing and explaining the causes of evolutionary variation, including constraints. I will pursue this point in one direction here, focusing on how demarcators provide a novel way of thinking about the control of inheritance. In particular, I show how using demarcators allows us to derive a version of Griesemer and Wimsatt's concepts of material overlap and scaffolding as pathways for heredity (Wimsatt and Griesemer 2007; Griesemer 2000a and b, 2002, 2014). In addition, I show how material overlap and scaffolding can apply to two different cases in the evolution of multicellularity, focusing on how a multicellular group can exert causal control over the functional states of its cellular parts. Although much work remains to be done to establish demarcators as an approach to individuality, in the conclusion I discuss the possibilities for a pluralist stance on biological individuality that is based on commensurate but distinct ways of defining the domain of biology.

The Multilevel Selection Framework

Given that evolutionary transitions pose deep problems for classical evolutionary theory, how can we go about explaining the evolution of the compositional hierarchy of biological individuals? Focusing on the process of a transition itself, Ellen Clarke has recently split this larger issue into three subproblems (Clarke 2014): How does the transition happen? Why does it happen? How is it maintained?[3]

The dominant framework for explaining evolutionary transitions, MLS theory, answers aspects of these subproblems using mathematical models from population genetics. I will not attempt a general review of MLS theory and its place in the larger controversies about group selection, since the general theory and debate have only limited relevance here and would take us too far afield. (For more discussion, see Okasha 2005; Leigh 2010.)

What questions, then, can multilevel selection theory answer about evolutionary transitions? Richard Michod's paradigmatic work on the evolution of multicellularity over the past two decades provides a convenient set of examples (e.g., Michod 1996, 1997; Michod and Roze 1999; Michod et al. 2003). The most basic question the model can address is the problem of maintaining the higher-level unit: whether evolved cooperation among single cells (the lower-level units) can become fixed in the population given loss of the cooperative trait through mutation and competition between cells (Michod 1997). The multilevel character of the model comes from specifying a life cycle structure, shown in Figure 3.1, which in this case features obligatory multi-cellular development from a single founder cell and reproduction through the dispersion of gametes. The multicellular individuals produce more gametes when they contain more cooperative cells, but the cooperative cells pay a price of slower reproduction within the group compared to defectors (free-riders or "cheaters" that benefit from the other cells' cooperation but do not contribute themselves). If the defectors come to dominate the group, then the cooperative trait is less frequent in the gametes produced and may be lost in future generations.

Variations on this model can also address other issues relevant to the sub-problems identified above, such as the benefits and costs of a distinct germ-line or unicellular genetic bottleneck (Grosberg and Strathmann 1998, 2007). The issue of the unicellular genetic bottleneck—whether it's easier to maintain cooperation when multicellular organisms reproduce using single-celled zygotes or multicellular propagules—addresses an important dimension of the "How?" problem. The origins of cooperation have received less attention in the literature (Calcott 2007), but MLS modeling can assess how strong a

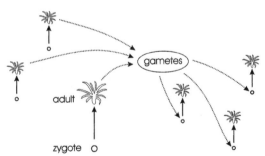

FIGURE 3.1. A multicellular life cycle. Illustration of a life cycle used by Richard Michod to specify a multilevel selection model. The cycle involves an obligatory stage as a zygote, which then develops into a multicellular adult. The adult produces gametes that generate new adults by a process that can vary in the mathematical model (not illustrated here). If reproduction is asexual, for example, these gametes are equivalent to the zygote, while if reproduction is sexual, two gametes would combine to form the zygote. During the life cycle, natural selection can operate at two levels, between single cells during the growth of the adult form and between multicellular adults. Modified from R. E. Michod, "Cooperation and Conflict in the Evolution of Individuality. I. Multilevel Selection of the Organism," *American Naturalist* 149, no. 4 (1997): 609, fig. 1.

source of beneficial cooperation would have to be to drive a transition, for example.

Another, more general function of MLS theory is as a tracking device for measuring the progress of a transition (Griesemer 2007). In other words, it helps us determine what stage a transition is in and how different events advance it as a process. Samir Okasha has shown the theory's utility in this regard using the multilevel Price equation, a formalism widely adopted in quantitative applications of MLS theory (Okasha 2006; also see Clarke 2014): given a choice of the lower- and higher-level individuals, the Price equation allows us to determine at which levels fitness is properly distributed. The idea, then, is that at the start of a transition the higher level is not a unit of selection that carries fitness, while at the end the lower level has lost most or all of its relevance for selection.

Individuation Mechanisms

However, considering Michod's models also illuminates what MLS theory cannot supply for the study of evolutionary transitions: the material structure of any specific case that serves to realize or specify the model. For example, we saw how Michod's model presupposed a particular life cycle structure. Moreover, the models don't tell us anything about the range of possible co-operative traits that might play a role in the transition, or the internal or external conditions under which different mechanisms for generating benefits are available. These are not weaknesses in the modeling work itself, but rather a reminder that MLS theory is not sufficient on its own to explain all of the questions we have about transitions. (For arguments related to the one I make here, see De Monte and Rainey 2014; and Winther 2006, 2009.)

In particular, MLS theory can tell us the degree to which fitness has transferred from one level of units to another, but it doesn't tell us which units to pick out in the first place (Clarke 2012, 2014). "Note that [the Price] equation itself does not tell us anything about how to choose these groups—they must be defined before the multilevel analysis can be applied" (Clarke 2014, 6). In order to interpret the abstract MLS framework and apply it to a concrete case, Clarke suggests that we look for causal mechanisms in the world that give objects the capacity to act as units of selection. Any mechanisms that contribute to an object's capacity in this way serve as "individuating mechanisms" for defining the units in an MLS model.

She goes on to provide criteria for what would qualify as an individuating mechanism based on Richard Lewontin's classic analysis of the units of selection (Lewontin 1970). "Selection can act on a collection to produce evolution only if its members vary heritably for some trait that affects their fitness," so individuating mechanisms define a collection's capacity for selection by "influencing the amount of genetic variance it contains; influencing the extent to which that genetic variance causes variance in fitness within the collection; [or] influencing the heritability of the genetic variance, or of the fitness effects" (Clarke 2013, 428). A similar set of functional criteria for individuating mechanisms would also apply for non-genetic inheritance processes. Examples of individuating mechanisms include genetic bottlenecks, separation of the germ line from somatic cells, fair meiosis, and the immune system (Clarke 2013).[4]

While the idea of an individuating mechanism is very useful in general, it faces several difficulties. In another paper, I argue that focusing on the capacity for selection as the only type of capacity for individuality is problematic (Sterner 2015). For example, a population of individuals may possess a high capacity for undergoing selection that nonetheless does not translate into adaptation because frequency-dependent effects block the trait from going to fixation.

Another option, which I will explore here, is to focus on the causal structure of the life cycle, including reproduction and development. This approach would draw on methods in molecular cell biology, developmental genetics, and comparative genomics, among others. The approach assumes that the living systems of interest share a general life cycle process whose causal structure can be analyzed without requiring the measurement or estimation of fitness.

Control of Life Cycles

I have argued so far that MLS theory is not sufficient in itself to explain all of the relevant questions scientists have about evolutionary transitions. What alternative to fitness could there be as a foundational concept for a level-neutral approach to defining and studying individuality? We would be looking for one or more functional capacities that can be realized in a variety of ways, such that its multiple realizability serves as its source of generality across levels. Moreover, there needs to be some objective criteria against which we can benchmark the degree of individuality ascribed to an object. The capacity of living things in a population to undergo selection, for example, tracks their importance for explaining and predicting evolutionary change over time.

Ideally, we would also have a precise theory for the degree of individuality that tells us how individuality changes as different mechanisms affecting the relevant functional capacity are gained or lost. Interaction effects between mechanisms would likely be important, and we should anticipate that non-linearity will be present as well.

In this section, I will start to develop an alternative theory of individuality using a functional-capacity-and-multiple-realizers schema that does not depend on fitness. The core concept that substitutes for a unit of selection in this manner is what I will call a demarcator, which distinguishes the parts of an individual from non-parts. The main significance of this distinction is that different classes of causal interactions will be possible among parts, among non-parts, and between parts and non-parts. Something will be a biological individual when it possesses at least one demarcator and this demarcator is a focal point for the causal control of key events in its life cycle. A material object or causal process is a focal point for control to the extent that the variation possible within a given life cycle can be explained by the following: a) the focal point's effects on what becomes a part of the object in question; b) its effects on what causal interactions are possible among parts, among non-parts, and between the two; and c) changes in the properties that underlie these two sets of effects that happen during the course of the life cycle. Causal processes will then count as individuating when they contribute to making the demarcator a focal point for control.[5]

My aim here is not to present a self-enclosed, purely theoretical definition of biological individuality. Rather, my goal is to show how the nature of individuality is an empirical problem that goes beyond ascertaining which entities can carry fitness or explaining how cooperation can be adaptive or maintained. I aim to show how one can theorize about biological individuality outside the domain of MLS theory by focusing on certain capacities of biological individuals that MLS theory presupposes but does not address.

In order to take this positive step, we will need to make the key assumption that a definition of biological individuality should allow us to say which things in the world are parts of an individual and which are not. Note that some ability to demarcate parts and wholes is essential to specifying any MLS model. Michod's model above, for example, depends on assigning cell lineages to particular multicellular groups in order to evaluate the effects of within-group conflict. More generally, evolutionary transitions produce new individuals that are composed of parts that were or still are individuals to some degree. Explaining how the higher-level individual evolved must involve some way of specifying the parts of which it is composed. A degree of fuzziness in distinguishing parts from non-parts is allowable so long as some things clearly do count as parts of a given individual and others do not, but I will set this issue aside for now for the sake of simplicity.

Instead of looking at the capacity of a kind of object to undergo selection over generations, we will examine the causes of variation for key events within the kind's life cycle. Reproduction is a particularly essential event in this regard. (I have something quite minimal in mind for the term reproduction here, so that it includes "making more of" an object by processes like fragmentation, similar to Maynard Smith's notion of multiplication [Griesemer 2000b].) Any new instance of a life cycle begins with a reproduction event that occurs during an already ongoing life cycle process. Generally, the entity we identify with the new life cycle won't yet be capable of reproducing. We can think of this period, between genesis and first reproduction, as the process of development. Griesemer, for example, defines development in terms of the acquisition of the capacity to reproduce (Griesemer 2002). The most familiar way for a life cycle to end is through death, for example by predation, disease, aging, or accident. A life cycle can also end in reproduction, however, if the parent is not preserved as an individual during the event. For example, when a cell divides symmetrically in half, biologists treat both of these cells as daughters of the parent rather than identifying one as the parent and the other as the offspring.

Consider the definition of life cycle the biologist John Tyler Bonner has given for multicellular organisms. Bonner describes four major stages: a single-cell stage, a period of growth and development, a period of maturity, and a period of reproduction (Bonner 1993, 17–18). The last two may coincide, but for many species the period of reproduction ends before the period of maturity, leading to senescence (i.e., aging).[6]

In order to investigate whether an object should count as a biological individual, we need to start by identifying a recurrent pattern of events that plausibly corresponds to a life cycle and then examine whether the pattern can be explained in terms of the effects of one or more demarcators. That is, we start with empirical observations of a repeating sequence of events that are connected together as a causal process. We then hypothesize that the observed phenomenon reflects the life cycle of some biological individual or group of individuals. If this is correct, then we can infer that one or more demarcators must exist for this individual. To get the research process moving, we then have to make a further hypothesis about what these demarcators are as material entities or causal processes.[7] As I'll describe further in a moment, any demarcator must play specific roles in explaining how, when, and where key events in the individual's life cycle occur. Determining whether these conditions hold for our proposed demarcators then becomes the major project for empirical research. As we come to understand what these demarcators do in the life cycle (and whether they are involved at all), we gain theoretical insight into the nature of the biological individual we proposed. If we can ultimately find no way to demarcate the supposed individual, then we are forced to abandon the original hypothesis and explain the data another way.[8]

What is involved in hypothesizing a demarcator for an individual and how do its properties relate to events in the life cycle? Since reproduction involves the making of more individuals, it must also involve the production of more demarcators, since each individual as a whole will differ in at least some of its parts. Moreover, the generation of new demarcators during reproduction will be essential to this process rather than incidental: whatever we have hypothesized the demarcator to be, it forms the causal basis for any new individual existing as a whole. How the demarcator changes during reproduction should therefore be central to explaining how the reproductive process happens. Plausible examples of demarcators would be the membrane of a cell, the immune system of an animal, successful interbreeding between members of a population, or the covalent bonds that hold together a plasmid as a single molecule.[9] Things that would probably not count as demarcators would be the process of fair meiosis or a unicellular bottleneck during reproduction.[10]

Another role for the demarcator is bringing about the actual specificity of the "contents" of the individual. The parts of the individual must stand in some, possibly complex relationship with the actual, concrete demarcator that marks them off as distinct from other objects in the world. Furthermore, the parts must differ in their capacities for causal interactions compared to things outside the whole. For example, the membrane of a cell is selectively permeable, affecting which molecules can interact inside the cell, and an animal's immune system selectively recognizes tolerable or beneficial entities, while rejecting or attacking potentially harmful ones. Any hypothesis of a demarcator, then, must serve to explain how certain things in the world become parts and how this status affects their causal capacities.

Lastly, there must be criteria for distinguishing between instances of a demarcator. This is a problem distinct from the individuation of biological individuals, since a given demarcator may be necessary but insufficient for characterizing that individuality. Prima facie it is not obvious that there is a general criterion or set of criteria which identify instances of every demarcator type. Plausible examples, though, would be the material continuity of a plasma membrane in a cell, or the contiguity of all covalent bonds in a molecule. Some sort of distinguishing criteria must be included, therefore, in putting forward a demarcator as an hypothesis.

The demarcator thus serves two necessary roles in a life cycle: an individual's demarcator must be able to change and multiply as part of reproduction, and the demarcator is also causally responsible for the specificity of parthood membership. Beyond this, the demarcator may have to undergo changes during the development of an individual from birth into reproductive maturity. It may also have to change or transform when the individual transitions to a new ecological niche or when the environment changes, for instance in metamorphosis during the complex life cycle of some insects. Additionally, as I'll discuss in more detail below, the demarcator is a crucial participant in the process of inheritance.

Each of these roles provides a dimension in which to evaluate the quality of the demarcator as an hypothesis. Without a demarcator that serves these two necessary roles, hypothesizing the existence of a biological individual as an explanation for the observed pattern of events must fail (or we have to abandon the demarcator framework). However, there is also another dimension to evaluating the quality of a demarcator: its importance to explaining how, when, and where key events in the life cycle occur. Recall from the earlier discussion that fitness, because it supports the prediction and explanation of demographic changes in a population, affords an objective basis for determining evolutionary individuality. The properties of a demarcator that influence life cycle control play an analogous role here by allowing one to predict and explain particular outcomes for a given life cycle as well as the observed range of variation within a population.

I will formulate this role for demarcators by requiring that they serve as focal points of control for events in the life cycle. By control, I don't necessarily mean to imply a centralized system for manipulating the life cycle.[11] More minimally, I use control to refer to the aggregation of all those causal processes that make a difference to how a life cycle happens over time, considered across the range of relevant conditions. For example, why did an individual reproduce sexually under these circumstances and asexually in another context? Alternatively, why did a free-swimming cell transform into a spore when the puddle it was living in started to evaporate? A material object or causal process is a focal point for control to the extent that the variation possible within the given life cycle can be explained by the following: a) the focal point's effects on what becomes a part of the object in question; b) its effects on what causal interactions are possible among parts, among nonparts, and between the two; and c) changes in the properties of the demarcator responsible for these two sets of effects during the life cycle. Note that the demarcator being a focal point of control does not say anything per se about whether the biological individual is a *locus* of control for its own life cycle (Bechtel and Richardson 1993). That is, the importance of the demarcator for the control of life cycles is neutral with respect to whether the origins of the control are internal or external.

With the basic properties of demarcators in hand, we can turn to analyze the process of an evolutionary transition in terms of the emergence of a higher-level demarcator that encompasses a set of parts that are or were biological individuals. One starting point for the transition is if preexisting demarcators for the lower-level individuals start to associate in some way, erasing the distinguishing features that established them as separate. For example, a group of cells might associate together by linking their cell membranes or walls through incomplete cell division or the creation of cytoplasmic bridges between membranes. Association between demarcators in this way would likely constrain the independence of events in each individual's life cycle, altering the process of reproduction as a result of the new linkages. At this point, evolution could then proceed to shift the focal point of control to the new, conjoined demarcator from the prior, separate demarcators.

A second pathway could be to evolve an entirely new demarcator without relying on the modification of preexisting ones. A group of cells might evolve a molecular signal they could emit to indicate membership in the group, for instance. This signal could serve as the basis of a new demarcator, one based on behavioral responses to the signal that would complement their cell membranes rather than merge them together.

Analyzing Inheritance in Terms of Demarcators

Although much more needs to be said about demarcators, I will focus for the rest of the paper on showing how they ground a novel perspective on biological heredity. One of the virtues of the demarcator approach is its potential to offer a level-neutral perspective on heredity that breaks free of the dichotomy between development and heredity put forward and entrenched by the Modern Synthesis. Nothing in the demarcator view of individuality presupposes the existence of DNA, genes, or other sorts of replicators as vehicles for heredity (Sterelny, Smith, and Dickison 1996), although it hardly denies their importance. This section and the next will show how we can use demarcators as a foundation for analyzing the nature of heredity and the tradeoffs between different kinds of inheritance processes.

The central insight is understanding the nature of heredity in terms of how one individual can exert causal control over another. Demarcators let us analyze the possible pathways of control into four categories:

1) control is exerted via some material part of one individual becoming part of the other;
2) control is exerted via a material entity that is not a part of either individual;

3) control is exerted via a mixture, where the material entity that starts as a part of the first individual does not become a part of the second;
4) control is exerted via a material entity that is not a part of the first individual but ends up becoming a part of the second.

We can think of heredity in the broadest possible sense as control exerted by one individual on another that causes the recipient to acquire one or more traits similar to the controller. This includes the familiar, vertical sense of heredity between parent and offspring, but also the horizontal transmission of traits between individuals. However, control exerted along the four pathways can also function in the other direction—that is, to produce dissimilarities between individuals. In the following section, for example, we will see how the asymmetric division of a cell into one larger and one smaller cell is essential for generating distinct cell types within a species of multicellular algae. As I will use the term here, then, the control of inheritance tracks both similarity and dissimilarity as outcomes, whereas heredity focuses solely on the extent of similarity between parents and offspring.

As I have defined them, the four pathways for control have a close relationship to the concepts of material overlap and scaffolding used by James Griesemer and William Wimsatt (Griesemer 2000a, 2002, 2014; Wimsatt and Griesemer 2007). Material overlap would correspond to the first category—that is, inheritance through the transfer of parts. Scaffolding would include the remaining three categories. Table 3.1 presents this relationship in a two-by-two matrix.

As an example, consider how one cell might exert control over another. If the controller passes some of its DNA, proteins, cell membrane, or other internal molecules over into the other, this counts as material overlap. Alter-

TABLE 3.1. Classification of mechanisms for the control of inheritance

		Controlled Individual	
		Internal	External
Controlling Individual	Internal	Material Overlap	Scaffolding
	External	Scaffolding	Scaffolding

NOTE. Four pathways along which one individual can exert causal control over the traits of another, based on whether the material entities used to exert this control are parts of either individual. When the material starts as a part of the controlling individual, for example, this counts as material overlap. The terms internal and external here are used simply as shorthand for being or not being a part of the relevant individual.

natively, there are three possible ways for it to use scaffolding. The controller might extrude an extracellular matrix that influences the controlled cell, for example (internal to external control). The controller could digest carbohydrate molecules in the environment that get taken up by the controlled cell (external to internal control). It could also act on an extracellular matrix connecting both cells to deform its shape and change the local pressure or other forces acting on the controlled cell (external to external).

It's worth pointing out, though, that if the controlled entity is actually a physical part of the controller, then the distinctions in Table 3.1 partially collapse. An example would be symbiosis through engulfment, such as mitochondria within eukaryotic cells. In this situation, the inside of the controller overlaps with the outside of the controlled, eliminating one kind of scaffolding. Nonetheless, the overall distinction between scaffolding and material overlap remains viable.

Griesemer and Wimsatt originally introduced material overlap and scaffolding in order to characterize inheritance from a developmental point of view. Griesemer in particular has argued that biological reproduction can be defined as a special kind of multiplication (i.e., making more of) that involves material overlap, which he calls progeneration (Griesemer 2000a). Griesemer suggests that material overlap is critical for biological reproduction because it transfers preexisting organizational structure to the new generation instead of attempting to form the offspring out of unorganized matter. The paradigmatic case would be how a daughter cell inherits one of the original DNA strands from its parent's double helix, along with a newly synthesized copy. (Note, though, that progeneration involves a particular kind of material overlap, in which the part transferred to the offspring produces similarity between parent and offspring because the causal dispositions of the part make the same difference to the offspring's traits as they did for the parent's.) While I don't presuppose his definition of reproduction here, his arguments are crucial to showing why material overlap has distinctive importance as a pathway for inheritance.

Scaffolding then serves as a complementary resource for the development of an individual. "Scaffolding refers to facilitation of a process that would otherwise be more difficult or costly without it, and which tends to be temporary—an element of a maintenance-, growth-, development-, or construction process that fades away, is removed, or becomes 'invisible' even if it remains structurally integral to the product" (Griesemer 2014, 26). Griesemer also writes, "More generally, scaffolds persist on different time scales than what they scaffold. Infrastructure can persist on very long time scales relative to individuals who use it and thus create correlated environments for organisms of different generations" (Griesemer 2014, 51).

While maintaining the crucial complementary role for scaffolding in relation to material overlap, I depart from Griesemer and Wimsatt's definition by privileging a spatial rather than temporal definition for the concept. This spatial approach also implies a strict dichotomy between material overlap and scaffolding that does not generally hold for Griesemer and Wimsatt's usage. We can see the difference by comparing Griesemer's emphasis on scaffolding as relative to the timescale of an individual and the version I give in Table 3.1, where scaffolding is defined in purely spatial terms. Additionally, Griesemer does not allow certain cases of the transfer of material parts from outside to inside an individual to count as scaffolding: "It would appear that a scaffold per se does not contribute material parts to the developing system, so it cannot count as food" (Griesemer 2014, 30). However, one can recapture some of Griesemer's key types of scaffolding—for example, the notion of infrastructure, by incorporating temporality as a secondary basis for categorization. Hence we could define infrastructure as including material objects that remain external to the controlled entities and that exist on much longer timescales than their lifecycles.

The major utility of material overlap and scaffolding for theorizing about inheritance follows the differing consequences of using one or the other as a pathway for control. As Griesemer has argued, material overlap is generally more reliable as an inheritance mechanism: the offspring acquires the traits of the parent because it has acquired some of the parts of the parent that causally produced those traits in the first place. "Material overlap can increase the robustness and reliability of transmission of capacities, compared to reliance on an unstable and uncertain environment to deliver components in suitable temporal order and spatial configuration, because complex organization can be preserved and propagated in material propagules" (Griesemer 2014, 26). Hence, "if material overlap is an efficient and effective way of propagating and producing developmental order and organization, then it should be favored, entrenched, conserved in evolution or else its absence should require special explanation" (Griesemer 2014, 28).

Indeed, material overlap is not always a possible or advantageous pathway for controlling inheritance. In some cases, for example, the inherited trait may need to vary with environmental circumstances that cannot be prespecified by the genome. The ability of some birds to recognize members of their own species, for instance, is acquired through interaction with other birds during development rather than hardwired into their genetics (Soler and

Soler 1999). Receiving material parts from another cell carries also risks as well as benefits: the sender could transmit its disadvantageous traits as well, or it could use the interaction as an opportunity for predation, parasitism, or competition. Additionally, material overlap requires the evolved capacity to generate and manage the transmission, which may limit the range of control a parent can exert if it requires close physical interaction.

In general, then, the value of scaffolding and material overlap lies in how they illuminate the potential costs and benefits of different combinations of control mechanisms for inheritance. Inheritance through material overlap may be most reliable, but if its ability to vary with environmental circumstances is constrained, then ecological specialization through scaffolding may be the best option. In the next section, I discuss two examples that illustrate how understanding inheritance using demarcators lets us evaluate different pathways toward higher-level individuality in evolutionary transitions. In particular, the demarcator view of individuality gives us a way to theorize about how different ways that lower-level individuals might associate together during a transition generates both affordances and constraints on the possibility for evolution of a new level of individuality.

Material Overlap and Scaffolding in the
Inheritance of Cell Differentiation

This section examines how life cycle control interacts with different ways of initiating multicellularity from lineages of bacteria or unicellular eukaryotes. Most work on the benefits and costs of multicellularity has focused on scenarios where multicellularity initiates through the division of a single cell and its descendants remain stuck together through adhesion or incomplete cytokinesis (the failure of cells to separate fully during division).[12] However, multicellularity can also initiate when multiple cells aggregate together, for example, in dictyostelid slime molds (Kessin 2001), myxobacteria (Pathak et al. 2012), and biofilms, which can include multiple species in the same aggregate (Claessen et al. 2014). Obviously, the availability of material overlap cannot be taken for granted when multicellularity starts by aggregation rather than cell division. Does this affect how life cycle control evolves during the transition? For instance, are there ways of policing cheating cells using scaffolding that provide alternatives to strategies based on material overlap, such as a genetic bottleneck? I will look at two cases of mechanisms for the inheritance of cell types, one based on material overlap and the other on scaffolding.

In general, cell differentiation depends on the ability of cells to occupy discrete overall physiological states in a stable manner. (For a philosophical discussion of cell types, see Slater 2013.) The key difference between cell types is not their DNA but how they use it. Controlling the inheritance of cell differentiation therefore goes beyond the material overlap involved in DNA replication. Although cell differentiation technically refers to the progressive specialization of cells within a multicellular organism, unicellular organisms also transition between discrete physiological states during their lifecycle. For example, many single-celled organisms alternate between states specializing in growth or reproduction, often driven by the presence or absence of nutrients in the environment.

Cell types can also have other crucial features. The type may be irreversible, in the sense that the cell possesses no internal capacity to undifferentiate. One type may also be capable of generating more specialized types. Totipotent cells can generate all other cell types in an organism, while multipotent cells are more limited. Figure 3.2 illustrates some of the various dynamic relationships that are possible between cell types.

MATERIAL OVERLAP: *VOLVOX CARTERI*

The process of cell division can cause material overlap between generations in two ways: first, through directly transmitting DNA and other cellular materials, such as cell membranes and protein complexes; second, by failing to achieve separation, such that partial connections, such as cytoplasmic bridges or conjoined cell walls, ensure ongoing overlap of material components between the cells. Asymmetric cell division—that is, where the two daughter cells receive different inheritances from their parent, appears to be a general mechanism for producing heritable differentiation through direct transmission. It is important for generating a split in germinal and somatic cell

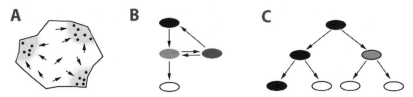

FIGURE 3.2. Networks between modular states. (A) Phenotypic modularity through forcing in the available space of physiological states. States of the cells in a group are represented by dots in the larger physiological space. Arrows could represent underlying dynamic forces in the gene network structure or environmental causes. (B) Four modular differentiated states. Notice that the transition from light gray to white is irreversible, while the other three form a cycle. (C) Cell differentiation during development. Black cells can differentiate into any other state (totipotent), while gray can only differentiate into white.

lineages in some species of the chlorophyte (green algae) Volvocales. However, incomplete cytokinesis is also crucial. In the species *Volvox carteri*, for
instance, it plays a central role in shaping the spatial development of the
multicellular organism (Kirk 1998).

The chlorophyte Order Volvocales has become a paradigm case study for
the evolution of eukaryotic multicellularity (Kirk 1998; Herron and Michod
2008; Michod 2007). The many species within the order are monophyletic
and exhibit a range of lifestyles from unicellular to complex multicellular
forms. The species *V. carteri* exhibits heritable differentiation into two cell
types: flagellated cells incapable of further reproduction that form a spherical
boundary around the organism and germ cells that divide within the organism to form small new individuals. (See Figure 3.3.)

V. carteri produces its separation of germ and soma cells through the combined effect of asymmetric cell division and the inhibition of growth. When
a germ cell divides, it splits into one large and one small cell. Through a still
unknown mechanism, cell size controls the expression of the gene that limits
chloroplast activity. As a result, the small cell can divide a few more times but
cannot grow and therefore represents a reproductive dead-end. The other

FIGURE 3.3. Cases of multicellularity. Images showing different varieties of multicellularity. (A) A group
of cells from *Gonium pectorale*, a colonial species in Volvocales. (B) The mature state of *Volvox carteri*,
a relative to *G. pectorale*. Each dot in the sphere is a somatic cell and the dark internal spheres are still-
developing, smaller versions of the adult organism. (A) and (B) from Kirk (2005), used by permission of
John C. Wiley and Sons. (C) Top-down view of a mature *Bacillus subtilis* biofilm. Reprinted by permission
of Macmillan Publishers Ltd.: *Nature Reviews Microbiology*, Hera Vlamakis, Yunrong Chai, Pascale Beauregard, Richard Losick, and Roberto Kolter, "Sticking Together: Building a Biofilm the *Bacillus subtilis*
Way," copyright 2013.

cell, however, stays large enough that its chloroplasts remain active, and can therefore continue to divide indefinitely. Heritable control of differentiation is thus produced by material overlap in bulk—that is, by the asymmetric distribution of cell volume—in combination with persistent negative regulatory feedback. Commitment to either trajectory—germ or soma—then invokes further physiological specializations. The somatic cells, for example, remain on the outside of the sphere and are responsible for controlling the locomotion of the group through the joint action of their flagella.

Interestingly, multicellularity in Volvocales appears to have initiated through clonal division and only later evolved incomplete cytokinesis. This suggests that the transition occurred through a sequence of modifications to the cells' demarcators. According to the most recent phylogeny, early colonial forms of multicellularity, lacking differentiation, began through the loss of individual cell walls and the growth of a shared extracellular matrix formed of carbohydrates and proteins (Herron and Michod 2008). At this early point, cell division proceeded to completion. However, one of the next major steps toward complex multicellularity involved the retention of cytoplasmic bridges between cells, effectively linking each cell's demarcator into a larger unit. One function of these linkages was to provide a more determinate geometric structure to the colony that was necessary for *V. carteri* to evolve its distinctive spherical shape. The bridges enable *V. carteri* to undergo inversion during development such that its gonidia (germinal cells) move from the outside to the inside of the sphere. This suggests one way in which the conjoined cell membranes serve as a focal point for control during the life cycle—that is, in order to develop a higher-order demarcator that establishes the gonidia as being "inside" the individual in a novel way.

SCAFFOLDING: *BACILLUS SUBTILIS*

Although material overlap may in general be a more reliable mechanism for controlling the inheritance of differentiation, in some contexts it can be unavailable or ineffective. In this subsection I describe a case of one-way (paracrine) signaling between cells in the bacterium *Bacillus subtilis* that also depends on the production of extracellular matrix to generate heritable differentiation. For a related debate about whether biofilms count as evolutionary individuals, see Ereshefsky and Pedroso (2012) and Clarke (2016).

B. subtilis forms complex multicellular biofilms through a combination of aggregation and clonal reproduction (Aguilar et al. 2007; Vlamakis et al. 2008).[13] Biologists have described a variety of cell types in the biofilms that are also spatially localized. Major types include motile cells, which are usually found at the bottom of biofilms, sporulating cells at the top, matrix-producing cells in clumps throughout the biofilm, and "miner" cells that secrete chemicals to break down nutrients for use by nearby cells (López and Kolter 2010). Biofilms in general exhibit a temporal program of development from initial formation to dispersal, and B. subtilis is a premier model organism in this regard (Monds and O'Toole 2009; Vlamakis et al. 2008). Mobile cells dominate early-stage biofilms in B. subtilis, followed by the formation of patches of matrix-producing cells. Matrix cells appear to enter spore formation soonest, although sporulation eventually spreads throughout most of the biofilm. Remaining cells may then disperse and return to a free-living state, while the spores spread from the elevated aerial structures created by the biofilm.

López et al. have found that biofilm development in B. subtilis depends on one-way signaling between cells (López et al. 2009; López and Kolter 2010). The sequence of steps goes as follows: cells initiating biofilm development generally produce a signaling molecule, comX. When comX exceeds a certain threshold, some cells begin producing a second signal called surfactin. Surfactin acts along a general pathway to disrupt the permeability of cell membranes in a way that stimulates the cell to differentiate and start producing extracellular matrix. Crucially, López et al. (2009) found that surfactin production and matrix production occurred in different populations of cells. This implied that matrix cells were not responding to comX, and López et al. found that the matrix surrounding these cells was sufficient to block the signal.

As Elizabeth Shank and Roberto Kolter point out in a recent review, "This unidirectional paracrine signaling, where one cell type produces a signal to which another cell type responds, allows the compartmentalization of cellular differentiation and permits cell-type status to be maintained over numerous generations" (Shank and Kolter 2011, 743). Descendent cells near the early matrix producers would inherit the extracellular matrix and its barrier against comX signaling. Hence we would expect that the matrix serves as scaffolding that constrains the range of types available to new cells by preventing their differentiation into surfactin-producers.

This case suggests an interesting alternative to policing mechanisms based on material overlap: when an early signal to initiate multicellularity also imposes a cost on cells that fail to cooperate. Although still speculative, surfactin could be playing both roles for B. subtilis. We do know that it causes stress to the membranes of any cells not protected by the extracellular matrix, and also that it activates a general signaling pathway that responds to many different sources, including environmental toxins or secretions from predators. It is conceivable, then, that *B. subtilis* might have appropriated surfactin from being an externally produced source of membrane stress to serving as an internally produced signal and policer of biofilm development. This dual role would effectively co-opt one of the basic benefits of life in a biofilm, protection from a harsh environment, to also serve as a mechanism for policing cells that don't cooperate in producing this benefit. It shows how an evolutionary transition that starts by aggregation could use scaffolding as an alternative control strategy instead of material overlap based on cell division. Furthermore, it points to how the disruption or breakdown of demarcators may be relevant to evolving a higher-level individuality in addition to their generation or conjunction.

Conclusion

Evolutionary transitions in individuality pose deep problems for biological theory. As a group, the transitions impose a dynamic chronological order to the forms of living things that have existed over time. Attempting to explain how and why the transitions occurred forces us to reckon with the messy and complex emergence of new forms of individuality over time rather than focusing primarily on paradigmatic kinds of individuals, such as sexually reproducing animals, which have already evolved. A successful answer to the problems raised by evolutionary transitions needs to at least address the three issues I raised in the introduction: it must offer a level-neutral account of what individuals are, provide a means for identifying whether some object is an individual, and address a variety of explanatory questions about the transition.

An evolutionary account of biological individuality, based on MLS theory, is arguably the most promising option we currently have for these first two challenges. However, it gains these merits in part by abstracting away from the material and causal properties of the systems under study, and I have argued that this limits its ability to address the full range of explanatory questions we have. This motivates the search for complementary accounts of individuality that are not based on fitness. I have presented one such alternative account here based on the concept of demarcators and argued for its potential—still in need of further development—to be level neutral and support identification of individuals.

Further development of the demarcator account could proceed in several directions. Demonstrating its level neutrality directly would require examining a number of the demarcators I proposed here across the different evolutionary transitions and showing that they have the correct functional roles and serve as focal points. Another possibility is to apply the concepts of material overlap and scaffolding to analyzing the affordances and constraints for evolution offered by different pathways to evolving higher-level individuality. In the transition to multicellularity, for example, there are a variety of ways for higher-level demarcators to emerge that would also have different tendencies to support the control of inheritance using material overlap or scaffolding—for example, compare aggregation via an extracellular matrix versus adhesion after cell division.

Additionally, the demarcator account, as grounded in the notion of causal control and the specificity of parts versus non-parts, has interesting connections to the topic of biological information. The notion of demarcator is defined here in terms of how much control is exerted via its properties rather than whether control originates among the parts it picks out. More broadly, however, the concept of control I have used here depends on the very sort of causal specificity that is fundamental to biological information (Sterner 2014). When life cycle control depends on environmental cues, inheritance, or communication, we could interpret the material structures carrying out control as information systems. This points to a way in which the comparative study of biological information (Jablonka and Lamb 2005) intersects with the comparative study of evolutionary transitions.

Finally, it may be possible to use a similar theoretical framework as I have here to develop analogous theories of individuality based on metabolism or cooperation. We could set the key functional capacity of a biological individual using an analysis of metabolic autonomy, for example. Mechanisms that contributed to this capacity would qualify as individuating mechanisms, and we would need to describe how the degree of individuality depended on the particular individuating mechanisms present. If this is possible, the domain of biological individuals—that is, the subject matter of biology—could then be defined according to a plurality of perspectives, each focusing on one aspect of the complex phenomena we traditionally group under the heading of "living things." What we commonly think of as competing approaches to individuality—metabolism, fitness, cooperation, life cycles—might then turn out to be epistemically complementary rather than ontologically exclusive.

Acknowledgments

This paper has developed and improved out of ongoing conversation and feedback from Scott Lidgard and Bill Wimsatt. My thanks to Scott and Lynn Nyhart for organizing the workshops on individuality that led to this volume, and also to Matt Herron, Alan Love, and Rick Grosberg for helpful comments and suggestions. Figure 3.1 is an adaptation of Figure 1 (p. 609) in R. E. Michod, "Cooperation and Conflict in the Evolution of Individuality. I. Multilevel Selection of the Organism," *American Naturalist* 149, no. 4 (1997), and is used by permission. This research was supported by NSF postdoctoral grant SES-1153114.

Notes to Chapter Three

1. By "thing" or "object" in this paper I mean simply what would be a plausible candidate for a biological individual.

2. The concepts of "part" and "whole" are complex technical terms in philosophy (Achille 2014), but I will not follow any particular philosophical account here. Instead, we can look to the local theories and practices of biologists to understand how the term should be applied for a particular phenomenon of study (see Winther 2006).

3. For the sake of completeness, I suggest that we also add, "How evolvable is the higher-order unit?" The algae species *Volvox carteri*, for example, evolved multicellularity 50–75 million years ago but has not developed further differentiation among cell types since then (Nedelcu and Michod 2004).

4. Fair meiosis occurs when there is an equal probability for each chromosome to end up in each haploid daughter cell. Fairness matters in situations such as the production of eggs in humans, where only one of the four haploid cells generated by meiosis matures to become a viable gamete.

5. The concept of a mechanism will not do any distinctive work in this paper in contrast to the more general concept of a causal process. Obviously the fact that mechanisms are recurring types of causal processes that reliably produce an effect will make them more interesting and tractable for biologists in general. To my knowledge, nothing in Clarke's account depends on choosing one notion of mechanism or another.

6. The philosopher Robert Wilson has also given a fairly general definition of a life cycle in the context of analyzing what it means to be an organism (Wilson 2005). Using the concept of a "living agent" that comes from a related part of his work, he defines a life cycle as being "comprised of a causal succession of entities, each a living agent, which themselves, together with the processes that mediate their succession, recur across generations" (Wilson 2005, 60).

7. In this paper, I will systematically avoid choosing between a substantive versus processual nature for demarcators. That is, I will allow demarcators to be material objects, enduring over time, and also sequences of events, connected by causal relations. Both ways of framing the underlying ontology behind life cycles have their own heuristic values for actual biological research, e.g., designing experiments or generating explanations.

8. Note that the sequence I describe here starts from a position of relative theoretical ignorance about what the individual is, so an initial empirical description of the target phenomenon is crucial for getting the investigation going.

9. I'm hedging here because being a demarcator is ultimately an empirical matter, as I'll describe in a moment, and I do not intend to stipulate that the concept must include these cases.

10. The demarcator must exist at the time that we wish to determine the parts of the individual. In the case of the unicellular bottleneck, what would matter for demarcation is whatever

generates cohesion among the descendants of the original cell over time, not the mere fact that they are descendants from a single individual. See above for a definition of fair meiosis.

11. I recognize that the word "control" has a number of potentially unfortunate connotations. My original inspiration for focusing on the control of life cycles came from Leigh Van Valen's classic idea that "evolution is the control of development by ecology" (Van Valen 1973). Other relevant sources are John Tyler Bonner's discussion of the control of pattern in development (Bonner 1974), and Bechtel and Richardson's discussion of a locus of control in a functional system (Bechtel and Richardson 1993).

12. Even here, there is a surprising diversity of alternatives to classical binary fission worth recognizing (Angert 2005).

13. In the wild, it's likely that many biofilms are multi-species and include a considerable range of genetic variation within species. Most laboratory cultures, however, focus on single-species biofilms grown from a single strain. That microbial culturing techniques now accommodate biofilm formation reflects a major advance, but the techniques still idealize away a large amount of ecological complexity and population structure present in nature.

References for Chapter Three

Achille, V. 2014. "Mereology." *Stanford Encyclopedia of Philosophy.* http://plato.stanford.edu / archives/spr2014/entries/mereology/.

Aguilar, C., H. Vlamakis, R. Losick, and R. Kolter. 2007. "Thinking about *Bacillus subtilis* as a Multicellular Organism." *Current Opinion in Microbiology* 10 (6): 638–43.

Angert, E. R. 2005. "Alternatives to Binary Fission in Bacteria." *Nature Reviews Microbiology* 3 (3): 214–24.

Bechtel, W., and R. C. Richardson. 1993. *Discovering Complexity: Decomposition and Localization as Strategies in Scientific Research.* Princeton, NJ: Princeton University Press.

Bonner, J. T. 1974. *On Development: the Biology of Form.* Cambridge, MA: Harvard University Press.

———. 1993. *Life Cycles: Reflections of an Evolutionary Biologist.* Princeton, NJ: Princeton University Press.

Buss, L. W. 1987. *The Evolution of Individuality.* Princeton, NJ: Princeton University Press.

Calcott, B. 2007. "The Other Cooperation Problem: Generating Benefit." *Biology & Philosophy* 23 (2): 179–203.

Calcott, B., and K. Sterelny, eds. 2011. *The Major Transitions in Evolution Revisited.* Cambridge, MA: MIT Press.

Claessen, D., D. E. Rozen, O. P. Kuipers, L. Søgaard-Andersen, and G. P. van Wezel. 2014. "Bacterial Solutions to Multicellularity: a Tale of Biofilms, Filaments and Fruiting Bodies." *Nature Reviews Microbiology* 12 (2): 115–24.

Clarke, Ellen. 2012. "Plant Individuality: A Solution to the Demographer's Dilemma." *Biology & Philosophy* 27 (3): 321–61.

———. 2013. "The Multiple Realizability of Biological Individuals." *Journal of Philosophy* 110 (8): 413–35.

———. 2014. "Origins of Evolutionary Transitions." *Journal of Biosciences* 39 (1): 1–14.

———. 2016. "Levels of Selection in Biofilms: Multispecies Biofilms Are Not Evolutionary Individuals." *Biology & Philosophy* 31 (2): 191–212.

De Monte, Silvia, and Paul B. Rainey. 2014. "Nascent Multicellular Life and the Emergence of Individuality." *Journal of Biosciences* 39 (2): 237–48.

Ereshefsky, M., and M. Pedroso. 2012. "Biological Individuality: The Case of Biofilms." *Biology & Philosophy* 28 (2): 331–49.

Griesemer, J. R. 2000a. "Reproduction and the Reduction of Genetics." In *The Concept of the Gene in Development and Evolution: Historical and Epistemological Perspectives*, edited by P. Beurton, R. Falk, and H.-J. Rheinberger, 240–85. Cambridge: Cambridge University Press.

———. 2000b. "The Units of Evolutionary Transition." *Selection* 1: 67–80.

———. 2002. "Limits of Reproduction: A Reductionistic Research Strategy in Evolutionary Biology." In *Promises and Limits of Reductionism in the Biomedical Sciences*, edited by M. H. V. Van Regenmortel and D. L. Hull, 211–32. West Sussex, UK: John Wiley & Sons.

———. 2007. "Tracking Organic Processes: Representations and Research Styles in Classical Embryology and Genetics." In *From Embryology to Evo-Devo: A History of Developmental Evolution*, edited by J. Maienschein and M. D. Laubichler, 375–435. Cambridge, MA: MIT Press.

———. 2014. "Reproduction and the Scaffolded Development of Hybrids." In *Developing Scaffolds in Evolution, Culture, and Cognition*, edited by L. R. Caporael, J. R. Griesemer, and W. C. Wimsatt, 23–56. Cambridge, MA: MIT Press.

Grosberg, R. K., and R. R. Strathmann. 1998. "One Cell, Two Cell, Red Cell, Blue Cell: The Persistence of a Unicellular Stage in Multicellular Life Histories." *Trends in Ecology & Evolution* 13 (3): 112–16.

———. 2007. "The Evolution of Multicellularity: A Minor Major Transition?" *Annual Review of Ecology, Evolution, and Systematics* 38 (1): 621–54.

Hendry, A. P. 2005. "Evolutionary Biology: The Power of Natural Selection." *Nature* 433 (7027): 694–95.

Herron, M. D., and R. E. Michod. 2008. "Evolution of Complexity in the Volvocine Algae: Transitions in Individuality through Darwin's Eye." *Evolution* 62 (2): 436–51.

Jablonka, E., and M. J. Lamb. 2005. *Evolution in Four Dimensions: Genetic, Epigenetic, Behavioral, and Symbolic*. Cambridge, MA: MIT Press.

Kessin, R. H. 2001. *Dictyostelium: Evolution, Cell Biology, and the Development of Multicellularity*. Cambridge: Cambridge University Press.

Kirk, D. L. 1998. *Volvox: Molecular-Genetic Origins of Multicellularity and Cellular Differentiation*. New York: Cambridge University Press.

———. 2005. "A Twelve-Step Program for Evolving Multicellularity and a Division of Labor." *BioEssays* 27 (3): 299–310.

Leigh, E. G., Jr. 2010. "The Group Selection Controversy." *Journal of Evolutionary Biology* 23 (1): 6–19.

Lewontin, R. C. 1970. "The Units of Selection." *Annual Review of Ecology and Systematics* 1: 1–18.

López, D., and R. Kolter. 2010. "Extracellular Signals That Define Distinct and Coexisting Cell Fates in *Bacillus subtilis*." *FEMS Microbiology Reviews* 34 (2): 134–49.

López, D., H. Vlamakis, R. Losick, and R. Kolter. 2009. "Paracrine Signaling in a Bacterium." *Genes & Development* 23 (14): 1631–38.

Maynard Smith, J., and E. Szathmáry. 1995. *The Major Transitions in Evolution*. New York: Oxford University Press.

Michod, R. E. 1996. "Cooperation and Conflict in the Evolution of Individuality. II. Conflict Mediation." *Proceedings of the Royal Society B* 263 (1372): 813–22.

————. 1997. "Cooperation and Conflict in the Evolution of Individuality. I. Multilevel Selection of the Organism." *American Naturalist* 149 (4): 607–45.

————. 2007. "Evolution of Individuality during the Transition from Unicellular to Multicellular Life." *Proceedings of the National Academy of Sciences* 104 (Suppl 1): 8613–18.

Michod, R. E., and D. Roze. 1999. "Cooperation and Conflict in the Evolution of Individuality. III. Transitions in the Unit of Fitness." *Lectures on Mathematics in the Life Sciences*: 47–92.

Michod, R. E., A. M. Nedelcu, and D. Roze. 2003. "Cooperation and Conflict in the Evolution of Individuality. IV. Conflict Mediation and Evolvability in *Volvox carteri*." *BioSystems* 69 (2–3): 95–114.

Monds, R. D., and G. A. O'Toole. 2009. "The Developmental Model of Microbial Biofilms: Ten Years of a Paradigm Up for Review." *Trends in Microbiology* 17 (2): 73–87.

Nedelcu, A. M., and R. E. Michod. 2004. "Evolvability, Modularity, and Individuality during the Transition to Multicellularity in Volvocalean Green Algae." In *Modularity in Development and Evolution*, edited by G. Schlosser and G. P. Wagner, 466–89. Chicago: University of Chicago Press.

Okasha, S. 2005. "Multilevel Selection and the Major Transitions in Evolution." *Philosophy of Science* 72 (5): 1013–25.

————. 2006. *Evolution and the Levels of Selection*. New York: Oxford University Press.

Orr, H. A. 2009. "Fitness and Its Role in Evolutionary Genetics." *Nature Reviews Genetics* 10 (8): 531–39.

Pathak, D. T., X. Wei, and D. Wall. 2012. "Myxobacterial Tools for Social Interactions." *Research in Microbiology* 163 (9–10): 579–91.

Shank, E. A., and R. Kolter. 2011. "Extracellular Signaling and Multicellularity in *Bacillus subtilis*." *Current Opinion in Microbiology* 14 (6): 741–47.

Slater, M. H. 2013. "Cell Types as Natural Kinds." *Biological Theory* 7 (2): 170–79.

Soler, M., and J. J. Soler. 1999. "Innate Versus Learned Recognition of Conspecifics in Great Spotted Cuckoos *Clamator glandarius*." *Animal Cognition* 2: 97–102.

Sterelny, K., K. C. Smith, and M. Dickison. 1996. "The Extended Replicator." *Biology & Philosophy* 11 (3): 377–403.

Sterner, B. 2014. "The Practical Value of Biological Information for Research." *Philosophy of Science* 81 (2): 175–94.

————. 2015. "Pathways to Pluralism about Biological Individuality." *Biology & Philosophy* 30 (5): 609–28.

Van Valen, L. 1973. "Review: Festschrift." *Science* 180 (4085): 488.

Vlamakis, H., C. Aguilar, R. Losick, and R. Kolter. 2008. "Control of Cell Fate by the Formation of an Architecturally Complex Bacterial Community." *Genes & Development* 22 (7): 945–53.

Vlamakis, H., Y. Chai, P. Beauregard, R. Losick, and R. Kolter. 2013. "Sticking Together: Building a Biofilm the *Bacillus subtilis* Way." *Nature Reviews Microbiology* 11 (3): 157–68.

Wilson, R. A. 2005. *Genes and the Agents of Life: The Individual in the Fragile Sciences: Biology*. Cambridge: Cambridge University Press.

Wimsatt, W. C., and J. R. Griesemer. 2007. "Reproducing Entrenchments to Scaffold Culture: the Central Role of Development in Cultural Evolution." In *Integrating Evolution and Development: From Theory to Practice*, edited by R. Sansom and R. N. Brandon. Cambridge, MA: MIT Press.

Winther, R. G. 2006. "Parts and Theories in Compositional Biology." *Biology & Philosophy* 21 (4): 471–99.

————. 2009. "Part-Whole Science." *Synthese* 178 (3): 397–427.

4

Discovering the Ties That Bind: Cell-Cell Communication and the Development of Cell Sociology

ANDREW S. REYNOLDS

Introduction

In 1838 the German cell theorist Matthias Schleiden remarked, "Each cell leads a double life": an independent one pertaining to itself and another pertaining to the larger organism of which it may be a part (Schleiden 1847 [1838], 231–32). Schleiden's observation highlights how thinking about cells is complicated by the fact that they may be at once a whole and a part. Consequently, the question of how to talk and think about cells has always been at least as important as the problem of how to observe and to physically study them. Current scientific discussions of cells are replete with mechanistic language and imagery. Cells are commonly described as chemical "factories," composed of "protein machines," regulated by genetic "programs" and so on. But there is another way of talking about cells that portrays them as minute agents making "decisions" about what to do and what type of cell to become (its "fate"), and living in rich social environments involving other cells. Philosophers call such talk the "agential perspective" (Godfrey-Smith 2009, 10). This paper considers how biologists have used a particular version of this agential perspective to think about cells and the multicellular organisms they compose as especially *social* entities.

While the cell theory of the early nineteenth century frequently portrayed cells as discrete "building blocks" (*Bausteine*) from which a more complex plant or animal is constructed, by the mid-nineteenth century cells were commonly regarded as autonomous "elementary organisms," and the plant or animal body was a "society of cells" or "cell-state."[1] The principle of the division of labor and its role in the organization of modern political states provided biologists an analogical model with which to think about the functional relationship between the organism as a whole and its constituent cell-parts. But how cells managed to arrange themselves into a hierarchical system of tissues and organ systems in which the physiological labor of life was divided among them was little discussed. Cells were said to "subordinate" their individual interests for the greater good of the organism or "cell society," but missing was an explicit account of why they would do this and what ties them together as a unified whole at the level of the organism.

This paper discusses two particular ties, first explicitly addressed in the latter half of the twentieth century, that work to bind cells together: one literal, one metaphorical. The first concerns the adhesion proteins and extracellular matrix glycoproteins that literally tie cells together, structurally, through various forms of direct contact and the creation of a shared environment. The second is the intercellular communication that allows cells to regulate one another's behavior, metaphorically tying their fates together. Some of the molecules and proteins relevant to the first function also have a part in the second, structurally ensuring that cells stick together and functionally communicate with one another so as to coordinate their behavior. Recognition of the importance of both these ties is the consequence of the tools and techniques of the molecular revolution of the twentieth century, which made their existence and significance evident. Cell theory has consequently undergone a significant revision since the 1970s with the inclusion of cell-cell communication and cell signaling theory, which describes how "signals" received at or through the cell membrane are internally processed to result in changes in cell behavior and morphology.[2] What effect has this new understanding of cells as communicative agents had on the earlier image of the body as a "society of cells"? For one, it helps explain how a clump of genetically and morphologically similar blastocyst cells, with no division of physiological labor, becomes differentiated and organized into a complex system of tissues and organs resulting ultimately in a highly integrated animal like ourselves. But it also provides further support for the view that cells and the higher-level organisms they create are the result of social forces and arrangements.

In what follows I illustrate how, despite the fact that understanding of the molecular details of cell communication is reliant on mechanistic metaphors (e.g., signal transduction, cell circuits, and wiring diagrams), what I will call the "social perspective" continues to play an important role for biologists working on the evolution, development, normal function, and pathology of cells and organisms. I focus on a key example of this approach, known as "cell sociology." This label has been used on several distinct occasions by different scientists sharing a particular interest in the *interactions* between cells within and among groups and their consequences. Cell sociology highlights the importance of what are known as "group" or "community effects," which are instances of emergent properties dependent upon the interaction of multiple cells. The lesson to be drawn from this is that cells are very much social be-ings, not just in the sense that they tend to engage in a lot of interaction with one another, but that they are *transformed* by these social interactions, so that a proper understanding of cells and the organisms they constitute requires thinking about cells as more than just inert "building blocks" or complicated machines.[3]

This social perspective constitutes a way of thinking about organisms and cells distinct from, but complementary to, the more dominant mechanistic and reductionist perspectives scientists use when thinking about the causal details of how cells manage to do what they do. On the whole it views cells as individuals—what Bechtel (2010) would call cells as the *object* of inquiry not just the *locus* of inquiry (where the real focus is on some sub-cellular part, typically a mechanism of some sort); but with the added proviso that cells must be understood within a larger social context, whether that context involves a dispersed community, a loosely organized colony, a tissue or organ system, or a tightly integrated multicellular organism as a whole.

Thinking about Parts and Wholes

Lynn Nyhart and Scott Lidgard (2011) have emphasized that questions about biological individuality and the relationship between parts and whole were central to much of biological theory in the nineteenth century, and James Elwick (2007) has shown how life scientists at this time used divergent styles of reasoning to understand the relationship between the biological organism and its component parts. That which he calls analysis:synthesis was associated with the idea that plants and animals are compound organisms, aggregates of more elementary organisms. Those advocating this view were prone to com-pare compound organisms to a human society and its component members. Herbert Spencer, for instance, described the animal body as a "common-wealth of monads" (Spencer 1868 [1850], 493; see Snait Gissis, this volume, for a different perspective on Spencer). Publication of Darwin's *On the Origin of Species* (1859) legitimized speculation about the origin of complex organisms from more ancient and simpler ones ("from monad to man" as the popular phrase puts it). Ernst Haeckel was one of the first to combine cell theory and the theory of evolution to search for clues to the origin of multicellular organ-isms from unicellular ones (those he called Protists). The cell theory (which Jan Sapp [1994, 36] has called a "social theory" at its core) made popular the idea that multicellular animals and plants are "cell-states" (*Zellenstaaten*) in which individual cells are analogous to the citizens of a higher level social unit.[4] This social metaphor was made plausible by Henri Milne-Edwards's observation that there is a "division of physiological labor" in plants and animals into specialized tissues and organs. Cells were construed as specialized individuals, arranged into various professions and classes within the larger cell society. Haeckel speculated on how complex cell-states like our own bodies might have evolved from more primitive sorts of cell societies. Long ago, he proposed, isolated hermit cells (*Einsiedler-Zellen*) may have gradually given up their solitary ways to form simple colonies similar to monastic communities.[5] Increased division of labor would result in more specialized and differentiated cells, eventually leading to the point that none could survive on its own, each having become dependent upon the specialized functions of its neighbors. In the end this resulted in the most complicated and centralized cell-states, animal bodies with a central nervous system—which Haeckel likened to his own late nineteenth-century German *Cultur-Staat* (Reynolds 2008). Haeckel also proposed a hierarchy of different levels of individuality—ranging from single cells to multicellular organisms to colonies of organisms like coral reefs and the elaborate Siphonophora, relatives of jellyfish and corals (Haeckel 1866).[6] But tied as this "colonial theory" of multicellular origins was to Haeckel's biogenetic law of recapitulation ("ontogeny recapitulates phylogeny"), in addition to a more mechanistic turn in twentieth-century biology generally, this social conception of the multicellular organism fell somewhat out of favor. The focus on genetics and then molecular biology relegated talk of the body as a "society of cells" to a lesser tributary, trickling alongside the more mainstream talk of mechanisms and information.[7]

The Rise of Cell Communication Theory

If political and economic metaphors (division of labor, cell state) dominated social metaphors of the body in the nineteenth century, communication metaphors emerged more slowly. By the mid-nineteenth century the nervous system was commonly compared to a system of telegraph wires carrying electrical signals throughout the animal body (Otis 2001), but that all cells in the body might be communicating with one another was hardly suspected. Even as late as 1906 it could only be hinted at that the cells of a developing embryo might be in "communication" with one another by means of the slender "protoplasmic bridges" seen connecting them (Shearer 1906). Experiments performed around the turn of the twentieth century showing the disruptive effects of killing or removing early blastula cells, and the inductive influence of tissues such as Spemann's organizer on developmental processes, all pointed to the existence of some kind of coordination between the developing structures within an embryo. How this was achieved was unclear, although some chemical basis seemed likely (Armon 2012). Physiologists rather than embryologists made more headway into the coordination of organ functions by means of chemical communication. In the first decade of the twentieth century Ernest Starling and William Bayliss collaboratively proposed that chemical molecules called "hormones" carried by the blood acted as "messengers" to coordinate the body's various functions, and fostered intense efforts to identify the body's chemical messengers (Henderson 2005). In the late 1950s Peter Karlson and Adolf Butenandt introduced the term "pheromone" to describe chemical messengers which act externally between individual organisms of various species of social insects (Karlson and Butenandt 1959).[8] Application of the techniques of molecular biology (e.g., x-ray crystallography, radioisotope and fluorescent labeling, and protein nuclear magnetic resonance) eventually provided insight into hormone activity at the cell level (Kohler 1982; Morange 1998). The theory of cell-cell communication and intracellular signalling emerged in the 1960s and '70s through the creative conceptual work of people like Earl Sutherland ("second messengers" [Sutherland et al. 1965]) and Martin Rodbell ("signal transduction" [Rodbell 1980]), who were trying to understand the activity of insulin hormone at the cell level. The discovery (Nealson et al. 1970) of the coordinated control of bioluminescence in colonies of bacteria by means of chemical signals (what later came to be known as "quorum sensing" [Fuqua et al. 1994]) kicked off a vibrant research program into the study of cell-cell communication in microbes. The theory of cell-cell communication entailed a significant amendment to classical cell theory. Cells were no longer mere "building stones," but gregarious social organisms in constant communication with one another by chemical and physical signals.[9]

Cell Sociology

While biochemists in the early twentieth century concerned themselves with understanding the internal chemical dynamics of cells as parts of larger tissue, organ, and organism systems, another group of biologists were using the newly developed techniques of cell and tissue culture to study the behavior of cells individually and in groups. Alexis Carrel in 1931 called for the creation of a "new cytology" that would leave behind the established methods of studying the corpses of dead cells with stains and dyes for the observation of live cells in culture, whose behavior could be captured with the new technology of micro-cinematography (Landecker 2007). This would allow for what Carrel called the study of "cell sociology" (Carrel 1931). "Cell colonies, or organs," he wrote, "are events which progressively unfold themselves. They must be studied like history. A tissue consists of a society of complex organisms" (ibid., 298); its physiological properties belong to "the supracellular order and are the expression of sociological laws" (Carrel 1931, 303). This idea was later repeated by the French pathologist Albert Policard, though without reference to Carrel (Policard 1964).[10] But it was not until the 1970s that an extensive attempt was made to develop a theory of "cell sociology"–and once again by a French biologist—the developmental biologist Rosine Chandebois. In a series of papers and a coauthored book in the '70s and early '80s Chandebois proposed a theory of development she called "cell sociology" (Chandebois 1976, 1977, 1980, 1981; Chandebois and Faber 1983).

As the name suggests, the focus of "cell sociology" is the interactions occurring within and between *groups* of cells. Chandebois based her theory on the phenomenon of *autonomous progression of differentiation.* This refers to the ability of a group of cells removed from an embryo to continue to develop as a particular structure in vitro in roughly the same period of time as it would have in its original in vivo environment. Since the environment in which the cell group is situated has no significant effect on its ability to differentiate to a particular stage, development, Chandebois argued, is in its most basic form an automatic or autonomous process. But autonomous progression is a density-dependent effect that only proceeds with a sufficient number of cells maintaining contact with one another. A similar phenomenon had been noted earlier by Albert Fischer, who in 1923 had failed in his attempts to get isolated fibroblast cells to divide and proliferate in vitro. Fischer found that proliferation in culture was dependent on his using a minimal number of cells that maintained close contact with one another (Fischer 1923). When only a few scattered cells are transplanted, not only do they not grow, they degenerate and die.[11] Noting that groups of cells have collective properties not seen in the individual cells alone, Chandebois insisted that development is a "social" phenomenon.

> A cell population is therefore much more akin to a human society than to a network of automata. A cell in isolation can neither maintain its activities unchanged nor respond to a stimulus that could transform *a group of cells of the same type.* In other words, cell individuality is not based on individual memory alone: we have to do, to different degrees in different tissues, with a "group effect" implying a "collective memory." And it is this aspect of the social behavior of cells that underlies the phenomenon of progression that we call development. Its course is in many respects comparable to the history of a civilization. And its study must be pursued from the viewpoint of a sociology of cells. (Chandebois and Faber 1983, 25)

Central to Chandebois's notion of "cell sociology" is the recognition that cells of the developing embryo are in constant communication with one another, and that this communication, by chemical and mechanical means, has an important influence on the cell's behavior and fate (its "individuality"), in a way analogous to how the social interactions experienced by a growing human influence his or her behavior and fate.

Evolutionary developmental biologist Brian Hall (Hall and Miyake 2000; Hall 2003) has applied Chandebois's notion of "cell sociology" to the issue of modularity in biological systems to explain that, while all cells in a developing embryo communicate, it makes a difference whether the signals in question are exchanged between cells *within* a similar group (a homotypic interaction in a localized condensation of cells) or *between* dissimilar groups (a heterotypic inductive signal between separate layers of cells, for example).[12] Signals exchanged among similar cells within a homotypic context make possible the emergent features known as group or community effects, such as the upregulation of certain tissue-specific genes (Chandebois and Faber 1983; Gurdon 1988; Gass and Hall 2007; Hall and Miyake 2000; Hall 2003). As these effects are not attainable by isolated cells or numbers of cells beneath a minimal density, they are akin to social phenomena. The adage "It takes a village" is also applicable, then, to the raising of specialized cells. Recently the idea of cell sociology has also been applied to the immune system to highlight the complex and dynamic interactions of the various types of cells in the body's defense system (Shirasaki et al. 2013). The communication between cells within a particular group (a condensation, a tissue, an organ, etc.) helps to explain the emergence and integratedness of these new levels of biological organization. The emphasis on group or community effects speaks to the organizational structure of the social interactions occurring between cells of a developing embryo, and reflects the significance of population structure for understanding transitions in evolutionary individuality more generally (see Huneman 2013 and others in the same volume).

In addition to this idea of cell "sociology" some recent writers have spoken of "cell sociobiology" (Keller 2012) or "socio-microbiology" (Strassman and Queller 2004; Parsek and Greenberg 2005; Velicer and Vos 2009). These sociobiological approaches to the study of cell behavior are distinguished by their preoccupation with the evolution of cooperative behavior among predominantly single-cell organisms such as bacteria and cellular slime molds, and an employment of inclusive fitness and evolutionary game theory, which are missing from Chandebois's strictly developmental focus.[13] This is already a well- established project operating under the label "evolutionary transitions in individuality" (Michod 1999), and it is unclear that the label "cell sociobiology" contributes anything more than a convenient shorthand description.

In all these various recent "cell sociologies," I would argue, cell-cell communication provides a significant explanatory principle missing from earlier discussions of a multicellular plant or animal as a "society of cells."

Communication, Evolutionary Transitions, and "Social Control"

"Communication" (from Latin "communicare" and "communis") incorporates the notion of community, of living in communion or in common. The verb "commune" ("to converse together, spiritually or confidentially") comes from old French, "communer," meaning to share (Macdonald 1966). Things in isolation of one another can share nothing in common except in a metaphorical sense—for example, a common shape or color. Communication, then, implies a common bond or link between two or more things, and it provides the means by which distinct agents may come together to form a coordinated whole at a "higher" level, and remain functionally integrated within a hierarchy of levels (e.g., cells, tissues, organs, organism, colony).

A honey-bee colony, for instance, is often referred to as a "superorganism." Its impressive cohesiveness is made possible because each bee is free to move about the nest and to exchange information with its hive-mates (Seeley 1989). So in this case a compound "organism" emerges not because its parts are tightly joined together but because they are able to exchange information within a shared environment. In biological systems, information may be communicated in two forms: signals and cues.[14] Signals, such as the famous honey-bee dance, have been naturally selected for their information-carrying function, whereas cues carry information only incidentally and are "by-products of behaviors performed for reasons other than communication" (Seeley 1989, 309). For example, a foraging bee gauges how well stocked in honey its colony is by the amount of time it waits for a storing bee to unload its nectar—a long wait means lots of bees bringing nectar and thus rich stores; a short wait means few bees bringing nectar and poor honey stores. Seeley predicts that in superorganisms like the honey-bee colony "the relatively subtle communication mechanisms of cues and the shared environment will prove even more important than the more obvious signals" (ibid., 311).

One might argue that communication constitutes a necessary (though not sufficient) condition not only for a superorganism such as the honey-bee hive but also for the existence of a multicellular individual (Knoll and Hewitt 2011). The authors of a recent book on evolutionary medicine go so far as to say "the essence of evolution can be reduced to forms of cell-cell communication" (Torday and Rehan 2012, 36, 42). The successive evolutionary transitions from single cells to cell colonies to proper multicellular organisms with physiological, behavioral, and genetic integration rely upon communication between the cells involved (Grosberg and Strathmann 2007). Since Buss (1987) it has been recognized that the history of life is a history of transitions in increasingly complex levels of selective units, and that the continued sustainability of a complex multicellular organism requires some means of mediating potential conflicts of interest between the various cell-lineages and those of the emerging colony or organism as a whole. Constant communication among the members of a specific group of cells within a larger cell population helps to define it as an integrated level within a developing hierarchy of similarly defined levels, between which there may be communication, but of lesser frequency, intensity, or impact. Cell-cell communication may also be viewed, then, as a critical means by which the "interests" of a higher-level entity, such as a multicellular organism, are asserted over the "interests" of the individual cell (cf. van Baalen 2013). The individual fates of the parts must be sufficiently tied to one another so that none stands a better chance of selfishly flourishing on its own at the expense of the whole (Maynard Smith and Szathmáry 1995; Michod and Nedelcu 2003).

Peter Godfrey-Smith (2009) suggests that the distinction between a clump of cells and a multicellular organism is not a categorical dichotomy but one best situated along a continuum. The tissue-cells of a tightly integrated animal, for instance, can in some cases be regarded as what he calls "Darwinian individuals," meaning that they can form populations capable of undergoing evolution by natural selection, as occurs in the case of cancer (Godfrey-Smith 2009, 84–86). But typically the cells of a higher level unit will be "marginal" Darwinian individuals that have been "de-Darwinized" by the organism so as to behave as subservient parts (ibid., 100–103).[15] Similarly, Queller and Strassman's (2009) "social definition" of an organism suggests that the essence of "organismality" lies in a shared purpose: "the parts work together for the integrated whole, with high cooperation and very low conflict" (Queller and Strassman 2009, 3144; see also Strassmann and Queller 2010). An organism, according to this proposal, is characterized by there being more cooperation than conflict among the parts. But as the authors stress, low conflict and high cooperation are not the same thing (ibid. 3151); both must be present for the parts to be *parts of an organism.*

The transition from a mere group of cells to a multicellular organism requires that heritable variation of fitness becomes possible at the new organismal level. This, Michod and Nedelcu (2003) propose, requires a reorganization of the basic components of individual cell fitness (survival and reproduction), the life-properties of the cells (immortality and totipotency), and a co-option of the lower-level cell processes for new functions at the higher organismal level. The fates of the individual cells must not only be bound together; they must be made to serve the interests of the new higher-level individual. Hence each cell's ability to replicate itself by division must be brought under control in such a way that, if it is allowed to replicate it does so in the interests of reproducing the whole organism and not just that particular cell. (This is achieved in many organisms by a separation between the germ and somatic cell-lines). Strict genetic control over each cell's ability to proliferate (i.e., its own division cycle) and even its ability to remain alive is the common way this has been done in animals. Each cell in a mammalian body, for instance, finds that control of its growth cycle and its ability to survive are tightly regulated by genetic programs carried in its nucleus, which are subject to close scrutiny and manipulation by other cells by means of cell signaling.

In animals and plants, one way such control is achieved is by each cell carrying its own "suicide program" that can be triggered by a number of factors: internal damage, communication from other cells of so-called suicide signals, or even failure to receive so-called survival signals. "Cell suicide" or "programmed cell death" (PCD) has been called a form of "social control" (Raff 1992) or "policing" (Michod 1999) exerted by the organism over its component cells. Mammalian cells are such highly socialized creatures and so reliant upon the company of their fellow cells that it seems about the only thing they can do on their own is die (Raff 1998). For mammalian cells, like Fischer's fibroblasts mentioned earlier, mere survival is a group or community effect monitored closely by cell-cell communication.[16]

Some (Ameisen 2002; Nedelcu et al. 2010) have proposed that the cell death program may have originally arisen when a bacterium infected another cell and used a plasmid-encoded toxin and its antidote (a so-called addiction module) as a means of rendering the host cell addicted to its presence, so that attempts to expel the intruder resulted in death. Gradually this death program may have been co-opted to establish a co-dependency among a colony of related cells—thereby enforcing a costly penalty to would-be defectors.

Given that components of the cell suicide program are found throughout the unicellular and multicellular eukaryotes (Huettenbrenner et al. 2003) and the recent discovery of key signaling molecules in unicellular microbes (King et al. 2003, 2008), it may be that the merger into multicellular bodies did not require new tools or mutations to become a possibility, but only the right circumstances. As Jean Claude Ameisen has proposed, "The emergence of what we call multicellularity may have only represented an extreme and irreversible manifestation of an ancestral feature on which single-celled eukaryotes have realized countless variations: the social control of cell fates *through intercellular signaling* at the level of a colony" (Ameisen 2002, 374, emphasis added).

PCD is one important means by which individual cells are led to "sacrifice" their own interests for the benefit of the society as a whole (Huettenbrenner et al. 2003). Because the elimination of unneeded cells is an important means by which the embryo is "sculpted" (for instance death of interdigital cells separates the fingers and toes in human embryos), PCD is commonly described as an "altruistic" act on the part of a cell. The highly regulated and orderly form of programmed cell death known as "apoptosis" is a clean and socially responsible means for a stressed or damaged cell to end its life. It has obvious benefits for the larger organism over a messy death (known as "necrosis") that involves spilling dangerous contents into the cell "neighborhood." Thus, in addition to signals and cues, we should add a third category of communication: coercion. Coercion is a communicative act by which one agent causes another to engage in a costly behavior benefitting the sender (Diggle et al. 2008). The death signals that result in another cell's programmed "suicide" might be construed as coercion. Signals or cues may result in the coercive death of a cell, and its suicide may still be fitness-enhancing (inclusive fitness), if it benefits closely related cells.

PCD is also an important component of the body's immune system response to invasion by foreign pathogens. Infected cells that destroy themselves along with an invasive virus, for instance, prevent it from commandeering the replicative machinery that would allow it to spread to neighboring cells. Pradeu (2012) provides a revised version of the immune theory definition of the biological organism that, rather than appealing to a problematic notion of the immune system as a biologically natural form of "self" confronting "non-self," defines an organism as "*a functionally integrated whole composed of heterogeneous components that are locally interconnected by strong biochemical interactions and controlled by constant systemic immune interactions of a constant average intensity*" (Pradeu 2012, 244). These interactions are forms of cell-cell communication and intracellular signaling. This account (further elaborated in Pradeu 2013), therefore, also relies upon cell-cell communication as the essential factor responsible for tying the organism's various components together.

Cell Contiguity and Communication

If the signal-induced initiation of programmed cell death metaphorically binds cells together, cell membrane proteins and other features of physical contiguity tie cells together more literally. These include the cadherins and other proteins that mediate cell adhesion and play an important role in the organization of cells into tissue structure. When cells share an environment by being bound together, communication becomes almost inevitable and can take many forms. It need not require diffusible chemical signals, nor even permeable gap junctions between adjacent cells. When cells are "joined at the hip," as it were, they can communicate through the brute exertion of mechanical pressures, twists, and shears. The cell's cytoskeleton itself has been identified as a possible communication network for intracellular signal transduction (Forgacs 1995). The cadherins and c-type lectin adhesion molecules are known to be present in choanoflagellates, which are thought to share important similarities with early unicellular ancestors of the metazoa (King et al. 2003, 2008). Cadherins are also thought to play a role in cell-cell communication (Yap and Kovacs 2003; Wheelock and Johnson 2003). They may in fact have originally served a cell communication and signal transduction role, before being repurposed for the job of creating cell junctions between animal tissue cells (Abedin and King 2010).

Cells physically bound together may also collaboratively create an extracellular matrix, a network of proteins and polysaccharides involved in cell communication and tissue organization (Alberts et al. 2007) or form, as many bacterial cells do, biofilm communities (Sterner, this volume). In the volvocine green algae, encapsulation of a colony of cells within a shared extracellular matrix binds the cells to a common fate (Herron 2009 and this volume). Cell-lines attempting to defect from the common good by selfishly reproducing appear to doom themselves and their peers to dysfunctional colony formation (Herron quoted in Akst 2011).

The recent experimental evolution of a multicellular form of yeast (Ratcliff et al. 2012) with a division of labor emerges from the death of specific cells forming the points of dispersal of multicellular daughter-propagules. The authors speculate that these "apoptotic cells," typically the oldest cells, may be "weak links" forming the adhesion between the younger propagule and the parent cluster. PCD is a feature of the original unicellular yeast in question (*Saccharomyces cerevisiae*), and while it is possible that the trigger for these cell deaths may simply be a stochastic event (Ratcliff, pers. comm.), it is worth considering whether the trigger for these cell deaths is perhaps a *mechanical cue* resulting from the dying cell's location tightly sandwiched within the center of a multicellular colony—an environmental or positional cue, which with time may become a signal.[17]

To sum up, I have discussed two sorts of "ties that bind" cells together: one that literally binds cells together—adhesion molecules and the proteins constituting the extracellular matrix; another that metaphorically binds cells together–the cell communication and signaling molecules responsible for such things as the cell death program. The two are not unrelated, however, for as mentioned above the cadherins and c-type lectins are known to play both roles in choanoflagellates and other basal metazoans (Yap and Kovacs 2003; Wheelock and Johnson 2003).

Conclusion

Early discussions of a compound organism as a "society of cells" were largely motivated by a division of labor among specialized cells. More recent biology accounts for the ties that bind the cells of a multicellular organism together. These are i) the cell adhesion and extracellular matrix molecules that literally tie cells together as colonies and bodies, and ii) the signaling molecules with which cells communicate to mutually control one another's growth and survival (e.g., PCD). Cell-cell communication also accounts for cell differentiation in development. What I have called the social perspective is a variant of the agential perspective that makes particular use of social metaphors. Does it lead (or rather mislead), as Peter Godfrey-Smith (2009, 142f) worries, to "Darwinian paranoia" —that is, the anthropomorphic attribution of goals and strategies to non-human entities, of which metaphorical talk of "selfish" genes is the best known example? "Cell sociology" does not involve the ascription of intentions or purposes to cells; it simply highlights the fact that groups of cells behave differently and have different capacities than individual cells on their own. Cell sociology makes the point that cells are not to the body as bricks are to a building, but as citizens are to a society. Just as it takes a society to create the special category of individuals known as citizens (or students, teachers, or scientists), so it takes a society or community of cells to create specialized types of cells. Cells may be parts of organisms, but they are not parts in the same way as grains of sand are parts of a sand heap, or lego blocks are parts of a lego-house, or even computers are parts of a computer network. Unlike these spatially and temporally autonomous units, cells are transformed by their social interactions; they become different types of thing, they become specialized or differentiated cells. If there is an increase in fitness (perhaps inclusively conceived), then this social perspective permits us to understand how a higher-level sort of individual emerges, as individual cells can achieve this greater fitness only as component parts of a larger group-entity. In this way a new level of *social* agency (the group or community effect) emerges, making possible the transition to a new level of individuality, at the tissue, organ, and ultimately organismal levels. Proper understanding of cellular conditions, such as development and cancer, requires paying attention to cell-cell interactions and tissue organization (Sonnenschein and Soto 1998), and talk of a "sociology of cells" is a useful heuristic emphasizing the importance of complex and structured social (communicative) interactions among populations and sub-populations of cells. The term "molecular sociology" has likewise been suggested for the study of protein interactions and functional organization within the cell environment (Robinson et al. 2007).

How far might we push this cell-sociological view of multicellularity? There are two chief means by which a multicellular individual can come about: i) by aggregation of autonomous free-living cells (e.g., cellular slime molds), and ii) by failure of daughter cells to separate after cytokinesis (Bonner 1998; Nedelcu 2012). But there is another scenario that doesn't quite fit into either of these: the case of "communities" of bacteria (e.g., biofilms) which display controlled morphologies and coordinated behavior similar to multicellular organisms. With a few exceptions (myxomycetes fruiting bodies, filamentous cyanobacteria) bacteria do not typically build bodies, yet some (Shapiro 1988, 1998) have argued that bacteria ought to be regarded as multicellular, for like other multicellular organisms, they display division of labor, phenotypic differentiation, and–more controversially perhaps–even rudimentary intelligence (Shapiro 2007). That bacteria achieve these communal benefits through intercellular communication (e.g., quorum sensing) in both specialized and universal chemical languages (Winans 2002) provides further support for Shapiro's contention that a proper appreciation of bacteria requires a "socio-bacteriological" approach. Others echo this view (Winans and Bassler 2002; Jacob et al. 2004; Parsek and Greenberg 2005; Nadell et al. 2008; Velicer and Vos 2009). Godfrey-Smith's framework allows those more reluctant to expand the notion of individuality to regard these bacterial communities as "marginal" Darwinian individuals (but see also Haber 2013).

What I hope to have shown is that understanding how new levels of biological individuality arise from previously existing types, from unicells to colonies, and from colonies to organisms and so on, requires attention to the communication going on amongst members of a similar level (homotypic) and between levels of dissimilar (heterotypic) kinds. It is this social communication that explains how cells can lead, as Schleiden said, a double life.

Acknowledgments

I am indebted to the participants of the two meetings in Philadelphia and Madison for helpful comments and questions, and to Lynn Nyhart and Scott Lidgard in particular, for inviting me to be part of the project and for their questions and suggestions, which greatly improved this paper. Thanks too to Will Ratcliffe, who read an earlier draft and provided insightful comments.

Notes to Chapter Four

1. In fact both Schleiden and Schwann, who are typically credited with founding the cell theory, described cells as little "organisms" in their pioneering publications. See Schwann (1847).

2. For whatever reasons, historians and philosophers of science have been slow to give the science of cell signaling the attention it deserves.

3. Typically when cell biologists speak of the transformation of cells they mean genetic changes that alter a cell's normal behavior and often lead to properties associated with cancer and tumorigenesis, unregulated cell division, or immortality. Here I mean more broadly changes in morphology and behavior of both individual cells and their cell-lineage descendants typically occurring during embryogenesis and development generally.

4. Rudolf Virchow, one of the chief proponents of the "cell state" metaphor, on the issue of individuality remarked that "the 'I' of the philosopher is a consequence of the 'We' of the biologist'" (Virchow 1958 [1858], 139). For more on the history and debates about the theory of the cell state see Reynolds (2007).

5. Haeckel called these simple cell communities "coenobia" after the term used to designate a monastic community. It is derived from the Greek "koinos" meaning "common" (*Concise Oxford English Dictionary*).

6. See Richards (2008), especially chapter 5, for a helpful discussion of these topics and see Rieppel (this volume) for discussion of later development of this idea of a nested hierarchy of individuality.

7. d'Hombres (2012) discusses the gradual decline in importance of the concept of the physiological division of labor as an explanatory principle in modern biological discussions.

8. They suggested "pheromone" to replace the previously used terms "ectohormone" and "social hormone." I am grateful to Hannah Landecker for drawing this research to my attention.

9. Ironically, though Rodbell thought of cell communication in social terms—"biological communication consists of a complex meshwork of structures in which G-proteins, surface receptors, the extracellular matrix, and the vast cytoskeletal network within cells are joined in a community of effort, for which my life and those of my colleagues is a metaphor" (Rodbell 1994, 221)—the theory of signal transduction which he helped create is firmly rooted in cybernetic and electronic engineering metaphors. Hannah Landecker is currently working on the history of how hormones underwent a conceptual metamorphosis from "messengers" to "signals" and the role of cybernetic and information theory in this story.

10. Policard described the attempt to understand the laws governing the function and behavior of tissue cells as "problems de sociologie cellulaire" (Policard 1964, 143).

11. Fischer and Bohus (1946) discussed the role of communication and "social forces" between tissue cells in this regard.

12. Cell condensations are a preliminary step in the creation of specialized tissues, the earliest stage at which tissue-specific genes are upregulated (Hall and Miyake 2000).

13. The use of the term "cell sociobiology" in Keller (2012) is actually misleading and "cell sociology" would have been more fitting, as the report actually discusses physical and mechanical interactions between developing embryo cells.

14. See Maynard Smith and Harper (2003) and several of the papers in D'Ettorre and Hughes (2008) for further discussion of the distinction in relation to different biological systems.

15. Godfrey-Smith distinguishes between the concepts of individual and organism. An organism such as a multicellular tree or human consists of multiple types of (Darwinian) individuals, for instance cells and genes.

16. See Reynolds (2014) for a discussion of the various metaphors that have been central to the science of cell death.

17. See van Baalen (2013, 124 and 128) for similar remarks on the significance of communication for aligning the interests of multiple partners.

References for Chapter Four

Abedin, Monika, and Nicole King. 2010. "Diverse Evolutionary Paths to Cell Adhesion." *Trends in Cell Biology* 20 (12): 734–42.

Akst, Jef. 2011. "From Simple to Complex." *Scientist* 25 (1). Accessed December 18, 2012. http://www.the-scientist.com/?articles.view/articleNo/30827/title/From-Simple-To-Complex/.

Alberts, Bruce, Alexander Johnson, Julian Lewis, Martin Raff, Keith Roberts, and Peter Walter. 2007. *Molecular Biology of the Cell*. 5th ed. New York: Garland Science.

Ameisen, Jean Claude. 2002. "On the Origin, Evolution, and Nature of Programmed Cell Death: A Timeline of Four Billion Years." *Cell Death and Differentiation* 9: 367–93.

Armon, Rony. 2012. "Between Biochemists and Embryologists—the Biochemical Study of Embryonic Induction in the 1930s." *Journal of the History of Biology* 45 (1): 65–108.

Bechtel, William. 2010. "The Cell: Locus or Object of Inquiry?" *Studies in History and Philosophy of Biological and Biomedical Sciences* 41 (3): 172–82.

Bonner, John T. 1998. "The Origins of Multicellularity." *Integrative Biology* 1: 27–36.

Buss, Leo W. 1987. *The Evolution of Individuality*. Princeton, NJ: Princeton University Press.

Carrel, Alexis. 1931. "The New Cytology." *Science* 73 (1890): 297–303.

Chandebois, Rosine. 1976. "Cell Sociology: A Way of Reconsidering the Current Concepts of Morphogenesis." *Acta Biotheoretica* 25 (2–3): 71–102.

———. 1977. "Cell Sociology and the Problem of Position Effect: Pattern Formation, Origin and Role of Gradients." *Acta Biotheoretica* 26 (4): 203–38.

———. 1980. "Cell Sociology and the Problem of Automation in the Development of Pluricellular Animals." *Acta Biotheoretica* 29 (1): 1–35.

———. 1981. "The Problem of Automation in Animal Development: Confrontation of the Concept of Cell Sociology with Biochemical Data." *Acta Biotheoretica* 30 (1): 143–69.

Chandebois, Rosine, and Jean Faber. 1983. *Automation in Animal Development: A New Theory Derived from the Concept of Cell Sociology*. New York: Karger.

D'Ettorre, Patrizia, and David P. Hughes, eds. 2008. *Sociobiology of Communication: An Interdisciplinary Perspective*. New York: Oxford University Press.

d'Hombres, Emmanuel. 2012. "The 'Division of Physiological Labour': The Birth, Life and Death of a Concept." *Journal of the History of Biology* 45 (1): 3–31.

Diggle, Stephen P, Stuart A. West, Andy Gardner, and Ashleigh S. Griffin. 2008. "Communication in Bacteria." In *Sociobiology of Communication: An Interdisciplinary Perspective*, edited by Patrizia D'Ettorre and David P. Hughes, 11–31. New York: Oxford University Press.

Elwick, James. 2007. *Styles of Reasoning in the British Life Sciences: Shared Assumptions 1820–58.* London: Pickering & Chatto.

Fischer, Albert. 1923. "Contributions to the Biology of Tissue Cells. I. The Relation of Cell Crowding to Tissue Growth in Vitro." *Journal of Experimental and Biological Medicine* 38 (6): 667–72.

Fischer, Albert, and Jensen A. Bohus. 1946. "Growth Limiting Factors of Tissue Cells in Vitro." *Acta Physiologica Scandanavica* 12 (1–2): 218–28.

Forgacs, G. 1995. "On the Possible Role of Cytoskeletal Filamentous Networks in Intracellular Signaling: An Approach Based on Percolation." *Journal of Cell Science* 108: 2131–43.

Fuqua, W. Clairborne, Stephen C. Winans, and E. Peter Greenberg. 1994. "Quorum Sensing in Bacteria: The LuxR-LuxI Family of Cell Density-Responsive Transcriptional Regulators." *Journal of Bacteriology* 176 (2): 269–75.

Gass, Gillian L., and Brian K. Hall. 2007. "Collectivity in Context: Modularity, Cell Sociology, and the Neural Crest." *Biological Theory* 2 (4): 1–11.

Godfrey-Smith, Peter. 2009. *Darwinian Populations and Natural Selection.* Oxford: Oxford University Press.

Grosberg, Richard K., and Richard R. Strathmann. 2007. "The Evolution of Multicellularity: A Minor Major Transition?" *Annual Review of Ecology, Evolution, and Systematics* 38: 621–54.

Gurdon, John B. 1988. "A Community Effect in Animal Development." *Nature* 336 (22/29): 772–74.

Haber, Matt. 2013. "Colonies Are Individuals: Revisiting the Superorganism Revival." In Frédéric Bouchard and Philippe Huneman, eds. 2013. *From Groups to Individuals: Evolution and Emerging Individuality*, 195–217. Cambridge, MA: MIT Press.

Haeckel, Ernst. 1866. *Generelle Morphologie der Organismen: Allgemeine Grundzüge der organischen Formenwissenschaft, mechanisch begrundet durch die von Charles Darwin reformirte Descendenztheorie.* Berlin: Reimer.

Hall, Brian K. 2003. "Unlocking the Black Box between Genotype and Phenotype: Cell Condensations as Morphogenetic (Modular) Units." *Biology & Philosophy* 18: 219–47.

Hall, Brian K., and T. Miyake. 2000. "All for One and One for All: Condensations and the Initiation of Skeletal Development." *BioEssays* 22 (2): 138–47.

Henderson, John. 2005. "Ernest Starling and 'Hormones': An Historical Commentary." *Journal of Endrocrinology* 184: 5–10.

Herron, Matthew. 2009. "Many from One: Lessons from the Volvocine Algae on the Evolution of Multicellularity." *Communicative & Integrative Biology* 2 (4): 368–70.

Huettenbrenner, Simone, Susanne Maier, Christina Leisser, Doris Polgar, Stephan Strasser, Michael Grusch, and Georg Krupitza. 2003. "The Evolution of Cell Death Programs as Prerequisites of Multicellularity." *Mutation Research* 543: 235–49.

Huneman, Philippe. 2013. "Adaptations in Transitions: How to Make Sense of Adaptation when Beneficiaries Emerge Simultaneously with Benefits?" In *From Groups to Individuals: Evolution and Emerging Individuality*, edited by Frédéric Bouchard and Philippe Huneman, 141–72. Cambridge, MA: MIT Press.

Jacob, Eshel Ben, Israela Becker, Yoash Shapira, and Herbert Levine. 2004. "Bacterial Linguistic Communication and Social Intelligence." *Trends in Microbiology* 12 (8): 366–72.

Karlson, Peter, and Adolf Butenandt. 1959. "Pheromones (Ectohormones) in Insects." *Annual Review of Entomology* 4: 39–58.

Keller, Ray. 2012. "Physical Biology Returns to Morphogenesis." *Science* 338: 201–3.

King, Nicole, Christopher T. Hittinger, and Sean Carroll. 2003. "Evolution of Key Signaling and Adhesion Protein Families Predates Animal Origins." *Science* 301 (5631): 361–63.

King, Nicole, Jody M. Westbrook, Susan L. Young, Alan Kuo, Monika Abedin, Jarrod Chapman, Stephen Fairclough, et al. 2008. "The Genome of the Choanoflagellate *Monosiga brevicollis* and the Origin of the Metazoans." *Nature* 451 (7180): 783–788.

Knoll, Andrew H., and David Hewitt. 2011. "Phylogenetic, Functional, and Geological Perspectives on Complex Multicellularity." In *The Major Transitions in Evolution Revisited*, edited by Brett Calcott and Kim Sterelny, 251–70. Cambridge, MA: MIT Press.

Kohler, Robert E. 1982. *From Medical Chemistry to Biochemistry: The Making of a Biomedical Discipline.* Cambridge: Cambridge University Press.

Landecker, Hannah. 2007. *Culturing Life: How Cells Became Technologies.* Cambridge, MA: Harvard University Press.

Macdonald, A. M. 1996. *Chambers's Etymological English Dictionary.* Edinburgh W. and R. Chambers.

Maynard Smith, John, and Eörs Szathmáry. 1995. *The Major Transitions in Evolution.* Oxford: Oxford University Press.

Maynard Smith, John, and David Harper. 2003. *Animal Signals.* Oxford: Oxford University Press.

Michod, Richard. 1999. *Darwinian Dynamics: Evolutionary Transitions in Fitness and Individuality.* Princeton, NJ: Princeton University Press.

Michod, Richard, and Aurora Nedelcu. 2003. "On the Reorganization of Fitness during Evolutionary Transitions in Individuality." *Integrative Comparative Biology* 43: 64–73.

Morange, Michel. 1998. *A History of Molecular Biology.* Translated by Matthew Cobb. Cambridge, MA: Harvard University Press.

Nadell, Carey D., Bonnie L. Bassler, and Simon A. Levin. 2008. "Observing Bacteria through the Lens of Social Evolution." *Journal of Biology* 7 (27). http://jbiol.com/content/7/7/27. doi: 10.1186/jbiol87.

Nealson, Kenneth H., Terry Platt, and Hastings J. Woodland. 1970. "Cellular Control of the Synthesis and Activity of Bacterial Luminescent System." *Journal of Bacteriology* 104 (1): 313–22.

Nedelcu, Aurora M. 2012. "Evolution of Multicellularity." In *Encyclopedia of the Life Sciences.* John Wiley and Sons: Chichester. doi:10.1002/9780470015902.a0023665.

Nedelcu, Aurora M., William W. Driscoll, Pierre M. Durand, Matthew D. Herron, and Armin Rashidi. 2010. "On the Paradigm of Altruistic Suicide in the Unicellular World." *Evolution* 65 (1): 3–20.

Nyhart, Lynn K., and Scott Lidgard. 2011. "Individuals at the Center of Biology: Rudolf Leuckart's *Polymorphismus der Individuen* and the Ongoing Narrative of Parts and Wholes. With an Annotated Translation." *Journal of the History of Biology* 44 (3): 373–443.

Otis, Laura. 2001. *Networking: Communicating with Bodies and Machines in the Nineteenth Century.* Ann Arbor: University of Michigan Press.

Parsek, Matthew R., and E. P. Greenberg. 2005. "Sociomicrobiology: The Connections between Quorum Sensing and Biofilms." *Trends in Microbiology* 13 (1): 27–33.

Policard, Albert. 1964. *Cellules vivantes et populations cellulaires: Dynamique et structure moléculaire.* Paris: Masson et Cie.

Pradeu, Thomas. 2012. *The Limits of the Self: Immunity and Biological Individuality*. New York: Oxford University Press.

———. 2013. "Immunity and the Emergence of Individuality." In *From Groups to Individuals: Evolution and Emerging Individuality*, edited by Frédéric Bouchard and Philippe Huneman, 77–96. Cambridge, MA: MIT Press.

Queller, David C., and Joan E. Strassman. 2009. "Beyond Society: The Evolution of Organismality." *Philosophical Transactions of the Royal Society B* 364: 3143–55.

Raff, Martin. 1992. "Social Controls on Cell Survival and Cell Death." *Nature* 356 (6368): 397–400.

———. 1998. "Cell Suicide for Beginners." *Nature* 396: 119–22.

Ratcliff, William C., R. Ford Denison, Mark Borrello, and Michael Travisano. 2012. "Experimental Evolution of Multicellularity." *Proceedings of the National Academy of Sciences USA* 109 (5): 1595–600.

Reynolds, Andrew S. 2007. "The Theory of the Cell State and the Question of Cell Autonomy in Nineteenth and Early Twentieth Century Biology." *Science in Context* 20 (1): 71–95.

———. 2008. "Ernst Haeckel and the Theory of the Cell State: Some Remarks on the History of a Bio-political Metaphor." *History of Science* 46: 123–52.

———. 2014. "The Deaths of a Cell: How Language and Metaphor Influence the Science of Cell Death." *Studies in History and Philosophy of the Biological and Biomedical Sciences* 48: 175–84.

Richards, Robert J. 2008. *The Tragic Sense of Life: Ernst Haeckel and the Struggle over Evolutionary Thought*. Chicago: University of Chicago Press.

Robinson, Carol V., Andrej Sali, and Wolfgang Baumeister. 2007. "The Molecular Sociology of the Cell." *Nature* 450: 973–82.

Rodbell, Martin. 1980. "The Role of Hormone Receptors and GTP-Regulatory Proteins in Membrane Transduction." *Nature* 284 (5751): 17–22.

———. 1994. "Signal Transduction: Evolution of an Idea (Nobel Lecture)." Accessed April 25, 2012. http://www.nobelprize.org/nobel_prizes/medicine/laureates/1994/rodbell-lecture.pdf .

Sapp, Jan. 1994. *Evolution by Association: A History of Symbiosis*. Oxford: Oxford University Press.

Schleiden, Matthias. 1847 [1838]. *Contributions to Phytogenesis*. Translation by Henry Smith. London: Sydenham Society.

Schwann, Theodor. 1847 [1839]. *Microscopical Researches into the Accordance in the Structure and Growth of Animals and Plants*. Translation by Henry Smith. London: Sydenham Society.

Seeley, Thomas D. 1989. "The Honey Bee Colony as a Superorganism." *American Scientist* 77 (6): 304–11.

Shapiro, James. 1988. "Bacteria as Multicellular Organisms." *Scientific American* (June): 82–89.

———. 1998. "Thinking about Bacterial Populations as Multicellular Organisms." *Annual Review of Microbiology* 52: 81–104.

———. 2007. "Bacteria Are Small but Not Stupid: Cognition, Natural Genetic Engineering and Socio-bacteriology." *Studies in History and Philosophy of Biological and Biomedical Sciences* 38: 807–19.

Shearer, Creswell. 1906. "On the Existence of Cell Communications between Blastomeres." *Proceedings of the Royal Society of London B* 77 (520): 498–505.

Shirisaki, Yoshitaka, Mai Yamagishi, Nanako Shimura, Atsushi Hijikata, and Osama Ohara. 2013. "Toward an Understanding of Immune Cell Sociology: Real-Time Monitoring of Cytokine Secretion at the Single-Cell Level." *IUBMB Life* 65 (1): 28–34.

Sonnenschein, Carlos, and Ana Soto. 1998. *The Society of Cells: Cancer and Control of Cell Proliferation.* New York: Springer-Verlag.

Spencer, Herbert. 1868 [1850]. *Social Statics: or, the Conditions Essential to Human Happiness Specified, and the First of them Developed, 2nd ed.* London: Williams and Norgate.

Strassman, Joan E., and David C. Queller. 2004. "Sociobiology Goes Micro." *American Society for Microbiology News* 70 (11): 526.

———. 2010. "The Social Organism: Congresses, Parties and Committees." *Evolution* 64: 605–16.

Sutherland, Earl, I. Oye, and R. W. Butcher. 1965. "Action of Epinephrine and the Role of the Adenylyl Cyclase System in Hormone Action." *Recent Progress in Hormone Research* 21: 263–46.

Torday, John, and V. K. Rehan. 2012. *Evolutionary Biology, Cell-Cell Communication, and Complex Disease.* Hoboken, NJ: Wiley-Blackwell.

Van Baalen, Minus. 2013. "The Unit of Adaptation, the Emergence of Individuality, and the Loss of Sovereignty." In *From Groups to Individuals: Evolution and Emerging Individuality*, edited by Frédéric Bouchard and Philippe Huneman, 117–40. Cambridge, MA: MIT Press.

Velicer, Gregory, and Michiel Vos. 2009. "Sociobiology of the Myxobacteria." *Annual Review of Microbiology* 63: 599–623.

Virchow, Rudolf. 1958 [1858]. "Atoms and Individuals." In *Disease, Life, and Man: Selected Essays by Rudolf Virchow*, translated and with an introduction by Lelland J. Rather, 120–41. Stanford: Stanford University Press.

Wheelock, Margaret J., and Keith R. Johnson. 2003. "Cadherin-Mediated Cellular Signaling." *Current Opinion in Cell Biology* 15 (5): 509–14.

Winans, Stephen. 2002. "Bacterial Esperanto." *Nature Structural Biology* 9 (2): 83–84.

Winans, Stephen, and Bonnie L. Bassler. 2002. "Mob Psychology." *Journal of Bacteriology* 184 (4): 873–83.

Yap, Alpha S., and Eva M. Kovacs. 2003. "Direct Cadherin-Activated Cell Signaling." *Journal of Cell Biology* 160 (1): 11–16.

5

Alternation of Generations and Individuality, 1851

LYNN K. NYHART AND
SCOTT LIDGARD

In the 1840s and 1850s, the nature of individuality was *the* fundamental underlying problem for biologists seeking to raise the theoretical level of their science. It was central to theoretical and practical discussions in cell theory, reproduction, morphology, physiology, classification, development, and evolution—virtually all the main areas of biological investigation in the period—and it tied them together (Nyhart and Lidgard 2011). In this chapter, we explore the understanding of biological individuality at a crucial moment in the middle of the nineteenth century, as illuminated by the phenomenon (and problem) dubbed "alternation of generations" by Johannes Japetus Steenstrup in 1842 (Geddes and Thomson 1902; Winsor 1976; Churchill 1979; Parnes 2007). Remarkably, the physiological zoologists Richard Owen, Thomas Henry Huxley, and Johannes Müller and the botanists Matthias Schleiden, Alexander Braun, and Wilhelm Hofmeister all wrote on alternation of generations in 1851. In tracing these responses to Steenstrup's proposed generalization, we gain new insight into the questions and assumptions surrounding biological individuality that animated nineteenth-century life scientists.

Ocean voyaging before 1818, the German naturalists Adalbert von Chamisso and Johann Friedrich Eschscholtz observed something strange about the small marine invertebrates called salps. There seemed to be two kinds. In one, "several individuals are born attached to one another," but the offspring differed entirely. "It swims free and solitary in the waters, and we have seen it give birth to numerous republics or confederations of beings resembling those from which it was born. Two generations unalike and alternating constitute the species" (Chamisso 1821, 207 [our translation]). These observations, along with the connection between "alternation" and "generations," were ridiculed or ignored.

But in 1842, the Danish naturalist Johannes Japetus Smith Steenstrup made the terms, and the idea, famous. He identified "alternation of generations" with the following phenomenon: one living being sexually produced an offspring that did not resemble itself. When that offspring, which he called the "nurse" form, reproduced again (asexually, through budding), its own progeny returned to the original form, if not in the next immediate generation then in a later one. Steenstrup focused less on reproductive mode than on the succession of dissimilar bodies, interpreting each one as a generation and a separate individual, rather than a metamorphic transformation of the same individual (Churchill 1979; Parnes 2007). He presented alternation of generations as a law-like principle of nature. Steenstrup's synthesis of alternating generations exposed semantic relations that were understood to hold between a part and the whole, and that identified what made a biological entity an individual, or not.

Then, as now, conflicts around these part-whole relations played out in the raw epistemological practice of scientists and the criteria developed for distinguishing individuals. Then, as now, scientists were concerned with epistemological and ontological questions surrounding unity, composition, and functional integration; with concepts of spatial and temporal boundedness and corporeal indivisibility; and in causal and processual explanations of biological individuality (Elwick, this volume). Different concepts of biological individuality derived from various combinations of these commitments, and vied for acceptance from the end of the eighteenth century through the late nineteenth century.

These many concepts became focused around 1851, like metaphorical grains passing through an hourglass. The neck of the hourglass, the forcing constraint, was the exposure of part-whole relations brought about by varied interpretations of alternation of generations. The collision of these individuality concepts with alternation of generations both revealed scientists' underlying assumptions and generated new ideas about levels of individuality above and below the level of the organism. These are well illustrated by the half-dozen important participants we focus on here. More would certainly be possible: we omit here Carl Nägeli, whose complex and important ideas deserve separate treatment. We also leave out Rudolf Leuckart, especially important for interpreting both alternation of generations and individuality in terms of the physiological division of labor (see Nyhart and Lidgard 2011; Leuckart 1851). Our historical actors' biological individuals are disparate, contextualized by different threads of reasoning and at times by different defining criteria. Foreshadowing today's discussions, one scientist might bundle several criteria together without disambiguating them; another scientist might propound a single absolute, universal criterion of individuality, disavowing the importance of others. Scientists also differed in their causal explanations for integration of subordinate individuals in relation to a whole: anatomical organization; efficient division of labor among (polymorphic) individuals for the well-being and perpetuation of the whole; similar developmental formation; the construction of higher-level individuals via transforming coalescence of lower-level ones; or an unseen "force" within germ-cells.

The controversies of the 1850s over alternation of generations did not result in a clear-cut outcome for definitions of individuality, beyond agreement that it was indeed a widespread phenomenon across the animal and plant kingdoms (see Elwick 2007 for a different claim for the British context). Nevertheless, the controversies were significant, for what they show about both the past and the present. They reveal that the scientists involved generally shared a broad commitment to a temporal perspective, an overall development toward "perfection" over time, and multiple levels of individuality. No one uses teleological terms like "perfection" anymore, but ideas of increasing complexity of life forms, both in the course of ("individual") development and over evolutionary history, persist. Biologists today still think of individuals in terms of development and life cycles, and integrate biological hierarchy into fundamental concepts such as modularity and levels of selection. The historical terms and circumstances may appear foreign to us, but we grapple in a similar way with the same phenomena.

Steenstrup: What Is Alternation of Generations?

Steenstrup was not the first to observe the pattern he called "alternation of generations." In addition to Chamisso's 1819 discovery about salps (Glaubrecht and Dohle 2012), Michael Sars (1841) had tentatively identified a seemingly parallel pattern in the jellyfish *Aurelia aurita*, in which the free-swimming medusa form alternated with a sessile polyp form. But Steenstrup showed that this pattern, previously thought to be an isolated phenomenon, appeared in a broad range of invertebrate groups.

In more complex life cycles, the different forms themselves often passed through their own dissimilar developmental stages. Some marine invertebrate polyp forms developed to their own mature level; the "strobila" that then budded off underwent development in turn before becoming the sexually reproducing, adult medusa (Fig. 5.1). Among genera of parasitic trematode worms or "flukes," the generation of "cercaria" (previously named as a

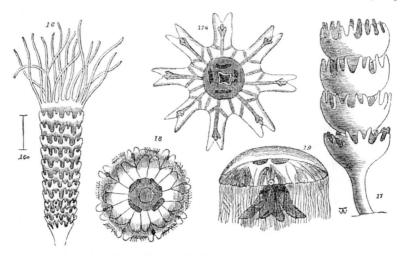

FIGURE 5.1. Alternation of polypoid and medusiform generations in the iconic *Medusa aurita* (now *Aurelia aurita*; Jones [1871], p. 85, fig. 49). 16 and 17 show the polyp form in the process of "strobilation" or detaching the next stage, which matures (17a, 18) into the medusa form (19). The latter sexually produces new polyps.

separate genus) was sometimes considered to pass through both "larva" and "pupa" stages before it produced the fluke (Steenstrup 1845, 52, 57–63; see Fig. 5.2). Steenstrup's insight was to discern in these unfamiliar, complex life cycles a common pattern that involved the production of two distinct forms rather than a single, continuous trajectory from embryo to adult. This was what he called "alternation of generations."

For Steenstrup, alternation of generations did more than describe a pattern; it was a unifying solution to several problems. Its most obvious benefit was to unite under a common rubric varieties of reproduction that had previously been seen as rare and anomalous. Yet it questioned the everyday notion—one associated with vertebrates—that the individual was an independent entity with spatial boundaries and organization (its form), a physiologically integrated whole that existed from birth (or egg), through sexual maturity and reproduction, to death. It also challenged those scientists who believed that the defining feature of animal individuality was the capacity for sexual reproduction. Christian Gottfried Ehrenberg (1851, 765–69) suggested that this was a legacy from Linnaean classification, which took as its object the "adult individual," defining both "adult" and "individual" by the ability to reproduce sexually. In decoupling these links, alternation of generations opened up individuality to a variety of interpretations.

Steenstrup's work exposed a broader comprehension of development in nature, by showing the actual means by which certain species attained their most perfect form—not through a linear development to maturity, nor through metamorphosis, but through a discontinuous process that took multiple generations. Steenstrup thought he had espied a further pattern as well. He posited a gradual rise in autonomy across groups, as expressed in both morphology and voluntary action—two different forms of a trend toward perfection. He even extended the theory to plants, referring to a plant or tree as a colony of individuals, unfolding over generations toward perfection. On this view, individuality could and did exist in a hierarchy of levels. In both plants and animals, moreover, the intermediate, vegetatively propagating generation(s) fostered the forms that would mature into the "perfect," sexually reproductive form, which Steenstrup rendered in language that ties his metaphysics to earlier *Naturphilosophen* (Churchill 1979). This "fostering" was the key function of his "nurse-generation" [*Ammengeneration*]. Indeed, for Steenstrup, the "nursing" function explained the significance of these non-reproductive beings' very existence. Even as he recognized their value, Steenstrup re-enshrined sexual reproduction as the most significant attribute of adulthood.

Crucially, Steenstrup's alternation of generations bore on individuality in distinguishing this process from metamorphosis: "a *metamorphosis* can only

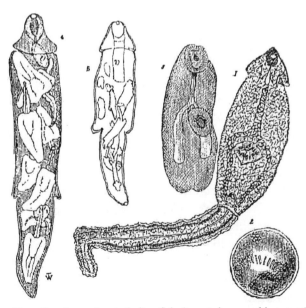

FIGURE 5.2. Alternation of generations in the liver-fluke (trematode worm of the genus *Cercaria*), emphasizing the physically nested nature of the different generations. (4) shows "a *nurse* containing fully developed *Cercariae*; (5) shows "a *parent-nurse* filled with partially-developed '*nurses*'" (Jones [1871], p. 158, fig. 110, after Steenstrup).

imply changes which occur in the *same* individual; but when from it other individuals originate, something more than a *metamorphosis* is concerned" (Steenstrup 1845, 6). An individual of a given generation might itself undergo metamorphosis during its development (as suggested by the cercaria "pupa"). But by this very definition, metamorphosis took place within one generation (that is, one individual). When development took place across individuals, it moved to another category; hence the idea "alternation of generations."

All this presupposed it was possible to know what an individual was. For Steenstrup, evidently, individuality was centrally defined by propagation of distinct bodies. Sexual reproduction was not the only option. Different modes of propagation and development—budding, fission and subsequent regeneration, metamorphosis, parthenogenetic embryo formation, sexual embryo formation by union of gametes—were all subsumed under the overarching principle of differing forms, irrespective of whether the mode of reproduction was asexual or sexual (Churchill 1979). The individual members of his different generations were separated or individuated from others partially by their autonomy and distinctiveness of form, but also temporally. Persistence was also part of this picture: whereas the caterpillar disappeared as it turned into the butterfly, the different developmental stages in the cases Steenstrup brought together could persist after budding off new individuals. Each phase therefore comprised multiple "individuals" and could be considered a "generation."

If Steenstrup was clear in distinguishing individual life cycles from generational ones, he also held to a yet higher level of individuality: the species. "Species individuality (if I may thus express it)" constituted the sum of the different forms entailed within the species (Steenstrup 1842b, 118). Whereas traditionally this was understood to encompass the adult male and female forms plus the developmental stages needed to reach them, Steenstrup argued that the forms of the alternating generations must be added to compose the whole species. By implication, the species was not defined by the sum of its individuals, but by the sum of its generations. Generations constituted the intermediate term between individuals and species.

The impact of Steenstrup's book was rapid, yet mediated. His original 1842 essay was published in Danish, translated into German that same year, and then translated into English in 1845 from the German (Steenstrup 1842a and b, 1845), gaining and losing meanings in translation. From Danish to German, it gained a title: the Danish title became the subtitle, and the title *Ueber den Generationswechsel* was added. This tipped the emphasis from the form of brood-care—Steenstrup's singular concept of the "nursing generation"—to an emphasis on the changing pattern—the succession of generations. The English version further metamorphosed the meanings, in particular muting the emphasis on development in nature. The title word "Generationswechsel" was translated into English as "alternation of generations," emphasizing two generations, rather than the more ambiguous "succession of generations" that could also have been meant by the German and Danish terms (Kramp 1943, 13; Cornelius 1990, 580). Steenstrup's terms would come under close scrutiny as scientists debated the nature of alternation of generations and its bearing on individuality.

Individuality and Causal or Explanatory Laws in Zoology

Steenstrup's generational conception helped to bring the biological problem of individuality into sharp yet chaotic focus: "ascribing generational identity to cells, germs, or buds was far from obvious in about 1850 and was often disputed" (Parnes 2007, 320). As researchers discovered more nuanced complexities of reproduction in many invertebrates and plants, they found themselves forced to make explicit their previously tacit assumptions about the defining characteristics of individuality (Thomson 1859; Phillips 1903; Elwick 2007).

The British anatomist Richard Owen both endorsed and criticized Steenstrup's concept. In *On Parthenogenesis* (1849), Owen praised the wealth of facts amassed as the *phenomena* of alternation of generations, and the "known" analogies of other embryonic developments or individualizations in animals and plants. Owen's critique began by dismissing Steenstrup's metaphor of the "nurse" as unsuitable. His main appraisal was still more severe—far from being a causal law of nature, Steenstrup's alternation of generations merely redescribed the phenomena and provided no real explanation or cause for it. Owen himself assumed that burden under his broad new term, "parthenogenesis," and his own theory of metagenesis. Unlike our current use of the term, Owen's parthenogenesis encompassed virtually all forms of development and propagation known at that time, thus subsuming alternation of generations (Fig. 5.3). Owen presented his theory of metagenesis as a law of nature, both causal and explanatory of the conditions constituting generation in plants and animals. His response to Steenstrup must be understood within this broader framework.

One important feature of Owen's theorizing about generation and, explicitly, organismal development is his hierarchy of individuality. In impregnation, the ovum receives "the matter of the spermatozoan," forming the "germ-cell," which propagates numerous "offspring," by growth, differentiation, and both asexual and sexual reproductive modes: "The progeny of the impregnated germ-cell . . . minister to the life of a being higher than themselves"

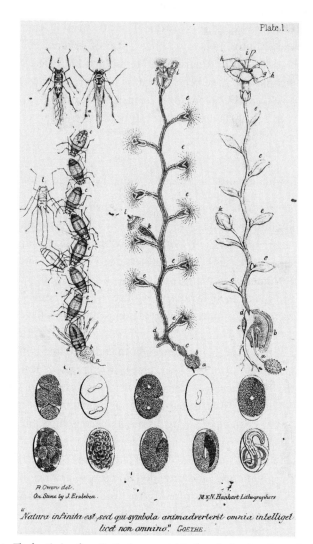

Plate.1.

R. Owen del.
On Stone by J. Erxleben.

M.&N. Hanhart Lithographers.

"Natura infinita est, sed qui symbola animadverterit omnia intelliget licet non omnino". GOETHE.

FIGURE 5.3. The frontispiece from *On Parthenogenesis* (1849) emphasizes Owen's perceived correspondence of life cycle stages across representative animals (aphids and hydroids) and a flowering plant, all of which show an alternation of generations. Near the bottom of the vertical life stages, the figures represent parallels of sexual reproduction for each organismal group: spermatic fertilization (a), and below this, cell cleavage divisions and embryogenesis. Rising vertically in the illustrated stages, asexual growth/reproduction iterates many individuals, some of which differ in form, eventually producing individuals that reproduce sexually (see also Churchill 1979).

(Owen 1849, 5). Owen thus recognized a hierarchy of biological individuality below the level of the organism, linked by developmental transformations. This view loosely recalls hierarchical individuality theorized by Leibniz, Kant, Goethe, and a host of earlier idealist naturalists (Nyhart and Lidgard 2011). Yet Owen's individuals begin at the cellular level: individual egg and sperm cells fuse into the primary germ-cell, which in turn undergoes fission into secondary germ cells—all are individuals. The secondary germ cells together form a "germ mass" that then produces new individuals at a higher hierarchical level via subsequent coalescence and dissolution of parts. These higher-level individuals, by growth, propagation, combination, and metamorphoses, yield the body of an organism, which is also an individual (Owen 1849, 4–5). Undoubtedly influenced by the new cell theory of Schleiden and Schwann, Owen developed his theory of metagenesis in a direction untouched by Steenstrup, who did not focus on germ cells or tissue formation.

Also far more than Steenstrup, Owen emphasized a cyclic perspective: "a species is represented by a series of individuals of different powers and forms succeeding each other in a cycle" (Owen 1849, 13). By 1851, Owen had replaced his use of "parthenogenesis" with the term "metagenesis," as in this passage about asexually reproduced individuals in *Aphis*: "Then recommences the cycle of change, which being carried through a succession of individuals and not completed in a single lifetime, is a 'metagenesis,' rather than a 'metamorphosis.' This phenomenon . . . now proves to be an example of a condition of procreation to which the greater part of organised Nature is subject" (Owen 1851, 271). Thus physically separate bodies in the life cycle, whether reproduced sexually or asexually, were of the same nature; the different forms that ensued were *necessary* to attain and understand the completeness or "perfection" of the species (Ogilvie 1861).

Owen's most serious criticism of Steenstrup was that he provided no causal explanation for the phenomena he identified. Owen (1849) argued that generation of life forms was explained by and contingent upon a "spermatic force" (Elwick 2007, 101–5) invested in the sexual union of the germinal vesicle (nucleated cell of ovum) and sperm-cell (nucleated spermatozoan). His appeal to such a force was apparent as early as his 1837 Hunterian Lectures (Owen 1992, 61), which were based partly on work by the anatomist Johannes Müller (Müller-Wille 2010). In 1837, Owen wrote, "The germ is the whole organism potentially; the different parts *actually* exist by its development. We recognize the power to produce the different parts of ye intire organism, as residing in, and operating from the germ, when we trace their *actual* formation in the incubating egg" (Owen 1992, 219). Owen's spermatic force (1849, 1851) was teleological in the sense that it explained the immanent course of morphological differentiation in a life cycle, not a pre-ordained structure of the great chain of being. (Sloan 2003 discusses Owen's complex developmental theories.) The processes by which physically separate bodies are generated differed "only in accessary and non-essential particulars; the nature of the essential spermatic force would be the same" (Owen 1849, 75).

Owen's famous conflict with Thomas Henry Huxley over the true meaning of biological individuality—and alternation of generations—has been well rehearsed elsewhere (Churchill 1979; Desmond 1984, 35–48; Elwick 2007, 138–48), but is illuminating for our story. Huxley's more materialist, anti-essentialist, and anti-idealist position clashed with Owen's views. Huxley commenced his siege in 1851, in a review of Müller's work showing a transition between different larval and adult forms in echinoderms. Huxley (1851a, 14) disparaged Müller, Steenstrup, and most especially Owen, who attempted to "unite all the aberrant generative processes . . . and to express them in its terms," associating these with a "psychical individuality," with which "the zoologist has no concern." Huxley and numerous others since that era painted Owen as a Platonic essentialist, yet, as we have seen, Owen's conceptual use of explanatory forces and their actions in organismal development is more complex (Sloan 2003). Opposing Owen, Huxley asserted his own definition of biological individuals and ranking of relevant criteria:

> Individuality has so long and so obviously, among the higher animals, been observed to be accompanied by independent existence, that the latter attribute has come to be considered as, conversely, an indication of individuality—to the neglect of the really characteristic attribute, which is—the circumstance of being the total result of the development of a single ovum. (Huxley 1851a, 15)

What then of Steenstrup's and Owen's other generative processes counting gemmation (budding) and fission, of resulting transitions in form, of separately bounded bodies that were generated asexually? Huxley offered the new term "zooids," that "are like individuals but not individuals." "Instead of saying then, that in a given species, there is an alternation of so many generations, we should say that the individual consists of so many zooids. Again, where no "alternation" takes place, the individual = the sum of its organs; where there is alternation, the individual = the sum of its 'zooids'" (Huxley 1851a, 15).

Huxley, too, endorsed an orientation toward life cycles, "beginning with an ovum, they end with an ovum; and thus, though there is a physical continuity between the progenitrix of a race and her latest progeny, yet the material substratum presents a series of distinct, though constantly recurring, cycles of form" (Huxley 1856, 481). But unlike Owen and Steenstrup, whose views of biological individuality might be construed as "bottom-up," wherein a succession of individuals transform or construct higher levels of individuality (and complexity) within cycles of development, Huxley's view seems decidedly "top-down." Huxley's metaphysics included a deep commitment to a fundamental unity of plan in animal organization. Alternation of generations and metagenesis were profoundly uncomfortable challenges to this commitment. How could "like beget like," if generation in aphids, jellyfish, and bees were utterly different from generation in higher animals? In several places, Huxley framed his view of individuality downward from the better-understood perspective of higher animals, especially vertebrates. In his analysis of different sexual and asexual forms in the life cycle of salps (pelagic tunicates), he wrote, "Not either of their forms, but both together, answer to the 'individual' among the higher animals" (Huxley 1851b, 52). And unlike Owen's epigenetic vision of individuals transforming (sometimes abruptly) and increasing in complexity, Huxley seems committed to the more linear development exhibited by higher animals. His lawlike epigenesis is simpler, merely the developmental appearance of parts from embryogenesis onward: "there is a kind of individuality which is constituted and defined by a fact or law of succession. Phenomena which occur in a definite cycle are considered as one in consequence of the law which connects them" (Huxley 1852, 185). We note that just as concepts of hierarchical or composite individuality had deep historical roots, so too did the view that individuality is defined by the fertilized ovum's course of development, with defenders such as the botanist Mirbel (1815), and attributions extending back even to Aristotle.

Some contemporaries questioned Huxley's neglect of boundedness and physiological unity as criteria of individuality (e.g., Lubbock 1857). Others went even further: "Hence if we would avoid the solecism of speaking of the *individual* as in some cases being *divided* into a number of parts, having all a general character of completeness, and use the word, as propriety seems to require, in its logical sense of *that which is individualized*, we must be prepared to admit as a consequence, that the individuals of one species are not always homologous with those of another" (Ogilvie 1861, 110). Still, many scientists sympathized in principle with Huxley's views (Quatrefages 1864). The zoologist William Benjamin Carpenter, in particular, presented largely the same case as Huxley in his review of then-current publications on development among zoophytes (1848), and savaged Owen's *On Parthenogenesis* a year later (Carpenter 1849; Churchill 1979; Elwick 2007).

The Huxleyite "victory" was less complete than has been suggested, however (Elwick 2007). Another salient feature of *On Parthenogenesis* is the idea of "germinal continuity" maintained through a genealogical succession of organisms, for which Owen is often credited (Stockberger 1913). He wrote here that not all the derivatives of the primary germ cell are required for the body's formation. Certain germ-cell derivatives remain "unchanged" and are retained in the body; these may give rise to the development of another body, an individual, which may or may not resemble the first body (Owen 1849, 4–6 and 72–73). In the wake of the publication of Weismann's germ plasm theory some fifty years later, older British scientists were quick to point out the strong resemblance to Owen's interpretation (Geddes and Thomson 1902; Collier 1997; Rupke 2009). Weismann (1893, 173–82) adopted alternation of generations and individualized germ and soma; new individuals might arise from each, although control of ontogeny ultimately resided in different kinds of particles in the germ plasm. Weismann's ideas contributed eventually to gene theory and the modern evolutionary synthesis. In the late twentieth century, however, reconsideration of "somatic embryogenesis" subtly echoed Owen and Weismann in an entirely new context: major transitions and the evolution of individuality (Buss 1983, 1987).

Disputes over the meanings and usefulness of different concepts of biological individuality continued over the remainder of the nineteenth century and through the twentieth (Nyhart and Lidgard 2011). Many scientists, often with opposing views on individuality, sought to generalize "alternation of generations" across major divisions of life (Carpenter and Dana 1851; Lankester 1857; Quatrefages 1864; Geddes and Thomson 1902; Svedelius 1927). Discoveries of new complex life cycles—especially in algae, lower plants, protozoans, and parasitic metazoans—continued to challenge the generality and even the applicability of different views. By the early 1900s, the enormous breadth of Owen's original "parthenogenesis" had narrowed in meaning (Phillips 1903; Lankester 1919), while the equivalence of "alternation of generations" in plants and animals, and its synonymy (or not) with "metagenesis," were still debated (Jones 1937; Boyden 1954). All these terms survive today, albeit with meanings altered by genetic discoveries. Perhaps most important for our story, however, the dominant biological focus shifted in Huxley's direction (Elwick 2007, 149–59), that is to say, toward elucidating sexual reproduction and transmission (Churchill 1987; Rheinberger 2008). At least in the English-speaking world and especially among zoologists, comparative morphology, development, and generation via asexual modes and their attendant consequences for individuality concepts were largely pushed aside for much of the next century (Boyden 1973).

Individuality, Alternation of Generations, and
Developmental Cycles of Plants

In the history of botany, 1851 is acclaimed as the year when Wilhelm Hofmeister "discovered" alternation of generations in plants. The historiographic emphasis on Hofmeister's genius, exhibited by his single-handed discovery (Morton 1981; Kaplan and Cooke 1996), has obscured the fact that its acceptance was not immediate, and that there was considerable earlier discussion of alternation of generations among botanists, spurred by previous zoological discoveries. This section situates Hofmeister's achievement within the broader botanical discourse on the nature of plant individuality and alternation of generations, focusing especially on Matthias Schleiden and Alexander Braun.

A critical component of the discussion among botanists (and some zoologists) concerned the legitimacy of analogies between plants and animals with respect to alternation of generations—analogies already suggested by Steenstrup himself (1845, 114–15). He urged that, like the animals he had discussed, the plant or tree (a "compound individual") "unfolds itself through a frequently long succession of generations, into individuals, becoming constantly more and more perfect, until, after the immediately precedent generation, it appears as *Calyx* and *Corolla* with perfect male and female individuals . . . and after, the fructification brings forth seed, which again goes through the same course" (Steenstrup 1845, 114–15). Here Steenstrup laid bare assumptions he shared with many contemporary Continental botanists. Over a succession of "generations," through vegetative propagation, the plant individuals moved toward their perfect state—the sexually reproducing male and female individuals; following sexual reproduction, the cycle of successions began again. Plants and animals thus exhibited a broadly similar pattern of vegetative and sexual propagation.

Yet plants and animals differed, and so, too, did the problem-complexes surrounding them. For zoologists, alternation of generations solved some problems about anomalous kinds of individuals, by asserting a common temporal pattern of animal generation and form-change that covered many zoophytes and parasites. But among botanists, relationships among alternation of generations and definitions of individuality and generation were profoundly shaped by ongoing debates over the nature of plant individuality in general, cell formation, and the identification of sexual structures (and sexual reproduction itself) in different kinds of plants, especially among the cryptogams (plants with "hidden sexual parts").

By standards promulgated by many German botanists since the early nineteenth century—whether followers of Schelling's *Naturphilosophie* or of Goethe's theory of plant metamorphosis and the looser "Romantic" style it conjured up—the determination of all of these questions would have to come via a developmental perspective. While scorning the speculative, obfuscatory style he attributed to the *Naturphilosophen*, Matthias Schleiden (1842) likewise advocated the study of development as key to understanding plant life. Indeed, he argued that the nature of plants mandated a particular kind of developmental perspective, one that emphasized the cyclicity of plant life. As he put it (1842–43, vol. 1: 31), "in nearly every moment of its life, the plant is only a part of itself." To reveal and understand the plant as a whole, then, required attention to all the moments in the plant's life, adding up all the parts that came and went (Schleiden 1842–43, vol. 1: 31–32). Using the simultaneously dying flower and emerging fruit as an example, Schleiden's contemporary Alexander Braun made a similar point in his 1851 monograph on "rejuvenescence": even as one part of the plant dies off, another begins anew.

Both Schleiden and Braun recognized that this cyclical pattern operated above the level of the most fundamental plant unit. But what was that unit? Goethe had followed earlier botanists in arguing that an idealized leaf-form was the fundamental unit of the plant (Arber 1950, 42–47; Nyhart and Lidgard 2011). Others, including Braun, argued that the shoot (comprising the stem, leaf, and bud together) was the basic unit (Braun 1851, 1853a; Arber 1950, 70–71). Schleiden (1838) famously argued that the cell was the fundamental unit—the individual—through which all the processes of life were sustained. But in his 1842–43 textbook, he was more nuanced. While positing the cell as the most fundamental level of individuality in plants, he cautioned against dogmatism. He viewed "individual" as an interpretation, its features variable depending on context and, especially, scale. Since, Schleiden thought, we lack agreement on what constitutes a "plant," then we can have only pragmatic definitions of plant individuals. He named a nested hierarchy of three orders of plants: simple plants of the first order (single cells), second-order plants that comprise a union of the first-order ones, and third-order plants that similarly unite second-order ones into composites (Schleiden 1842–43, vol. 2: 4–6).[1] Schleiden thus made explicit notions of hierarchical levels of plant individuality that had been mooted for many years.

In 1851 Braun, too, entered into the discussion of individuality, its levels, and alternation of generations in his book *Betrachtungen über die Verjüngung in der Natur* (translated [1853a] as "Considerations on the phenomenon of rejuvenescence in nature"); he would elaborate further two years later (Braun 1853b, translation 1855). Like many German biologists (Rieppel 2011), Braun believed that a nested hierarchy of individuality extended from the lowest level all the way up to Nature as a whole. Whereas Schleiden saw the expression of plant life as inhering in the first instance in the cell, and higher orders of individuals as built up though aggregation, for Braun, individuals could be comprehended only when seen as members of a higher-level temporal entity, the species. When the mature fruit produced a new seed, it did more than simply "start" a life. It formed a link in a chain of generational succession, which itself constituted a higher level of individuality—the species. This pattern repeated, and at each higher level, "the individual is realized through a chronological succession of formations and material division into subordinate links . . . realized in a complex of a higher order," all the way up to the "organism" of Nature as a whole (1853a, 323). Ultimately, for Braun, it was God who organized this great multileveled system of development such that at each level development adhered to a plan generated "internally."

Within this broad framework, Braun considered alternation of generations at the levels of leaves, cells, and shoots—the chief contested lowest-level units of plant individuality. In a discussion of the leaf that cited Steenstrup, Braun wrote (1853a, 106) that some botanists had viewed the plant "as a series of generations, the individuals of which are represented by the leaves, and the metamorphosis of the plant as an alternation of generations, through which, after numerous preparatory and asexual individuals, the generation finally arrives in the flower at the formation of the sexual individuals closing the series." Here we have alternation of generations specified to the leaf as individual. However, Braun rejected the leaf-as-individual idea, arguing that the stem was always a prerequisite component (107)—the relevant unit being the stem, leaf, and bud together, developmentally conjoined as the shoot.

Braun also considered that alternation of generations might occur at the level of the cell, as represented by certain unicellular algae. Here the cells divided vegetatively, "but the plant is incapable of keeping intimately connected the generations of cells thus produced . . . ; it breaks up more or less completely into the individual cells, which thus, although members of a definite cycle of vegetation, lead a separate individual life. We see that which is completed by the higher plants in an organically connected series of cells, represented here by an alternation of generations of unicellular vegetable individuals" (Braun 1853a, 126). Braun defined cell "generations" by the formation of new cells; "alternation of generations" again meant the alternation of vegetative cell division and the fructification that "closes" the cycle. Here Braun brought Steenstrup's idea down to the level of the cell, but again unified by reference to a higher level of organization—the "cycle of vegetation."

At the end of this work Braun generalized his image of the composite biological individual "breaking up into a limited or unlimited series of subordinate (morphological) individuals" that exhibited various degrees of autonomy (Braun 1853a, 322–23n). He argued that this pattern, which encompassed alternation of generations, held across numerous groups in both the plant and animal kingdoms. Thus, he presented an image of top-down organization that differed from Huxley's in its much more deeply layered teleology. Ultimately, in Braun's system, both individuality and alternation of generations were explananda, but at different levels: the latter was encompassed by the former, and explained within a developmentalist framework of the succession of forms, dependent on a multileveled concept of individuality.

Hofmeister came to alternation of generations from a perspective different from either Braun or Schleiden. Remarkably, in his famous 1851 work, *Vergleichende Untersuchungen der Keimung, Entfaltung und Fruchtbildung höherer Kryptogamen und der Samenbildung der Coniferen* (Comparative investigations on the germination, development, and fruit-formation of the higher cryptogams and of seed-formation in the conifers), he used versions of the term "individual" only eight times in 142 pages, and he was completely unconcerned to define "individuals," or even generations. Indeed, he barely used the term "alternation of generations." Perhaps this was just part of his laconic style, which reflected a strongly empiricist bent he shared with Schleiden. It may also have been a deliberate effort to sidestep what may have seemed the fruitlessly philosophical tenor of discussion surrounding individuality and alternation of generations.

Alternatively, Hofmeister's reticence with respect to these issues may simply reflect his primary interest in a different pair of hotly contested botanical concerns: the identification of homologies among the sexual parts in different plant types, and the detailed processes by which cells form and multiply. Hofmeister's achievement, which would ultimately draw together the entire plant kingdom into a common life-cycle pattern, lay in the way he combined these two issues in a study that swept across the higher cryptogams. First, he identified the egg- and sperm-producing organs (the archegonia and antheridia) with slightly different structures than had his predecessors, which allowed him to draw new homologies across different plant types—especially mosses and ferns. He derived the alternating generations from two differing modes of cell formation, distinguished by the botanist Carl Nägeli in 1844 (tr. Nägeli 1846).

Schleiden held that all new cells formed by coalescing around a nucleus within an existing cell ("free cell formation"). While accepting this as one mode of new cell formation, Nägeli argued that cells could also multiply by division. In the latter case, the original cell dissolved into its offspring cells, marked by the creation of new cell walls between them. This mode he called "parietal" (*wandständige*) cell formation. As the historian John Farley has shown (1982, 86–97), Hofmeister's innovation was that only a cell arising through free cell

formation marked a true new individual and thus a new generation; multiplication through division was just part of a plant's growth and did not signify
the origination of a new individual. Free cell formation occurred just twice in
the life cycle of the higher cryptogams: once in producing the "mother-cell"
from which developed the sexual bodies (the archegonium and antheridium),
and once in producing the spore-bearing body. Like Braun and Steenstrup,
then, Hofmeister also saw the alternation of generations as residing in the alternation of sexual and asexual modes of generation (and also in the clearly
distinct forms produced by these two modes). But the reasoning underlying his assertion was novel, and the extent of empirical observation running
across the higher cryptogams rendered the reasoning persuasive to many.
Only Schleiden and his closest followers, such as Hermann Schacht (1852),
remained unmoved. While the chief reason may well have been Hofmeister's
rejection of the universality of free cell formation, Schleiden offered other ob-
jections: as late as 1861, in the fourth edition of his textbook, he harrumphed
that the very concept of alternation of generations depended on a comparison
between animals and plants that was illegitimate in principle (xiii).

Parasitism and Evolution

On October 23, 1851, Braun wrote to his friend Carl Theodor Ernst von
Siebold. "I've just come from the meeting at the Academy [of Sciences] and
want to tell you, while it's still hot, the extraordinary thing I've heard there.
Johannes Müller . . . held a lecture that kept me in the greatest suspense until the end, which brought me some peace" (Braun to Siebold in Mettenius
1882, 541–42). Müller had found many instances of a species of holothurian
echinoderm, the sea cucumber *Synapta* (now *Labidoplax*) *digitata*. Most developed normally, but occasionally he found one with an "abnormal sexual
apparatus," the eggs of which developed not into other sea cucumbers but
into a brood of creatures resembling pectinibranchiate snails. Could this be
some kind of alternation of generations? The snails certainly could not be
expected to complete the cycle by producing sea cucumbers in their turn.
This would go against everything scientists knew about molluscs. But maybe
it was a novel kind of alternation of generations—a true "heterogony," as
Müller called it, in which one kind of organism could truly produce another
kind. This seemed far-fetched, but how else was this phenomenon to be
interpreted?

Müller's struggles over interpreting the sea cucumber and its snail progeny allow us to examine how scientists studying alternation of generations
ca. 1851 engaged with another facet of the problem of biological individual-

ity, namely *individuation*. By this we mean how scientists addressed continuity, discontinuity, and transformation in nature, whether at the level of the common-sense individual organism, the species, or higher taxa. Whereas Braun and Huxley were concerned in their own ways to instill continuity in the face of the evident anatomical discontinuities in alternation of generations, Müller's case illustrates the attention given to discontinuity and continuity as simultaneous features of individuals and species, in particular as they wove through three different problems confronting naturalists: parasitism, classification, and evolution.

First, alternation of generations potentially contributed to solving the problem of the origins, life cycles, and classification of parasites, whose confusing life histories stubbornly resisted analysis. Parasites had often seemed to appear spontaneously in the bodies of their hosts. Many scientists had held that parasites were born through spontaneous generation, though that view was well on the decline by 1851 (Farley 1972). The life history of the parasitic trematode worm had been interpreted at length by Steenstrup as a case of alternation of generations. He identified distinct forms as belonging to different generations of a species, where these forms were previously understood as separate parasites of two different hosts. Siebold (Braun's correspondent above) soon identified another trematode stage, in which the cercaria became encysted within yet another intermediate host (Farley 1972, 119–20). More generally, alternation of generations showed that different generations occurred as free-living forms and as parasitic ones, connected through ingestion by a host or burrowing into it. One form entered the host; another appeared from within it.

The most likely explanation of Müller's mysterious mollusc was that it was a parasite. Certainly Müller rejected spontaneous generation as antithetical to science (1851, 645). But in the printed version of his talk (1851, translated 1852a), he observed that neither it nor a progenitor snail was eaten by the sea cucumbers, which in any case were filter feeders, and that no parental snail had crawled into the ovarian capsule (Müller 1852b, 24). So maybe this anomalous capsule, the "molluskigerous sac" (schneckenbildender Schlauch), was not an ovarian capsule at all, but was itself the wormlike parasite—"a vermiform metamorphosis of a mollusk, as it were" (Müller 1852a, 34; note mollusk = mollusc in modern terminology). This faltered on the observation that the "knob-like end" of the molluskigerous sac was not merely attached to the sea cucumber by a sucker. Rather, "the vessel of the *Holothuria* embraces and is grown to the knob of the sac" (Müller 1852a, 34). Based on this developmental criterion, the sac could not be a parasite, but had to be part of the sea cucumber. (See Figure 5.4.)

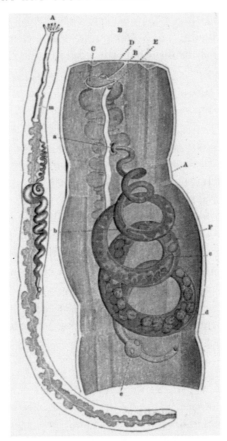

FIGURE 5.4. The sea cucumber *Synapta digitata* with "molluskigerous sac" that generated the snail eggs (Schmidt 1878, 322). (A): *Synapta digitata* with the parasitic molluskigerous sac. Natural size. (B): Close-up of the central portion showing the molluskigerous sac. Illustrations derive from Baur (1864), figs. 23 and 24.

Perhaps the sac formed as a "bud" from the parental holothurians (Müller 1852a, 34). Müller's next possibility was a case of alternation of generations, with the sea cucumber producing the mollusc, which would at some point, as yet unobserved, again produce a sea cucumber. Would the holothurians and molluscs thus form a single larger taxonomic group, like medusae and polyps? It seemed "highly improbable that the alternation of generations ever goes so far" (1851, 644; 1852a, 34). As Müller later elaborated (1852c, 23), even if polyps and medusae had been considered separate species before, at least they were in closely related classes, whereas the echinoderms and molluscs were far more distant branches of the animal kingdom.

What had excited Braun, though, was not primarily the implausibility of lumping together holothurians and molluscs under the rubric of alternating generations. It was, rather, Müller's new extension of alternation of generations, "heterogony."[2] As Müller put it, "the essence of the alternation of generations is, that the form B, produced from, and dissimilar to, A, reproduces the form A. How would it be, however, if B propagated itself as B, and A as B, but also as A?" (Müller 1852a, 34). "It could be that we must distinguish a heterogony via alternation of generations from another kind of heterogony, which does not necessarily alternate but which continues the produced form, a heterogony with similar continuation" (Müller 1852b, 31–32). In other words, in the second case, one kind might produce a completely different kind, which would then perpetuate itself. It was thus a form of instant speciation—and evolution.

At Muller's 1851 lecture, Braun told Müller that the new finding would shock naturalists. As for himself, "I've long been unable to think otherwise about the origin of organic beings. Geology clearly shows that a higher law of organic continuity must rule in the successive creation of organisms; but in our stationary times to have found an example of this—that I never expected" (Braun to Siebold in Mettenius 1882, 542–43). Müller seems to have shared his amazement, for it was exactly catching nature in the act of making new species that held profound meaning:

> The origin of different species in the creation is an indubitable fact of palae-ontology, but it remains in the region of the supernatural so long as the very act of origin is not seen, and traced out into its elements. As soon as this is possible, it ceases to be supernatural, and falls among the higher order of phenomena, whose laws also observation will one day discover. (Müller 1852a, 34–35)

Müller held fast to the conviction that once observed, the phenomenon had to be comprehensible within the realm of science. Yet his tentative heterogonic solution drew the skepticism of his English translator, who was "anything but convinced" by it. The novelty did not lie entirely in the invention of a saltationist version of transmutation; earlier scientists, such as Étienne Geoffroy St. Hilaire, Robert Chambers, Owen, and Arthur Schopenhauer had proposed speculative theories of saltational evolution involving developmental ruptures or teratologies (Lovejoy 1911; Richards 1987, 1994). However, none of these was grounded, as Müller's was, in alternation of generations and consequent individuation of different forms, and none seemed to have a basis in direct observation. Still, "so novel and startling a theory should at least be based upon very strong evidence," which was lacking; in addition, the alternative explanation of parasitism lay at hand (Müller 1852a, 38). As Müller continued to puzzle over the molluskigerous sac, he grew more inclined toward the solution of parasitism, in part swayed by the English commentator (1852c, 25). He never did resolve the question for himself before his death in 1858 (du Bois-Reymond 1860, 132–33; see also Baur 1864; see also Osborne, this volume, for later developments concerning parasitism).

Perhaps the most significant larger result of the interest in the sea cucumber and the snail was the later revival of Müller's "heterogony" (so quickly buried in 1851–52), in response to Darwin's 1859 *Origin of Species*. Both Richard Owen and the anatomist Albert von Kölliker (among others) found Darwin's theory unpalatable. Kölliker disliked Darwin's reliance on accident; Owen similarly opposed Darwin's unconstrained variation of lineages (though he also opposed Darwin's Lamarckian mechanism of use and disuse in generating species transmutation). Both men sought instead to account for the origin of new species through general "laws of nature" that would be more, well, lawful in character. Kölliker (1864, 235) modeled his law explicitly on alternation of generations. Here, one form might produce another very different from itself, which then might reproduce that second kind and never return to the first. One of the natural laws Owen proposed as alternatives to Darwin's transmutationism was "heterogeny," which referred to the very same thing as Kölliker (Rupke 1994, 226–30). Both laws, in fact, are indistinguishable from Müller's theory of heterogony.

Kölliker's scheme, built on Müller's, neatly solved the problem of species individuation—at least as a useful model. The connections across these classificatory links were simultaneously continuous and discontinuous: they provided a direct, continuous genetic link among species but also a clear shift from one species to another across generations. But whereas in 1851, Müller's theory appeared radical, in the new context of Darwin's theory, Kölliker's and Owen's looked conservative.

Concluding Remarks

Let us return to where we started. Today is hardly the first time that scientists have questioned what constituted an individual. The linkage of alternation of generations to the problem of individuality, the relationships between the animal and plant kingdoms, the individuation of species, the relations of parts and wholes in the organic world, and developmental concepts at different hierarchical levels—all these infused the conflict between physiological mechanists, who sought to reduce life to physics and chemistry, and comparative morphologists, who defined organisms through their part-whole relations. The story told here might seem stuck in the nineteenth century. In the succeeding 150 years, evolutionary biology and genetics have dominated biological research in an increasingly reductionist framework. Yet recent attempts at synthesis have made new comparisons with the past relevant for today's historians, philosophers, and biologists (Gilbert 2003; Love 2003).

Among the diversity of criteria defining individuality in the mid-nineteenth century, a few significant themes emerge. First, deciding what constituted a generation entailed deciding what constituted an individual: that preeminent link bound alternation of generations to the problem of individuality. Many of our actors—Steenstrup, Owen, Schleiden, Braun, Müller (also Leuckart and Nägeli)—resolved discrepancies among logic, observation, and intuitions about individuals and generations in part by resorting to nested levels of individuality, whether understood structurally or temporally. This solution extended back much earlier, especially in botany, whose practitioners often viewed the plant as a colony of lower-level individuals (see, e.g., Geddes 1878, 841; Rieppel 2011). Today, multiple units and levels of selection (Godfrey-Smith 2009; Clarke 2012; Bouchard and Huneman 2013; Sterner, this volume), major transitions in evolution and individuality (Grosberg and Strathmann 2007; Sapp 2009; Calcott and Sterelny 2011; Herron, Sterner, this volume), and the growing empirical support for modularity, symbiotic transitions, environment-epistatic developmental, and functional integration (Winther 2001a and b; Bouchard 2010; Gilbert and Epel 2015; Gilbert, this volume) continue to disrupt our orthodox notions of biological individuality and just what a part and an organism are (McShea and Venit 2001; Pepper and Herron 2008; Clarke 2011; Gorelick 2012; Brigandt, this volume; Love and Brigandt, this volume), especially when these are contextualized in a biological hierarchy (Bouchard and Huneman 2013; Lidgard and Nyhart, Love and Brigandt, and Rieppel, this volume). Looking back to around 1851, the challenges to orthodoxy seem remarkably similar.

Second, nearly all our scientists (with the prominent exception of Schleiden) sought to generalize their own interpretations about both alternation of generations and individuality. To what degree did plants provide a model for animals, and vice-versa? Could alternation of generations unite the plant and animal kingdoms through general laws? And if so, what was the nature of such laws? Mid-nineteenth century botanists and zoologists, including Steenstrup, Owen, and Braun, perceived certain parallels in plant and animal alternation of generations and their associated views of individuality. However, they and later zoologists and botanists frequently did not follow parallel courses as they incorporated aspects of both ideas into their respective canons (Allen 1937; Jones 1937; Walbot and Evans 2003). As we have indicated, the "Huxleyan individual," which treated "zooids" as mere parts within the true individual, prevailed in zoology, whereas aspects of hierarchical individuality have endured in certain realms of botany (Bailey 1906; Classen-Bockhoff 2001; Barthélémy and Caraglio 2007) and among certain zoologists concerned with clonal animals (Child 1915; Boardman et al. 1973; Hughes 1989). The mid-nineteenth century differences over lawlike generalizations covering alternation of generations and individuality in animals and plants continued in many quarters into the next century (Chamberlain 1905; Causey 1935; Vergara-Silva 2003). Alternation of generations in other major groups of organisms—fungi, algae, and protozoa—was not well-established until the late nineteenth and early twentieth century (Calkins 1901; Johnson 1915). Comparison of ideas about individuality and alternation of generations across zoology, botany, and perhaps mycology and microbiology thus opens a promising new avenue of historical inquiry.

The same can be said for looking across the living world and its scientific investigators today. Discoveries of complex reproduction and life cycles (Bell 1994; Achtman and Wagner 2008; Herron et al. 2013) continue to challenge earlier, simpler views of generation, sex, and individuality. The notion of the organism as a life cycle, which became paramount in both the metaphysics and epistemology of biology in the mid-nineteenth century, continues to test our understanding of individuality and its causal explanations. Stem cells, first discovered in hydroids by Weismann, now provide central problematics in embryonic determination, developmental plasticity, and regulation of complex life cycles (Frank et al. 2009). Of this plasticity, even the ontogenetic *reversal* of life cycles and alternate "generations" has been documented (Piraino et al. 1996)! And as we struggle to incorporate microbes—with rampant horizontal gene transfer, indistinct genealogies, and pervasive symbioses—into our more conventional understanding, generations and individuals become ever more perplexing (Gilbert et al. 2012; Doolittle 2013; Gilbert, this volume). A century and a half after Steenstrup, we are still trying to "explain" the ontogeny of organismal form (Love 2008) while contending with spatial and temporal hierarchies.

Acknowledgments

We thank the participants in the two workshops that preceded this volume for discussion of numerous ideas presented here. Ned Friedman inspired us to look more deeply into the alternation of generations story, especially in botany. Phillip Sloan's helpful review of the manuscript aided us with a better understanding of Richard Owen's complex ideas on individuality and development. We also gratefully acknowledge support from the Negaunee Foundation and the Graduate School of the University of Wisconsin-Madison.

Notes to Chapter Five

1. This will have a familiar ring to those who know Ernst Haeckel's (1866) six orders of individuals. Though it appears that historians have not drawn the connection before, it seems plausible that Haeckel drew upon Schleiden here, as the latter was one of the scientists Haeckel had most admired as a student.

2. The terms homo- and heterogonic continued to be used into the twentieth century in works on parasitic worms and plants, where they refer to particular kinds of alternating dimorphism.

References for Chapter Five

Achtman, Mark, and Michael Wagner. 2008. "Microbial Diversity and the Genetic Nature of Microbial Species." *Nature Reviews Microbiology* 6 (6): 431–40.

Allen, Charles E. 1937. "Haploid and Diploid Generations." *American Naturalist* 71 (734): 193–205.

Arber, Agnes R. 1950. *The Natural Philosophy of Plant Form*. Cambridge: Cambridge University Press.

Bailey, L. H. 1906. "The Plant Individual in the Light of Evolution. The Philosophy of Bud-Variation, and Its Bearing upon Weismannism." In *The Survival of the Unlike: A Collection of Evolution Essays Suggested by the Study of Domestic Plants*, 5th ed., 81–106. New York: Macmillan.

Barthélémy, Daniel, and Yves Caraglio. 2007. "Plant Architecture: A Dynamic, Multilevel and Comprehensive Approach to Plant Form, Structure and Ontogeny." *Annals of Botany* 99 (3): 375–407.

Baur, Albert. 1864. *Beiträge zur Naturgeschichte der Synapta digitata: drei Abhandlungen*. Novorum Actorum Academiae Caesariae Leopoldino-Carolinae naturae curiosorum 31. Dresden: E. Blochmann.

Bell, Graham. 1994. "The Comparative Biology of the Alternation of Generations." In *The Evolution of Haploid-Diploid Life Cycles: 1993 Symposium on Some Mathematical Questions in Biology, June 19–23, 1993, Snowbird, Utah*, edited by Mark Kirkpatrick, 1–26. Providence, RI: American Mathematical Society.

Boardman, Richard S., Alan H. Cheetham, and William A. Oliver. 1973. *Animal Colonies: Development and Function Through Time*. Stroudsburg, PA: Dowden, Hutchinson & Ross.

Bouchard, Frédéric. 2010. "Symbiosis, Lateral Function Transfer and the (Many) Saplings of Life." *Biology & Philosophy* 25 (4): 623–41.

Bouchard, Frédéric, and Philippe Huneman, eds. 2013. *From Groups to Individuals: Evolution and Emerging Individuality*. Cambridge, MA: MIT Press.

Boyden, Alan. 1954. "The Significance of Asexual Reproduction." *Systematic Zoology* 3 (1) 26–47.

———. 1973. *Perspectives in Zoology*. Vol. 51. International Series of Monographs on Pure and Applied Biology: Division, Zoology. Oxford: Pergamon Press.

Braun, Alexander. 1851. *Betrachtungen über die Erscheinung der Verjüngung in der Natur*. Leipzig: W. Engelmann.

———. 1853a. "On the Phenomenon of Rejuvenescence in Nature." In *Botanical and Physiological Memoirs*, 1–342. London: Ray Society.

———. 1853b. *Das Individuum der Pflanze in seinem Verhältniß zur Species. Generationsfolge,*

Generationswechsel und Generationstheilung der Pflanze. Berlin: Königlichen Akademie der Wissenschaften.

———. 1855. "The Vegetable Individual in Its Relation to Species." *American Journal of Science and Arts* 19–21. 2nd Series: 19: 297–318, 20: 181–201, 21: 58–79.

Buss, Leo W. 1983. "Evolution, Development, and the Units of Selection." *Proceedings of the National Academy of Sciences USA* 80 (5): 1387.

———. 1987. *The Evolution of Individuality.* Princeton, NJ: Princeton University Press.

Calcott, Brett, and Kim Sterelny. 2011. *The Major Transitions in Evolution Revisited.* Cambridge, MA: MIT Press.

Calkins, Gary N. 1901. *The Protozoa.* London: Macmillan.

Carpenter, William B. 1848. "On the Development and Metamorphoses of Zoophytes." *British and Foreign Medico-Chirurgical Review* 1: 183–214.

———. 1849. "Owen and Paget on Reproduction and Repair." *British and Foreign Medico-Chirurgical Review* 4: 409–49.

Carpenter, William B., and James D. Dana. 1851. "On the Analogy between the Mode of Reproduction in Plants and the 'Alternation of Generations' Observed in Some Radiata." *Edinburgh New Philosophical Journal* 50: 266–68.

Causey, David. 1935. "Individuality in the Animal Kingdom." *Biologist* 17 (1): 9–16.

Chamberlain, Charles J. 1905. "Alternation of Generations in Animals from a Botanical Standpoint." *Botanical Gazette* 39 (2): 137–44.

Chamisso, Adelbert von. 1821. "Lettre écrite à M. le comte de Romanzoff, par M. de Chamisso, naturaliste français, qui a fait le voyage autour du monde, avec M. de Kotzebue, sur le brick russe le Rurick." In *Journal des voyages, découvertes et navigations modernes; ou Archives géographiques et statistiques du XIXe siècle,* 2nd ed. 1: 201–8. Paris: Colnet, Roret et Roussel.

Child, Charles M. 1915. *Individuality in Organisms.* Chicago: University of Chicago Press.

Churchill, Frederick B. 1979. "Sex and the Single Organism: Biological Theories of Sexuality in Mid-nineteenth Century." *Studies in History of Biology* 3: 139–77.

———. 1987. "From Heredity Theory to 'Vererbung': The Transmission Problem 1850–1915." *Isis* 78: 337–64.

Clarke, Ellen. 2011. "Plant Individuality and Multilevel Selection Theory." In *Major Transitions in Evolution Revisited,* edited by Brett Calcott and Kim Sterelny, 227–50. Cambridge, MA: MIT Press.

———. 2012. "Plant Individuality: A Solution to the Demographer's Dilemma." *Biology & Philosophy* 27 (3): 321–61.

Classen-Bockhoff, Regine. 2001. "Plant Morphology: The Historic Concepts of Wilhelm Troll, Walter Zimmermann and Agnes Arber." *Annals of Botany* 88 (6): 1153–72.

Collier, J. R. 1997. "The Localization Problem: A Review." *American Zoologist* 37 (3): 229–36.

Cornelius, Paul F. S. 1990. "Evolution in Leptolid Life-cycles (Cnidaria: Hydroida)." *Journal of Natural History* 24 (3): 579–97.

Desmond, Adrian J. 1984. *Archetypes and Ancestors: Palaeontology in Victorian London, 1850–1875.* Chicago: University of Chicago Press.

Doolittle, W. Ford. 2013. "Microbial Neopleomorphism." *Biology & Philosophy* 28 (2): 351–78.

Du Bois-Reymond, Emil H. 1860. *Gedächtnissrede auf Johannes Müller.* Berlin: Königlich Akademie der Wissenschaften.

Ehrenberg, Christian G. 1851. "Ueber die Formbeständigkeit und den Entwicklungskreis der organischen Formen. Ein Bild der neuesten Bewegungen in der Naturforschung." *Berichte über die zur Bekanntmachung geeigneten Verhandlungen der Königlichen Preussischen Akademie der Wissenschaften zu Berlin*: 761–95.

Elwick, James. 2007. *Styles of Reasoning in the British Life Sciences: Shared Assumptions, 1820–1858*. London: Pickering and Chatto.

Farley, John. 1972. "The Spontaneous Generation Controversy (1859–1880): British and German Reactions to the Problem of Abiogenesis." *Journal of the History of Biology* 5 (2): 285–319.

———. 1982. *Gametes and Spores: Ideas about Sexual Reproduction 1750–1914*. Baltimore: Johns Hopkins University Press.

Frank, Uri, Günter Plickert, and Werner A. Müller. 2009. "Cnidarian Interstitial Cells: The Dawn of Stem Cell Research." In *Stem Cells in Marine Organisms*, edited by Baruch Rinkevich and Valeria Matranga, 33–59. Dordrecht: Springer.

Geddes, Patrick. 1878. "Morphology." *Encyclopaedia Britannica*. 9th ed. New York: C. Scribner's Sons.

Geddes, Patrick, and J. Arthur Thomson. 1902. *The Evolution of Sex*. Rev. ed. New York: Walter Scott.

Gilbert, Scott F. 2003. "The Morphogenesis of Evolutionary Developmental Biology." *International Journal of Developmental Biology* 47 (7–8): 467–77.

Gilbert, Scott F., and David Epel. 2015. *Ecological Developmental Biology: Integrating Epigenetics, Medicine, and Evolution*. Sunderland, MA: Sinauer.

Gilbert, Scott F., Jan Sapp, and Alfred I. Tauber. 2012. "A Symbiotic View of Life: We Have Never Been Individuals." *Quarterly Review of Biology* 87 (4): 325–41.

Glaubrecht, Matthias, and Wolfgang Dohle. 2012. "Discovering the Alternation of Generations in Salps (Tunicata, Thaliacea): Adelbert von Chamisso's Dissertation 'De Salpa' 1819—Its Material, Origin and Reception in the Early Nineteenth Century." *Zoosystematics and Evolution* 88 (2): 317–63.

Godfrey-Smith, Peter. 2009. *Darwinian Populations and Natural Selection*. Oxford: Oxford University Press.

Gorelick, Root. 2012. "Mitosis Circumscribes Individuals; Sex Creates New Individuals." *Biology & Philosophy* 27 (6): 871–90.

Grosberg, Richard K., and Richard R. Strathmann. 2007. "The Evolution of Multicellularity: A Minor Major Transition?" *Annual Review of Ecology, Evolution, and Systematics* 38: 621–54.

Haeckel, Ernst. 1866. *Die Generelle Morphologie der Organismen*. 2 vols. Berlin: Georg Reimer.

Herron, Matthew D., Armin Rashidi, Deborah E. Shelton, and William W. Driscoll. 2013. "Cellular Differentiation and Individuality in the 'Minor' Multicellular Taxa." *Biological Reviews* 88 (4): 844–61.

Hofmeister, Wilhelm. 1851. *Vergleichende Untersuchungen der Keimung, Entfaltung und Fruchtbildung höherer Kryptogamen und der Samenbildung der Coniferen*. Leipzig: Friedrich Hofmeister.

Hughes, Roger N. 1989. *A Functional Biology of Clonal Animals*. New York: Chapman & Hall.

Huxley, Thomas H. 1851a. "Report on the Researches of Prof. Mueller into the Anatomy and Development of the Echinoderms." *Annals and Magazine of Natural History*, 2nd Series 7 (43): 11–19.

———. 1851b. "Observations upon the Anatomy and Physiology of *Salpa* and *Pyrosoma*." *Philosophical Transactions of the Royal Society of London* 141: 567–93.

———. 1852. "Upon Animal Individuality." *Proceedings of the Royal Institution of London* (1851–1854) 1: 184–89.

———. 1856. "Lectures on General Natural History: Lecture II." *Medical Times and Gazette*, New Series 12 (307): 481–84.

Johnson, Duncan S. 1915. "History of the Discovery of Sexuality in Plants." In *Smithsonian Report for 1914*, 383–406. Washington, DC: Smithsonian Institution.

Jones, Roy W. 1937. "Alternation of Generations as Expressed in Plants and Animals." *Bios* 8 (1): 19–29.

Jones, Thomas R. 1871. *General Outline of the Organization of the Animal Kingdom, and Manual of Comparative Anatomy*. 4th ed. London: J. Van Voorst.

Kaplan, Donald R., and Todd J. Cooke. 1996. "The Genius of Wilhelm Hofmeister: The Origin of Causal-Analytical Research in Plant Development." *American Journal of Botany* 83 (12): 1647–60.

Kölliker, Albert. 1864. "Über die Darwin'sche Schöpfungstheorie." *Zeitschrift für wissenschaftliche Zoologie* 14: 174–86.

Kramp, P. L. 1943. "On Development through Alternating Generations, Especially in Coelenterata." *Videnskabelige meddelelser fra Dansk naturhistorisk Forening* 107: 13–32.

Lankester, E. Ray. 1857. "Alternation of Generations and Parthenogenesis in Plants and Animals [Reports of Societies (Continued)]." *British Medical Journal* 2 (40): 833–37.

———. 1919. "Memoirs: The Terminology of Parthenogenesis." *Quarterly Journal of Microscopical Science*, 2nd Series 63: 531–36.

Leuckart, Rudolf. 1851. *Ueber den Polymorphismus der Individuen oder die Erscheinungen der Arbeitstheilung in der Natur. Ein Beitrag zur Lehre vom Generationswechsel*. Giessen: Ricker.

Love, Alan C. 2003. "Evolutionary Morphology, Innovation, and the Synthesis of Evolutionary and Developmental Biology." *Biology & Philosophy* 18 (2): 309–45.

———. 2008. "Explaining the Ontogeny of Form: Philosophical Issues." In *A Companion to the Philosophy of Biology*, edited by Sarkar Sahotra and Anya Plutynski, 223–47. Malden, MA: Blackwell.

Lovejoy, Arthur O. 1911. "Schopenhauer as an Evolutionist." *Monist* 21 (2): 195–222.

Lubbock, John. 1857. "An Account of the Two Methods of Reproduction in *Daphnia*, and of the Structure of the Ephippium." *Philosophical Transactions of the Royal Society of London* 147: 79–100.

McShea, Daniel W., and Edward P. Venit. 2001. "What Is a Part?" In *The Character Concept in Evolutionary Biology*, edited by Günter P. Wagner, 259–84. San Diego: Academic Press.

Mettenius, Cecile. 1882. *Alexander Braun's Leben, nach seinen handschriftlichen Nachlass*. Berlin: G. Reimer.

Mirbel, Charles François Brisseau de. 1815. *Élémens de physiologie végétale et de botanique*. Vol. 2. Paris: Magimel.

Morton, A. G. 1981. *History of Botanical Science*. London: Academic.

Müller, Johannes. 1851. "Die Erzeugung von Schnecken in Holothurien." *Bericht über die zur Bekanntmachung geeigneten Verhandlungen der Königlichen Preussichen Akademie der Wissenschaften zu Berlin*: 628–55.

———. 1852a. "Upon the Production of Mollusks in Holothuriae." *Annals and Magazine of Natural History*, 2nd Series 9: 22–39.

———. 1852b. "Ueber die Erzeugung von Schnecken in Holothurien." *Archiv für Anatomie, Physiologie und wissenschaftliche Medizin*: 1–36.

———. 1852c. *Ueber Synapta digitata und über die Erzeugung von Schnecken in Holothurien*. Berlin: G. Reimer.

Müller-Wille, Staffan. 2010. "Cell Theory, Specificity, and Reproduction, 1837–1870." *Studies in History and Philosophy of Science Part C: Studies in History and Philosophy of Biological and Biomedical Sciences* 41(3): 225–31.

Nägeli, Carl. 1846. "Memoir on the Nuclei, Formation, and Growth of Vegetable Cells." In *Reports and Papers on Botany*, 213–92. London: Ray Society.

Nyhart, Lynn K., and Scott Lidgard. 2011. "Individuals at the Center of Biology: Rudolf Leuckart's *Polymorphismus der Individuen* and the Ongoing Narrative of Parts and Wholes. With an Annotated Translation." *Journal of the History of Biology* 44: 373–443.

Ogilvie, George. 1861. *The Genetic Cycle in Organic Nature: Or, the Succession of Forms in the Propagation of Plants and Animals*. Aberdeen: A. Brown.

Owen, Richard. 1849. *On Parthenogenesis; or, The Successive Production of Procreating Individuals from a Single Ovum*. London: J. Van Voorst.

———. 1851. "Professor Owen on Metamorphosis and Metagenesis. Being Abstract of a Lecture Delivered by Him to the Royal Institute of Great Britain in February 1851." *Edinburgh New Philosophical Journal* 50: 268–78.

———. 1992. *The Hunterian Lectures in Comparative Anatomy, May–June 1837*, edited by Phillip R. Sloan. Chicago: University of Chicago Press.

Parnes, Ohad S. 2007. "On the Shoulders of Generations: The New Epistemology of Heredity in the Nineteenth Century." In *Heredity Produced: At the Crossroads of Biology, Politics, and Culture, 1500–1870*, edited by Staffan Müller-Wille and Hans-Jörg Rheinberger, 315–46. Cambridge, MA: MIT Press.

Pepper, J. W., and M. D. Herron. 2008. "Does Biology Need an Organism Concept?" *Biological Reviews* 83 (4): 621–27.

Phillips, Everett F. 1903. "A Review of Parthenogenesis." *Proceedings of the American Philosophical Society* 42 (174): 275–345.

Piraino, Stefano, Ferdinando Boero, Brigitte Aeschbach, and Volker Schmid. 1996. "Reversing the Life Cycle: Medusae Transforming into Polyps and Cell Transdifferentiation in *Turritopsis nutricula* (Cnidaria, Hydrozoa)." *Biological Bulletin* 190 (3): 302–12.

Quatrefages, A. de. 1864. *Metamorphoses of Man and the Lower Animals*. Translated by Henry Lawson. London: R. Hardwicke.

Rheinberger, Hans-Jörg. 2008. "Heredity and Its Entities around 1900." *Studies in History and Philosophy of Science Part A* 39 (3): 370–74.

Richards, Evelleen. 1987. "A Question of Property-Rights—Owen's Evolutionism Reassessed." *British Journal for the History of Science* 20 (65): 129–71.

———. 1994. "A Political Anatomy of Monsters, Hopeful and Otherwise: Teratogeny, Transcendentalism, and Evolutionary Theorizing." *Isis* 85 (3): 377–411.

Rieppel, Olivier. 2011. "Species Are Individuals—The German Tradition." *Cladistics* 27 (6): 629–45.

Rupke, Nicolaas A. 1994. *Richard Owen: Victorian Naturalist*. New Haven: Yale University Press.

———. 2009. *Richard Owen: Biology without Darwin*. Chicago: University of Chicago Press.

Sapp, Jan. 2009. *The New Foundations of Evolution: On the Tree of Life*. New York: Oxford University Press.

Sars, M. 1841. "Ueber die Entwicklung der Medusa *Aurelia aurita* und der *Cyanea capillata*." *Archiv für Naturgeschichte* 7: 9–34.

Schacht, Hermann. 1852. *Physiologische Botanik. Die Pflanzenzelle, der innere Bau und das Leben der Gewaechse*. Berlin: G. W. F. Mueller.

Schleiden, Matthias J. 1838. "Beiträge zur Phytogenesis." *Archiv für Anatomie, Physiologie und wissenschaftliche Medizin* 13: 137–76.

———. 1842–43. *Grundzüge der wissenschaftlichen Botanik, nebst einer methodologischen Einleitung als Anleitung zum Studium der Pflanze.* 2 vols. Leipzig: Wilhelm Engelmann.

———. 1861. *Grundzüge der wissenschaftlichen Botanik, nebst einer methodologischen Einleitung als Anleitung zum Studium der Pflanze.* 4th ed. Leipzig: Wilhelm Engelmann.

Schmidt, Oscar. 1878. Die niederen Thiere. In: *Brehms Thierleben.* 2nd ed., Vol. 10 (4. Abt., Bd. 2). Leipzig: Verlag des Bibliographischen Instituts.

Sloan, Phillip R. 2003. "Whewell's Philosophy of Discovery and the Archetype of the Vertebrate Skeleton: The Role of German Philosophy of Science in Richard Owen's Biology." *Annals of Science* 60 (1): 39–61.

Steenstrup, Johannes J. S. 1842a. *Om Forplantning og Udvikling gjennem vexlende Generationsrækker, en særegen Form for Opfostringen i de lavere Dyrklasser.* Copenhagen: Lunos.

———. 1842b. *Ueber den Generationswechsel, oder die Fortpflanzung und Entwickelung durch abwechselnde Generationen, eine eigenthümliche Form der Brutpflege in den niederen Thierclassen.* Copenhagen: C. A. Reitzel.

———. 1845. *On the Alternation of Generations; or the Propagation and Development of Animals through Alternate Generations: A Peculiar Form of Fostering the Young in the Lower Classes of Animals.* London: Ray Society.

Stockberger, W. W. 1913. "A Literary Note on the Law of Germinal Continuity." *American Naturalist* 47 (554): 123–28.

Svedelius, Nils. 1927. "Alternation of Generations in Relation to Reduction Division." *Botanical Gazette* 83 (4): 362–84.

Thomson, Allen. 1859. "Ovum." In *The Cyclopædia of Anatomy and Physiology,* edited by Robert Bentley Todd, Vol. 5 (Supplementary Volume), 11–142. London: Longman, Brown, Green, Longmans & Roberts.

Vergara-Silva, Francisco. 2003. "Plants and the Conceptual Articulation of Evolutionary Developmental Biology." *Biology & Philosophy* 18 (2): 249–84.

Walbot, Virginia, and Matthew M. S. Evans. 2003. "Unique Features of the Plant Life Cycle and Their Consequences." *Nature Reviews Genetics* 4 (5): 369–79.

Weismann, August. 1893. *The Germ-Plasm: A Theory of Heredity.* New York: Charles Scribner's Sons.

Winsor, Mary P. 1976. *Starfish, Jellyfish, and the Order of Life: Issues in Nineteenth-Century Science.* New Haven: Yale University Press.

Winther, Rasmus G. 2001a. "August Weismann on Germ-plasm Variation." *Journal of the History of Biology* 34 (3): 517–55.

———. 2001b. "Varieties of Modules: Kinds, Levels, Origins, and Behaviors." *Journal of Experimental Zoology* 291 (2): 116–29.

6

Spencer's Evolutionary Entanglement: From Liminal Individuals to Implicit Collectivities

SNAIT GISSIS

> But I had undertaken to set forth a general theory of Evolution as exhibited throughout all orders of existence. . . . On the whole the result seems to have shown that the attempt was not unwarranted.
>
> In one respect, indeed, I had, as an outsider, studying the phenomena of organic life as phenomena of Evolution at large, a certain kind of advantage over specialists, dealing after the ordinary manner with their respective separate subjects—plant-life and animal-life. The man of science who limits himself to a department, is apt to overlook, or else not sufficiently to appreciate, those most general truths which the phenomena he studies display in common with other groups of phenomena. . . . thus conducing to a more philosophical conception of biological facts.
>
> HERBERT SPENCER, *An Autobiography*, vol. 2, 1904, 103–4[1]

Introduction

The works of Herbert Spencer (1820–1903), the deeply influential Victorian polymath, spanned many fields: philosophy, psychology, biology, sociology, political theory, and ethics. In all his works from the early 1850s on, Spencer was struggling to refashion the human sciences. He sought to base them on a theory of evolution that would make humans one component of a natural evolutionary continuum and naturalize human behavior by putting it within an evolutionary framework. In this way, it would be on par with the behavior of any other living organism. In the mid-1850s Spencer crystallized his view that Evolution (with a capital e) is the all-encompassing framework of living nature, a general process and principle whose main feature on all levels was a growing complexity of relationships between environments and organisms and their evolving effects. Spencer enunciated this radically innovative stance when introducing evolution to the analysis of the mental apparatus in the 1855 first edition of his *Principles of Psychology* (*PPi*). He there articulated what I consider to be his basic epistemological and methodological frame of analysis of living nature, namely, *the changing interrelations of the changing interactions between organism and environment, and the changing effects they induce.*

Spencer's work was carried out participating in the contemporaneous controversies, not only using the conceptual tools available at the time, but also forging new ones that were used and criticized within the next half century. In fact, the problématique he grappled with, and the answers he came up with, resonate with present-day questions and concerns (as discussed in more detail in the conclusion). He was concerned with structures of dynamic interactions between organisms, between organisms and their environments, between various components of organisms, and with the nature of function within organisms. Similarly, he repeatedly addressed the issue of the collectivity perspective and its roles in the constitution of individuality. And more generally, answering the question whether there could be evolutionary developmental systems of rising complexity is the thread that ties together all his works. He indeed well deserves a hearing.

Spencer worked over 50 years on these issues. In the present paper I analyze its initial phase: the principal works Spencer wrote before as well as between the publication of the first edition of *Principles of Psychology* (*PPi*, 1855) and the first edition of the *Principles of Biology* (*PBi*, 1864–67),[2] with the crystallization of the notion of biological individuality as the focus of my study. The context in which these works were written has been addressed by his biographers.[3] Here I merely want to point out that during the period under consideration the sociological point of view was beginning to emerge, and by the time Spencer was writing the many volumes of his *Principles of Sociology* (*PS*, 1874–96), the struggles to establish social science, also called "sociology," as an autonomous discipline were widespread in Western Europe and in the USA.

Spencer's work exemplifies a particular mode of transferring notions of individuality and collectivity between the social and the biological spheres. In contrast to both Lamarck (1809) and Darwin (1964 [1859]), who channeled their approach from the non-human to the human, Spencer arrived at the problems of biological individuality by initially addressing questions regarding human individuals and human sociality. This perspective guided his inquiry on the potentialities of complex relations of partial or full interdependence, of independence among parts and wholes, and more specifically, on the feasibility of cohesive, non-compulsive, collaborative structural forms of biological individuals and collectivities. It served in his time as a paradigm for a biological-social approach toward constituting social science; this irrespective of whether Spencer started with ethical-social motivations and looked for answers through the biological or the other way around.

In order to understand Spencer's weighty apparatus, his innovative stance, as well as the unresolved inner tensions within his theoretical framework, one has to relate his expository efforts to the overall conceptual scheme he had devised. This will help clarify the presuppositions and theoretical constraints with which he struggled. One also has to relate Spencer's framework to the prevalent contemporaneous conceptualizations of physics and chemistry. In Great Britain in Spencer's time the Newtonian tradition in the physical sciences served both as a basic conceptual scheme and as a paradigmatic model of how to do science. Spencer not only had read and was familiar with some of the major physical and chemical scientific and philosophical writings of the first half of the nineteenth century, but he also was friends with some of the leading British physicists and chemists, in particular John Tyndall and Edward Frankland.

Two features of the Newtonian tradition are relevant: 1) the scientific search was for *immutable laws of nature* that had universal extension and universal applicability, and were to account most generally for the totality of phenomena. Science was supposed to move through series of abstractions. One of these abstractions, theory formation, made possible predictions and rules. These in turn led back to observations. An implicit assumption of that procedure was the possibility of going from the more concrete and material, macroscopic description, to the more abstract, specifically, to a microscopic description involving ever smaller—possibly unobservable—entities. The direction from the parts—the microscopic entities—to the whole—the macroscopic, concrete entity—was that of bottom to top causal explanations. 2) The *particles*, the basic entities—the "individuals" in the Newtonian approach—had no history and carried no history. These *particles* were regarded as simple and unchanging, whether considered in the past, present, or future. For Newton—and for Maxwell—they were God-created; the dynamics describing the world made up from them were deterministic, and their causal explanations invoked immutable, unchanging interactions. Furthermore, the dynamical laws were assumed to hold at both the microscopic and the macroscopic levels (see Merz 1976).

Spencer's overarching framework was also influenced by another mode of conceptualizing the material world, namely that offered by chemistry, and in particular that advanced by organic chemistry at midcentury in Britain. For chemists, atoms were not conceived as the ultimate ontological units, but rather as components of their models. This conceptualization had a significant impact on Spencer, partly because of his relationship to Frankland, an outstanding organic chemist who was a member of the powerful "X Club" of liberal scientists, founded in the 1860s, and so important to Spencer in keeping him updated. Frankland was a close friend of Tyndall, another member of the club, who was close to Spencer from the late 1850s on (see MacLeod 1970;

Barton 1990, 1998; Desmond 2001). Organic chemistry at this time was becoming an important separate sub-discipline of chemistry, with its own concerns, methods, and approach. Organic chemists strove to understand the dynamics of matter through questions with a rather different emphasis on reactivity and change than the inorganic chemists, and in particular, through the deployment of analogies. Theirs was a search for descriptions and theoretical explanations in terms of forms or *structures* and *functions*, which paralleled the biological approach. Chemistry was (and is) a practical, pragmatic science which, unlike physics, did not look for ultimate explanations; and organic chemists accepted multiple explanations rather than an ideal, single one in terms of Newtonian or Laplacian corpuscles, motion, and forces. In organic chemistry a molecule was perceived as an entity responsive to its surrounding, and thus closer in its behavior to an organism as conceived at the time; "For most chemists of Hofmann's and Frankland's generation analogies between molecules and natural organisms seemed more appropriate than analogies between molecules and billiard balls or falling bodies" (Nye 1996, 127; see also idem, 1993; Rocke 2002; Schütt 2002).

Many among Spencer's contemporaries (but not some Germans) perceived the individual not only as an entity in terms of which all else had to be described, analyzed, and explained, and as such a focus of causation, but also as a substantive *spatio-temporal* basic entity: an entity whose coming into being had *to be explained*. By this decision, I argue, "the individual" became naturalized. And the more "the individual" became entwined into theories referring to levels above and below it, the more "the individual" gained in naturalness and robustness.

In the corpus of Spencer's work that I deal with, as well as in his later writings on sociality, Spencer was groping towards an evolutionary framework that first and foremost would provide a rich and detailed account of the complexity of the living world and its biological, psychological (mental), and social diversity, and give meaning to structure, functionality, adaptability, intentionality, and other aspects of life as these evolved. All these were characterized by interactions, openness, and directionality; all these were time-and-history bound. Concurrently, Spencer wanted this evolutionary framework also to encompass the non-organic, the subject matter of the physical sciences, and yet to make this exceedingly complicated (and unrealistic) endeavor scientific in the eyes of his contemporaries.

Isolatable individuals that could be decoupled from collectivities were generally looked upon by many among his contemporaries as the most fundamental elements of any scientific endeavor. However, while working his way through the psychological, the biological, and the social, Spencer encountered mounting difficulties in maintaining this position when attempting to account for the interactive patterns of evolutionary processes, for the complexified features of evolved living systems, and more generally for the emergence of novel features. I believe that one reason for the unresolved inner tension stemmed from the conflicts between the contemporaneous conceptual scheme in the physical sciences and his evolutionary theoretical framework. These differences are addressed in the various sections of the paper. The first section deals with Spencer's conception of complexity, the second with his views on biological individuality, and the third with his treatment of individuals and collectivities. In the final section, *Spencerian problématique*, I deal briefly with some contemporary problematizations of the concept of biological individuality.

Complexity

In the mid-1850s Spencer crystallized his view that Evolution is the all-encompassing framework of living nature. He then also enunciated his basic epistemological and methodological frame of analysis of living nature. This will be the main topic of the first subsection below, while I reserve the clarification of Spencer's coupling of complexity and heterogeneity for the beginning of the next subsection.

ORGANISM-ENVIRONMENT-EFFECTS

Commenting on Spencer's evolutionary views, John Tyndall, in a *BAAS* address delivered in Belfast in 1874, pointed to the crucial element of Spencer's conceptualization: "There are two obvious factors to be . . . taken into account—the creature and the medium in which it lives, or as it is often expressed, the organism and its environment. *Mr Spencer's fundamental principle is, that between these two factors there is incessant interaction* . . . Life he defines to be 'a continuous adjustment of internal relations to external relations'" (Tyndall 1892, 183, my emphasis).

Tyndall's "incessant interaction" between a "creature" and its environment was, in fact, a characterization of the Spencerian *three-place relationship*—his "*fundamental principle*"—expressed in the terms "creature" (i.e., organism), "medium" (i.e., environment), and their mutual interactions and the ensuing effects of these interactions. Spencer's "fundamental principle" could be applied to other entities taken from other fields of knowledge, with these entities fulfilling the place and role of "organism," "environment," and "interactions and ensuing effects." This "fundamental principle" or "basic frame of analysis," I argue, served as a basic model that he eventually sought to apply to the mental, social, and anthropological realms as well as the biological one. Furthermore, I will argue that the role played by "organism" in Spencer's three-place model did not have to be an individual, but that this role could be fulfilled by both individuals and collectivities, and indeed became so. This assumed that when asking specific questions from specific perspectives one could draw a boundary delineating what entities and/or objects were of relevance for the system under consideration, which entities and/or objects could be considered irrelevant when constituting the "environment," and which kinds of interactions were relevant by virtue of temporal, spatial, or other factors when specifying "interactions." This relegation into relevance and irrelevance could readily be made in the physical sciences. Drawing boundaries and justifying irrelevance was much more difficult with entities in living and social systems. There the dependence of the "organism" on the "environment" is crucial, and the specificity of the questions and of the perspectives much more limited and temporary.

A word on the evolution of Spencer's model: a preliminary basic frame of analysis began to be formulated in the Spencerian corpus with his carefully considered efforts during the 1850s relating to biological, social, and psychological development. It came to the fore again in two essays: "The Social Organism" and "Bain on the Emotions" (1860a and b; reprinted in Spencer 1891, vol. 1). In an attempt to broaden the scope of his analytical framework, Spencer elaborated its application to the physical realm in an article on the Nebular Hypothesis (1858; reprinted in Spencer 1891, vol. 1) and also in his article, "Progress its Law and Cause" (1857a; reprinted in Spencer 1891, vol. 1). However, his extension might not have seemed wide-ranging enough given the very far-reaching, universalized claims of contemporaneous physics (see Tyndall 1892 [1874], 192–93; Spencer 1904, 2: 12–16, 20, 67–86). In 1862 Spencer published his *First Principles* (*FPi*), in which the widest application of a generalized evolutionary model of processes of change was constructed (see preface). In it the concept of Force—what we now call "energy" (Cohen 1981)—and its conservation, in parallel with the law of conservation of matter, as well as the notion of equilibrium, played central roles.[4] I believe that in the next book, the *Principles of Biology* (*PBi*), Spencer found this broadening of the model to be counterproductive. He had to put constraints and modifications to a model which in *FPi* was conceived as uniformitarian and universal, in order to make it applicable to much more complex systems, namely living ones (and in the 1870s, social ones). The modifications gave the model a specific plasticity that was of particular relevance to the entities it was applied to. Furthermore, a close reading of *FPi* makes it clear that Spencer had given up the hope of answers to "ultimate questions" and was limiting himself to offering explanations of networks of *relations,* to investigating not the nature of the acting forces themselves but of their evolving effects (e.g., *FPi,* 86).

As conceived by Spencer, interactions changed over time, resulting, as their most significant feature, in there being greater complexity when going from some earlier time to some later time. Complexification was the assumed result of the dynamics of "organism-environment-interactions" and their evolving effects in time. It was also the explanatory justification for the dynamics, because Spencer could not provide sufficient specifications of the interactions.[5] And for Spencer, complexification was also the measure of progress. Within the Spencerian theoretical framework the description of the process, which would apply at all levels and in all fields (but primarily in biology, psychology, and sociology), was "the movement from homogeneity to heterogeneity." This, for Spencer, was a description of evolution, with his three-place model playing the role of mediator between the assumption of the dynamics engendering "complexity" and the description of "heterogeneity." Thus "heterogeneity" came to stand for "complexity" and vice versa. Within his theoretical framework the hierarchical feature of living nature was evidenced by increased heterogeneity, by progressing from a lesser to a greater complexity, particularly in the *relations* among the tripartite elements of the model. The connections between "the internal and the external," and in psychology between physiological and mental states, were encompassed under the same conceptualization.

In both *PPi* and *PBi,* Spencer often called these integral features of the model "correspondence" (see *PPi,* 373, on the inaptness of the term). He devoted nearly all the third part of *PPi* to a discussion of the concept of correspondence and its correlates under the general heading of "General Synthesis." He there stressed that "the vast array of phenomena . . . form in reality, one general, continuous and inseparable evolution. . . . The correspondence between the organism and its environment, while becoming in each higher phase more specialized and heterogeneous, must ever remain, as it has been from the beginning, one and indivisible" (*PPi,* 482–83, 485–86). I look upon this quote as corroborating my argument.

Within Spencer's framework, "physiological division of labor"—that is, the combined processes of "differentiation," "specialization," and "integration"—functioned as a generalized appellation and feature of mechanisms that were to explicate the process whereby complexity increased (see already in Spencer 1857b). The same applied to "use and disuse" and "inheritance of acquired modifications." They provided what Spencer wanted: change and continuity, especially trans-generational continuity. These features of mechanisms exemplified the effects brought about by those interactions on individual organisms and correspondingly, on the biological and ecological environment an organism finds itself in.

Given the existing level of knowledge, and the available technical and conceptual tools, the model Spencer was enunciating could not be mathematized, nor even quantified.[6] Nevertheless, Spencer believed that his model did incorporate a fundamental intuition expressed by some contemporaneous naturalists and philosophers regarding living nature and social nature that rejected the metaphysics of isolatable, unchanging, objects and/or entities and emphasized the evolvement of an ever-changing, increasingly complex order within these realms (e.g., *PPi*, 368, 381, 382). In his earlier writings and still in *PPi*, Spencer, in a Lamarckian fashion, attributed to organisms an internal drive that would account both for the presumed responsiveness of organisms to challenges from their environment and for a necessarily growing complexification of that responsiveness through functional or behavioral changes in the organisms.[7] In *PBi*, for reasons described by his biographers as social, scientific, personal-psychological, and/or political (e.g., Turner 1985; Peel 1992 [1971]; Francis 2007), Spencer tried to rid his theoretical framework of any allusion to such an internal drive. This is found more explicitly in *PBi*, vol. 1: 405–8. Nonetheless, throughout both volumes there were explicit and implicit traces of the difficulties he coped with in his efforts to eliminate attribution of an organismal internal force without giving up either his dynamics, or complexification, or the description of evolution—on all levels—as a movement from the homogeneous to the heterogeneous.

When Spencer's views were becoming crystallized, there were a number of extant modes of viewing the relations between living entities, whether human or not, and the conditions, circumstances, surroundings, or milieu in which they were located[8] (this within the context of explaining change and innovation as well as continuity and persistence and/or stability). In some, "environment" was depicted as passive, since any mechanisms of change as well as those of stable continuity were contained within the organism and worked from the inside out. (See John Knox as discussed in Desmond 1989, esp. 73–74; E. Richards 1989.) In others, "organism" was posited as passive and subjected to acting external forces, mostly physical but also social. In yet others—as with Lamarck—a more elaborate attempt had been made to produce a combination of the internal and the external, whereby the organism (often considered as an individual, but not necessarily so) could be interpreted as responding to the external in a differential manner. For a portion of the living entities it could be described as direct external influence, and for others as a chain of interactions, which was viewed as a mode by which the organisms, altered by the environment, also altered it (Barsanti 2000).

For Spencer, organisms could be considered only within some environment; thus, rather than positing a dichotomy, his theoretical efforts were directed at establishing a strong coupling of the two. Within Spencer's system, in order for "environment" to have reference there had to be an "organism-environment-evolving effects" frame of analysis. Spencer considered "environment" as the sum of all—physical, chemical, biological, social—external conditions and circumstances in which organisms found themselves, including the existence of other organisms. A significant question is whether Spencer thought of "environment" in a global manner, as existing "out there" irrespective of organismal perspectives on it—in a manner analogous to the way one would assume the independent existence of the external world—or whether he thought of "environment" as a specific environment for specific organisms. I believe the latter was the case. Spencer's treatment of nutrition as a process of environment-organism relations is an illuminating example. He considered nourishment an environmental action that sparked internal processes of physiological differentiation, promoting complexity of the organism (*PBi*, vol. 2, 378–83; compare with Landecker, this volume, on metabolism).

Spencer commented in various places that the effects of the environment on the lowest and on lower organisms were limited, and that these effects would become more diversified and more complex as the organisms themselves became more complex. Conversely, organisms impacted differentially on their environment (e.g., *PBi*, vol. 1, 426): the lowest organisms did so to a minimal degree and the most complex organisms to a maximal degree, in accordance with the hierarchies of individuation. However, the boundary between environment and organism did not signify zones of internal cooperation versus zones of external competition.

In the Spencerian framework, then, an environment was considered as an *environment for* an organism, and thus local and not rigidly circumscribed, and there was no way to conceive of organisms independently of their environment, nor of environment-less organisms. Therefore, some of an organism's changes could be looked upon as being passed on trans-generationally by virtue of the changed organisms and the generalized mechanisms of heredity Spencer offered. Similarly, the organism's changes could be looked upon as "embedded" in their environment and thus passed on to their organism's environment trans-generationally. Spencer posited that the changing states of organisms in time, by virtue of their interactions (which were also changing), reached differing states of *balanced adaptation* with their respective environments. He viewed these states as reflecting active processes taking place both inside organisms and within their "organism-environment-interactions and evolving effects," and named them "equilibrations" or "moving equilibrium" (see *FPi*, chapter 16, 440–48; *PBi*, vol. 1, part 3, chapter 11, and also chapter 12: natural selection as an "Indirect equilibration"; see also *PBi*, vol. 2, part 5).

In Spencer's framework these equilibrations had possible alternative trajectories in the future with differential effects on the components of the framework. Because the reciprocal effects were not fully predetermined there was latitude; the framework contained a measure of openness toward the future. This allowed for "emergence"—for emergent new structures, and emergent new means of interaction. This is particularly conspicuous in Spencer's sociology.

To sum up: Spencer's basic frame of analysis methodologically and epistemologically posited a three-fold relation of organism, environment, and their interactions, and their evolving effects in time. Contrary to some readings of Spencer, both contemporaneous and contemporary, I believe the following:

a. The relationship posited was not a dichotomy but rather an interrelated three-way dependency;

b. None of the posited elements by this frame of analysis was viewed as totally passive;

c. "Environment" and "organism" were explicated through a perspectival positioning, and their actions depicted earlier in my exposition as "differential"; and therefore

d. The frame of analysis was reciprocal and looping;

e. "A movement from the homogeneous to the heterogeneous" was presented as existing on all levels of the process of *progressive evolution*, the result being increasing complexity;

f. Complexification was assumed to be indicative of evolution and advanced as an explanation of that process.

THEORY OF COMPLEXITY

Peel (1992 [1971]) and Bowler (1989) have argued that British Victorian thinkers exhibited a tension deriving from the adoption of the "progress" discourse and the growing significance attributed to the laws of thermodynamics. William Thomson's formulation of the second law of thermodynamics concerning the heat death of the universe was soon interpreted as applying also to organisms. Many of his contemporaries believed in a general tendency to decomposition (Thomson 1862; Smith and Wise 1989). By contrast, in *FPi* Spencer strove to provide a universal growing complexity *and* decomposition, in a uniform manner in all areas. He thought that states of homogeneity were inherently unstable, and that states of heterogeneity produced webs of inter-dependencies, rather than relations marked by dominance and subordination. He also conceived of changes of characteristics rather than of accumulations of variations. Into this framework he incorporated Henri Milne-Edwards's "division of physiological labour" (Milne-Edwards 1834, 1851), and Karl Ernst von Baer's concept of directional development of the embryo from the homogenous to the heterogeneous. Moreover, Spencer's notion of recapitulation added further complexity to living systems. Spencer had decided to adopt from among the developmental models then available that of the development of the embryo.[9] Embryogenesis offered insights into the general processes of development, including openness, constraints, and limits of development (Churchill 1991), as well as indicating irreversibility: organisms could not revert to an original state from a more evolved, hetero-geneous state. The embryonic model also made it possible to talk about the micro-level of heterogeneity (Gould 1977; Adam 1988; Hopwood 2009). But note that Spencer consistently refrained from discussing an originating unit of life in either *PBi* or in *PS*.

Spencer mapped onto human society a generalized notion of biological organisms that enhanced features, which would enable him to claim that so-cieties and biological organisms were different from any other kind of enti-ties. Two fundamental features were key: the continual increase in complex-ity of their structure and the functional interdependence of their component parts. Thus at that generalized level of resolution, society could be said to be an organism (a superorganism).

Spencer's commitment to his basic tripartite frame of analysis and to com-plexification implied a hierarchical view of individuation and of living enti-ties at large. It was *hierarchical* in that the *progressive* development Spencer was committed to required stable component entities, stable (on a specifiable timescale) but also endowed with a measure of plasticity that would allow for some change in structure, properties, or functions. Examples would be "evo-lution of the embryo" at the component level of organisms, polymorphism at the level of individual organisms, and varieties at the collectivity level. This prescription served as a common ground that he considered would allow for continuity between the physiological, psychological, and social forms of liv-ing, and within the field of psychology between the simplest activities and those of reasoning and inference.

Spencer's theory of complexity did not match his practice. He declared that methodologically one should start from complex systems to investigate hypotheses regarding simple ones, and similarly, that in the social realm one should proceed from the present to the past (*PPi*, 71; Spencer 1873, 109, 320–

22). However, in both *PBi* and *PS* Spencer's path of analysis (and in fact a reductive one) started from the simple, considered the most elementary, and followed its compounding, re-compounding, and re-re-compounding, to produce multiplicity, variety, and complexity.

Biological Individuality

The mere concept of individual was a given for Spencer, as he had available to him a great deal of what was known on the living world. His concern was focused on the *evolutionary relationships* among the *diversity* of individuals, within a framework that was *progressivist*. For Spencer, an individual was a consequence of evolutionary processes, and not a foundational unit in a description of living nature. Stability did not inhere in particular components of biological individuals, but in the sameness of mechanisms and patterns and in the continuity of their workings. Viewed from Spencer's evolutionary perspective the entities of his wide-ranging investigations manifested—to put it anachronistically—extensive phenotypic plasticity, tangled hierarchies, complicated classifications, and controversial developmental histories. Yet despite this overwhelming diversity, Spencer attempted to arrive at a notion of individuality, and tried to offer a systematic review of examples across the gamut of living nature, so as to anchor his theorizing in interpretation of well-accepted empirical findings. Spencer presented the evolvement of individuation in *PBi* as a broad-ranging time-dependent panorama. Each aspect was delineated through the widest range of living entities, from unicellular animals and plants to vertebrates. The style of the descriptions was as if they were field reports or at times concrete experimental reports. One could view the book as a collection of evolutionary-developmental narrations by virtue of which some were also stories of individuation. Although the developmental narrative related to groups of living entities (rather than to species), it treated the members of the group as homogenous as far as the mechanisms of individuation were concerned. Change, activating individuation, was explained through the basic frame of analysis. At one end of the range of possible mechanisms he placed amalgamation of material components, understood as a process of becoming part of something more complex. At the other extreme, he referred to intelligence, reasoning, and information exchange and storage as integration of change (e.g., *PBi*, vol. 1, 62–63). Let me now turn to the differential repertory of patterns that was constitutive of the process of individuation of organisms.[10] Inheritance of specific patterns of the repertory—an inheritance which encompassed both development and biological heredity—assumed a central role in the claim that the change was an evolutionary one, which could be anachronistically called a phylogenetic change.

Individuation began with a gradually evolving material partitioning of internal from external and definiteness of form, and Spencer described its subsequent evolvement using an array of categories taken from the major controversies of the period.

In *PBi* Spencer started with a *liminal individuality* (in fact, the single-celled organism) in which the categories of differentiation from the environment operated at a lowest degree of bounding. The further development of individuality followed a path, charted by the basic framework, which most of the time was posited as necessary. An inherently unstable homogenous state developed to a more stable heterogeneous one, by virtue of temporary balanced states inside the organism, and temporary equilibria between the organism and its specific environment. Spencer characterized this process of temporal evolution as a "moving equilibrium" (e.g., *PBi*, vol. 1, chapter 5, part 2, "adaptation," in the discussion of a social analogy, 193–99, and in ibid., chapters 11-12, part 3). Notions of equilibrium permeated expositions both in physics and in the life sciences during the nineteenth century. At issue for Spencer was how to construct mechanisms that would allow for change and stability as compatible notions within the description of systems, with or without evolution. The hypothesis, which Spencer had discussed in a long article in 1858 (reprinted in Spencer 1891, vol. 1), was an example wherein a mechanism—gravitational attraction—had been adduced to give rise evolutionarily to the observed stability and equilibrium of the solar system. When dealing with living systems within an evolutionary framework, one had to account not only for continued change over time but also for emergence. Spencer sought to provide for processes that would ensure that similarly the system could overcome perturbations, be stable enough in their presence that an equilibrium state would be re-gained and re-established, and thus account for robustness under conditions of seeming instability as "a signature" of living systems.

The patterns arising out of the above movement were responsible for Spencer's criteria of individuality, according to similar functional and morphological patterns, rather than common ancestry. The only clue in *PPi* to a "paradigmatic individual" is Spencer's suggestion to consider the graded continuity from the lowest forms of irritability and contractility to the complex processes of an "advanced man of science," as falling under the same dynamics.[11] "There is, indeed, as already implied, no definition of individuality that is unobjectionable. All we can do is to make the best practicable compromise. . . . A biological individual is any concrete whole having a structure which enables it, when placed in appropriate conditions, to continuously adjust its internal relations to external relations, so as to maintain the equilibrium of its functions" (*PBi*, vol. 1, 206–7). Spencer depicted individual organisms as having some bounding material that delimits and differentiates them from their environment. They thus had a definite shape. These two characteristics—material bounding and shape—were the most basic conditions for being a liminal living individual (*PBi*, vol. 2, 78). In addition, they possessed material living-tissues that exhibited continuity with that of lower forms.

However, Spencer's categories of individuation reflected the difficulties he found in contemporaneous investigations on how to provide an encompassing definition of individuality when confronted with the baffling diversity and symbiotic relationships of living entities. Lower-level organisms were ordinarily more homogenous both in functional specializations and in their inner structures: there was little or no division of labor, no systems of dependence among parts, and barely a differentiation between the inner and the outer (as in his designation of protozoa as lowest organism). A discussion of heterogeneity inside organisms thus also marked levels in evolutionary organizational hierarchies.

Convergence of particular correlations or features produced not just a simple uni-linear hierarchy but hierarchies reflecting various levels of complexity. Thus the components of a higher-level living system could themselves be elements found at lower levels of the living entities. The components of a lower-level entity would possibly reflect only the mechanisms of aggregation or material compounding. Still-lower-level organisms would supposedly be made up of apparently liminal-individuals as components. Earlier organisms that formed quasi-stable components of a higher-level organism in the hierarchy could be autonomous when considered as living entities at a lower level. An example would be organisms with forms of reproduction that encompassed multiple or single life cycles and rhythms with both difference and similarity when looked upon inter-generationally (e.g., Nyhart and Lidgard 2011).

For Spencer these hierarchies were feature-dependent, pragmatic, and not "part of nature."

In Spencer's evolutionary theory new types of individuals were not viewed as marking evolutionary transitions, and thus did not call for new evolutionary-developmental mechanisms. In fact, rather the opposite: the variety of individual types was viewed within a historical and methodological continuum.[12]

For Spencer liminal individuals—whose boundaries were analyzed earlier—were those that could be pointed to qua individuals, due to those boundaries. But within the hierarchy of organisms of central significance in Spencer's theorizing was the distinction among individuals in accordance with progressive levels of complexity. Thus I conclude that for Spencer "biological individuality" was graded, wherein there was a continuity from the bottom liminal individuals to the most complex ones, at the uppermost, abstract, boundary. Considerations of speciation were secondary to him, and phyletic evolution would be the result of individual changes.

Individuals and Collectivities

Spencer struggled to put limits on the concept of biological individuality. The phenomena he encountered in his attempt to both describe and analyze objects that were perceived as separate, such as living entities, societies, and "the mental," did not seem to him to fit any available definition of "individual." The main reason was that most of them exhibited various or plural modes of collective existences. Perceiving any living entity as an individual for some purposes and as a collectivity for others would have greatly simplified his work. As other papers in this volume attest, it is possible to analyze a living entity independently and separately from its full evolutionary account, from its place in the basic frame of analysis, and to set aside its level of organization and degree of complexity. However, this requires viewing the entity from a particular, limited, chosen perspective, because specific questions are being asked, and only with a specific purpose in mind. I believe that Spencer was acutely aware of the tensions produced by the constraints of his framework of individuality, and that the categories of organizational grades and levels of individuals in fact constituted his conceptual gesture towards a hybrid notion of a collective individuality. His framework contained multiple modes of relating, as well as hierarchies of organizational complexity of relations. It encompassed a collectivity of diverse kinds of individuals, differentially related to each other and to the collectivities whose members they were. I find implicit support for this suggestion in the role of collectivities in the construction of a repertory of mental activity patterns in *PPi* and in the construction of the history of "evolutionized individuals"—"wholes"—in *PBi*, vol. 2.

In *PPi* Spencer argued for a developmental view of various procedures of cognition, such as classifying, reasoning (e.g., induction and deduction), and language itself. He usually drew his examples from the process of development "in children" or in (evolutionarily elementary) "savages," and used habits formed as a result of repetition as an explanatory tool. Habituation allowed the reasoner to perform significant shortcuts, because stages in the process "merged" and "sank to the unconscious," and the "shorter version" became the usual, ordinary one. He defined consciousness itself as an emergent consequence of experiencing changes, rather than its precondition. Later on Spencer explained in great detail how experiencing produced neural changes through the mechanism of accumulative-repetitive experience, which he regarded as developmental-evolutionary. These changes were to be passed on hereditarily—that is, as a biological transmission of habituated psychological patterns of experiencing, acquired through the life of an individual. In *PPi* (1855) and *PPii* (1870–72) human competences, "tendencies," or "latencies" were neither solely biological nor solely cultural-mental. They were pre-organized molds, inherited as neural modifications functioning as neural-psychological patterns for organizing experience. By that experiencing they were open at the same time to further functional-adaptational modification, which in turn would become biologically inherited. In order to use these hereditarily transmitted biological patterns to explain a gamut of psychological-mental activities such as instinct and reflex, these activities had to acquire a universally necessary and permanent character. Spencer achieved this by making them *collectively* held molds, which modified the modes of experiencing of the members of human collectivities (e.g., *PPi*, 188–89, 566–72; *PBi*, vol. 1, part 3, chapter 5). In a somewhat similar fashion, he argued in various places in *PBi* that ancestral paths of heredity, particularly in higher (complex) individuals, affected the way they were modified by contemporaneous action of evolutionary mechanisms. The impact of collective past generations—forcefully present in *PPi* in the notion of collectivities inhering within individuals through the back and forth mechanism between biological heredity and cultural-psychological inheritance of acquired characteristics—was similarly employed in *PS*. This was beautifully expressed in a critical comment on *PPii* by Douglas Spalding in *Nature*, February 20, 1873, who wrote of "the doctrine that the brain and nervous system is an organized register of the experiences of past generations, that consequently the intelligence and character of individuals and of races depend much more on this, on the experiences of their ancestors, than on their individual experiences" (299). See, for example, *PPi*: "Every one of the countless connections among the fibres of the cerebral masses, answers to some permanent connection of phenomena in the experience of the race" (581), and then Spencer goes on to say "*These pre-established internal relations* [i.e., the nervous molds—my addition], *though independent of the experiences of the individual, are not independent of experiences in general; but that they have been established by the accumulated experiences of preceding organisms. The corollary from the general argument that has been elaborated, is, that the brain represents an infinitude of experiences received during the evolution of life in general: the most uniform and frequent of which, have been successively bequeathed, principal and inter-est"* (583, my emphasis; see also ibid., 539–53). In Spencer's view (and that of some of his contemporaries), these molds fulfill the function of the Kantian transcendental apparatus, and yet are themselves a product of experiencing.

The commonly accepted view has been that the Spencerian narrative was one of individuals, though some would add groups, and I would agree that Spencer declared himself an "individualist" in matters political. Nevertheless, his basic evolutionary unit in *PS* was the collectivity, presented as a special kind of individual. As he argued in "The Social Organism" (1860b), societies and organisms grew both by an increase of structural complexity, and by an increase of the mutual interdependence of parts, resulting in the "whole" being more stable and longer lasting than its component parts (see also Spencer 1874, 330).

For most of his contemporaries, individuals were conceived to be epistemologically the real, observable entities. The reality of the collectivity was often envisaged as secondary, deriving from the reality of the individuals and thus unobservable on its own. By deploying the analogy to the individual biological organism, Spencer allowed the social collectivity in its interrelations to become epistemologically real, observable, perceivable, and accessible. One could thus construct its visualization at a time when the "visibility of Nature" had become a significant scientific issue (Gissis 2002).

In all fields Spencer held a hybrid notion of individuality, which could be viewed either as "collective individuality" or as "collectivity of individuals." Spencer's use of this hybrid category allowed him to use either individuals or hybrid individuals or collectivities in the role of the "organism" within his basic frame of analysis, and thus stretch both the extension of the model and the extension of possible interactions. This fitted well with his agenda in the *Principles of Ethics* (1892–93) to explicate the deep roots of human morality through evolution—that is, through a reconstruction of the history of evolvement of living entities as an evolutionary development of moral actions— starting with lower life forms and ending with humans.

In all the fields in which he theorized, Spencer made explicit his underlying assumption that it was conceptually impossible to separate and isolate individuality from collectivity. And this conceptual impossibility is crucial for understanding his totalizing endeavor.

The Spencerian *Problématique*—Is It Relevant?

Spencer's wide influence waned in the 1890s and all but died out after the turn of the century, particularly in Britain and continental Europe. However, in recent years there has been a renewed interest in some of his writings; I would like briefly to indicate why.

Relating to Spencer does not entail a commitment to his specific positions nor to the specific mechanisms he envisaged, but rather attentiveness to his *problématique* and its topography, and to the *kind* of issues he was struggling with. Thus, what is significant is not the fact that Spencer has been mentioned more often in recent years in articles written by biologists and philosophers of biology, or is being interpreted as "a forefather" of one position or another. In fact, some of these readings of his work are highly controversial. What are relevant are his questions, the tensions within his theoretical framework, the trove of theorizing and conceptualizing possibilities proffered by the models and metaphors that he deployed. These have resonated with present-day efforts to broaden, extend, revise, change, or revolutionize the theoretical evolutionary framework that came to predominate in the 1940s–1980s. And they could be generative.

The present-day efforts at making sense of life at the molecular level face problems similar to those that Spencer faced. Organisms interact with their surroundings and make use of their interactions with their environments to survive and reproduce. Function is thus inherent in life. But organisms are made up of molecules that are "dumb" and have no particular function when wrenched out of a particular context. How does one relate these two ways of analyzing the world?

I have tried to present Spencer's works as the evolution and unfolding of a theoretical framework that attempted to explain how function, purpose, and decision-making came into being in living organisms. To work out his viewpoint, Spencer used materials from countless past and contemporaneous sources. Some, such as those he took from the chemical and the biological discourses of the time, were deployed with a keen awareness of their embeddedness in other conceptual frames. He appropriated other less technical materials and used them with the complacency that accompanies that which was considered well known or self-evident. His endeavor should be looked upon as an attempt to answer questions posed within the then-existing framework. The conceptual constraints and the state of contemporaneous knowledge could not provide satisfactory answers to these questions within the terms of that framework. Thus he hypothesized a new tripartite frame of analysis of nature, and conjectured (imagined, speculated on) mechanisms that would substantiate this new foundational (basic) frame. The then-available knowledge could not provide the justification of his assumptions, neither those he made at the microscopic level, nor those which were to relate that level to the macroscopic level. This new framework and its conceptual and terminological apparatus, as well as the interrelationship between the "old" and this new framework, contributed both to the growing popularity of Spencer's work and to the ways it was misunderstood.

Present-day (evolutionary) biology faces analogous problems. There is a renewed interest in Spencer's works, especially in those related to the problematical and controversial notions of "individuals," "complexity," and "life." Thus one could take Clarke's recent enumeration of possible "in-dividualities" (2010) as a point of departure to think about contemporary discussions on the notions of organism, environment, and their interrela-tions, and inquire about the relevance and possible value of Spencer's work to these issues. Strands of Spencer's work relate to organism-environment relations, as found in niche construction theory, the notion of an active or-ganism shaping the environment and shaping itself, as well as the causal role of the environment (e.g., Santelices 1999; Lewontin 2000; Odling-Smee et al. 2003, R. Wilson 2007); to plasticity as response to environmental change (West-Eberhard 2003) and the importance of "behavior" in evolution, as in developmental systems theory (the whole corpus of the Developmental Sys-tems Theory group: see Oyama, Griffiths, and Gray 2001). Spencer's views on heterogeneity and systems interactions may be thought-provoking comple-ments to discussions of progressive and non-progressive complexity (Mc-Shea and Brandon 2010; Ruse 2004, 2013), to systems biology on one hand and the debate on levels of selection on the other (e.g., already Needham 1937; Buss 1987; Maynard Smith and Szathmáry 1995; Gould and Lloyd 1999; Gerson 2007; Okasha 2007; Folse and Roughgarden 2010; Godfrey-Smith 2011). Finally, Spencer's challenges and solutions regarding collectivities of-fer an oblique and interesting perspective of temporal change in part-whole relations, in the transition from uni- to multicellularity, of symbiosis, of the role of holobionts as an encompassing theoretical framework, and of the evo-lution and role of the immune system (e.g., D. Wilson 1989; Margulis and Fester 1991; Sapp 1994, 2005; Michod and Roze 1997, 2001; Gilbert and Sarkar 2000; Gayon 2008; Dupré and O'Malley 2009; Pradeu 2010; Gilbert et al. 2012; Bouchard and Huneman 2013).

Although I have but adumbrated Spencer's views of heredity in the pres-ent article, I have previously detailed Spencer's use of Lamarckian hereditary notions concerning the significance of development and of the inheritance of patterns acquired throughout the life of the organism in its interactions with its environment (Gissis 2005). The Spencerian *problématique* resonates with the burgeoning field of epigenetics. It would be interesting and valuable to clarify the connection.

I end by repeating once again that in spite of inner tensions Spencer was an emergentist, trying to convey within nineteenth-century conceptual frameworks certain insights on diversity and complexity, and on individuals and collectivities. These frameworks were rather inhospitable to such views. I believe that his endeavors to posit and generate a synthetic all-encompassing evolutionary world view merit respect and reconsideration. In fact, this may be part of the reason for the renewed (if still limited) present-day interest in his work.

Acknowledgments

It was a privilege to take part in the two workshops on biological individuality organized by Lynn Nyhart and Scott Lidgard during 2012. I learned much from the questions and critical comments by the participants and am very grateful for the detailed, extremely helpful, questions and editorial comments on my paper by Lynn Nyhart and Scott Lidgard. The criticism of, and discussions and disagreements with, Sam Schweber while preparing the final version did wonders in clarifying, sharpening, and compacting my views. I am deeply indebted to him for his involvement with this paper.

Notes to Chapter Six

1. Spencer wrote about his work on the first edition of his *Principles of Biology* (*PBi*, 1864–1867) and the significant help proffered to him by his friends in two chapters of his *Autobiography* (vol. 2, chapters 39, 41).

2. In particular Spencer 1851, 1852a and b, 1853, 1857a and b, 1859, and the second edition of the *Principles of Biology* (1898).

3. I have used principally three detailed groups of interpretation since the 1970s: 1) Carneiro (1973), Peel (1992 [1971]), and Young (1967, 1970), which focused on Spencer's concepts of "evolution" and "progress" and their relation to those of Spencer's contemporaries, including Darwin and somewhat earlier writers; 2) Haines (1988), Jacyna (1981, 1984), Perrin (1993), R. Richards (1987), C. Smith (1982), and Turner (1985), who grappled with Spencer's multidisciplinary network of concepts, the centrality of "evolution," and "complexity." Indirectly, I have also drawn on Bowler (1989), Desmond (1989, 1997, 2001), Limoges (1994), Rylance (2000), Schweber (1980), and Stocking (1987); 3) Barton (1990), Dixon (2008), Elwick (2003, 2007, 2014), Francis (2007), Godfrey-Smith (1996), Gray (1996), LaVergata (1995), Offer (2000, 2010), Pearce (2010), and Taylor (2007), who situated Spencer within historical, philosophical, scientific, and biographical contexts and contemporaneous controversies, and offered insights on the primacy of the ethical, social or scientific in his work. Indirectly, I have also drawn on d'Hombres (2010), Nyhart (1995), Nyhart and Lidgard (2011), Renwick (2009), Reynolds (2007), R. Richards (2008), and R. Smith (2010).

4. In "Filiation of Ideas" (1899), Spencer indicated that the conceptualization came after a talk with Tyndall in 1858. Also Francis 2007, 372n27.

5. In *FPi* Spencer admitted that he could give a general causal explanation of the relations

between forces (interactions) and their evolving effects when dealing with living entities, but not a quantitative one (e.g., *FPi* 1862, 265).

6. At the time both Spencer and Huxley, when discussing the methodological aspects of physical laws vs. biological laws, were critical of mathematics as a tool for describing change in living nature (e.g., see Desmond 1997, 2: 377).

7. For example, as he argued in "Theory of Population" (1852a). The article is to be read keeping in mind that for Spencer evolution was progressive, and was evidenced in the "internal coordination" of organisms generally, and more specifically in humans. This was noted in the first part of the article, until p. 8, as well as in the very last pages.

8. Trevor Pearce in his erudite article (2010) argued that unifying the extant modes into one term "environment" and positing organism and environment as dichotomous had been Spencer's critical innovation, though qualified to the British context. I disagree with this, both on the issue of historical beginnings and on the attribution of significance to the passage from "circumstances" to "environment."

9. It is interesting to note a criticism referring to the significance of embryology to developmental explanations was given by the sociologist Frederic Harrison. He claimed that Spencer dealt with beginnings—primitive man, embryos, invertebrates—and with the mature forms such as contemporary society, but did not provide the actual process of development (1905, 102).

10. For example, rhizopods, elaborated upon by Huxley (1898 [1852]). Spencer (1904), chapters 35, and 37.

11. Following Malthus, but as an answer to his scenario in *An Essay on the Principle of Population* (1798), Spencer elaborated in "Theory of Population" (1852a) on the differential relations between preserving biological individuality and reproduction. Spencer there posited the question: "to what extent does reproduction weaken the individual?" and his answer was that there was a direct relation all along the evolutionary progressive hierarchy between the preservation of biological individuality and a low rate of reproduction. By incorporating parts of this essay into his *PB* he held on to this position there, too. However, the relation between preservation of individuality and reproduction was in accordance with the organism's degree of complexity. Thus only humans were to be exempt from it, this both in his early essay and in the much later *PB* (*PBi*, vol. 1, part 2, chapter 7; vol. 2, part 6, "Laws of Multiplication").

12. One should add that in a couple of places he tentatively offered to look upon species as some such individual, e.g., *PBi*, vol. 1, 458 ff, vol. 2, 422 (to some extent).

References for Chapter Six

Adam, Alison E. 1988. *Spontaneous Generation in the 1870s: Victorian Scientific Naturalism and Its Relationship to Medicine.* PhD diss., Sheffield City Polytechnic. http://shura.shu.ac.uk/2695/.

Barsanti, Giulio. 2000. "Lamarck: Taxonomy and Theoretical Biology." *Asclepio* 52 (2): 125.

Barton, Ruth. 1990. "'An Influential Set of Chaps': The X-Club and Royal Society Politics 1864–85." *British Journal for the History of Science* 23 (1): 53–81.

———. 1998. "Huxley, Lubbock, and Half a Dozen Others": Professionals and Gentlemen in the Formation of the X Club, 1851–1864." *Isis* 89 (3): 410–44.

Bouchard, Frédéric, and Philippe Huneman, eds. 2013. *From Groups to Individuals: Evolution and Emerging Individuality.* Cambridge, MA: MIT Press.

Bowler, Peter. 1989. *The Invention of Progress: The Victorians and the Past.* Baltimore: Johns Hopkins University Press.

Buss, Leo W. 1987. *The Evolution of Individuality.* Princeton, NJ: Princeton University Press.

Carneiro, Robert L. 1973. "Structure, Function and Equilibrium in the Evolutionism of Herbert Spencer." *Journal of Anthropological Research* 29 (2): 77–95.

Churchill, F. B. 1991. "The Rise of Classical Descriptive Embryology." In *Conceptual History of Modern Embryology,* edited by Scott Gilbert, 1–29. New York: Plenum.

Clarke, Ellen. 2010. "The Problem of Biological Individuality." *Biological Theory* 5 (4): 312–25.

Cohen, I. Bernard, ed. 1981. *The Conservation of Energy and the Principle of Least Action.* New York: Arno Press.

d'Hombres, Emmanuel. 2010. "The 'Division of Physiological Labor': The Birth, Life and Death of a Concept." *Journal of History of Biology* 45: 3–31.

Darwin, Charles. 1964 [1859]. *On the Origin of Species.* A Facsimile of the First Edition. Cambridge, MA: Harvard University Press.

Desmond, Adrian J. 1989. *The Politics of Evolution: Morphology, Medicine and Reform in Radical London.* Chicago: University of Chicago Press.

———. 1997 [1994]. *Huxley: From Devil's Disciple to Evolution's High Priest.* Reading, MA.: Addison-Wesley.

———. 2001. "Redefining the X Axis: 'Professionals,' 'Amateurs' and the Making of Mid-Victorian Biology—A Progress Report." *Journal of the History of Biology* 34: 3–50.

Dixon, Thomas M. 2008. *The Invention of Altruism: Making Moral Meanings in Victorian Britain.* Oxford: Oxford University Press for the British Academy.

Dupré, John, and Maureen O'Malley. 2009. "Varieties of Living Things: Life at the Intersection of Lineage and Metabolism." *Philosophy and Theory in Biology* 1. http://quod.lib.umich.edu/p/ptb/6959004.0001.003/—varieties-of-living-things-life-at-the-intersection?rgn=main;view=fulltext.

Elwick, James. 2003. "Herbert Spencer and the Disunity of the Social Organism." *History of Science* 41: 35–72.

———. 2007. *Styles of Reasoning in the British Life Sciences: Shared Assumptions, 1820–1858.* London: Pickering & Chatto.

———. 2014. "Containing Multitudes: Herbert Spencer, Organisms Social, and Orders of Individuality." In *Herbert Spencer: Legacies,* edited by Mark Francis and M. W. Taylor, 89–110. Abingdon, Oxon: Routledge.

Folse, H. J., III, and Joan Roughgarden. 2010. "What Is an Individual Organism? A Multilevel Selection Perspective." *Quarterly Review of Biology* 85 (4): 447–72.

Francis, Mark. 2007. *Herbert Spencer and the Invention of Modern Life.* Chesham: Acumen.

Gayon, Jean. 2008. "Les espèces et les taxons monophylétiques sont-ils des individus?" In *L'individu, perspectives contemporaines,* edited by Pascal Ludwig and Thomas Pradeu, 127–76. Paris: Vrin.

Gerson, Elihu. 2007. "The Juncture of Evolutionary and Developmental Biology." In *From Embryology to Evo Devo: A History of Developmental Evolution,* edited by Manfred D. Laublicher and Jane Maienschein, 435–63. Cambridge, MA: MIT Press.

Gilbert, Scott, and Sahotra Sarkar. 2000. "Embracing Complexity: Organicism for the 21st Century." *Developmental Dynamics* 219: 1–9.

Gilbert, Scott, Alfred Tauber, and Jan Sapp. 2012. "A Symbiotic View of Life: We Have Never Been Individuals." *Quarterly Review of Biology* 87 (4): 325–41.

Gissis, Snait B. 2002. "Late Nineteenth Century Lamarckism and French Sociology." *Perspectives on Science* 10 (1): 69–122.

———. 2005. "Herbert Spencer on Biological Heredity and Cultural Inheritance." Paper presented at the third workshop on "The Cultural History of Heredity," Berlin, January, Max Planck Institute for the History of Science; subsequently published in a temporary preprint volume edited by Hans-Jörg Rheinberger and Staffan Müller-Wille. 137–52.

Godfrey-Smith, Peter. 1996. *Complexity and the Function of Mind in Nature.* Cambridge: Cambridge University Press.

———. 2011. "The Evolution of the Individual." Lakatos Award Lecture, London School of Economics, June 2011. http://petergodfreysmith.com/wp-content/uploads/2013/06/Evo _Ind_PGS_Lakatos_2011_C.pdf .

Gould, Stephen J. 1977. *Ontogeny and Phylogeny.* Cambridge, MA: Belknap.

Gould, Stephen J., and Elizabeth Lloyd. 1999. "Individuality and Adaptation across Levels of Selection: How Shall We Name and Generalize the Unit of Darwinism?" *Proceedings of the National Academy of Sciences USA* 96 (21): 11904–9.

Gray, Tim S. 1996. *The Political Philosophy of Herbert Spencer: Individualism and Organicism.* Aldershot: Avebury.

Haines, Valerie A. 1988. "Is Spencer's Theory an Evolutionary Theory?" *American Journal of Sociology* 93 (5): 1200–1223.

Harrison, Frederic. 1905. *The Herbert Spencer Lecture.* Oxford: Clarendon Press.

Hopwood, Nick. 2009. "Embryology." In *The Cambridge History of Science Volume 6: Modern Life and Earth Sciences,* edited by Peter J. Bowler and John V. Pickstone, vol. 6, chap. 16, 285–315. Cambridge: Cambridge University Press.

Huxley, Thomas. H. 1898 [1852]. "Upon Animal Individuality." In *The Scientific Memoirs of Thomas Henri Huxley,* edited by Michael Foster and E. Ray Lankester, vol. 1: 146–51. London: Macmillan.

Jacyna, L. S. 1981. "The Physiology of Mind, the Unity of Nature and the Moral Order in Victorian Thought." *British Journal for the History of Science* 14 (2): 109–32.

———. 1984. "Principles of General Physiology: The Comparative Dimension to British Neuroscience in the 1830s and 1840s." *Studies in History of Biology* 7: 47–92.

Lamarck, Jean Baptiste Pierre Antoine de Monet de. 1809. *Philosophie Zoologique.* Paris: Dentu. Accessed at Site Lamarck: www.lamarck.cnrs.fr.

LaVergata, Antonello. 1995. "Herbert Spencer: Biology, Sociology, and Cosmic Evolution." In *Biology as Society, Society as Biology: Metaphors,* edited by Sabine Maasen, 193–229. Dordrecht: Kluwer Academic.

Lewontin, Richard. 2000. *The Triple Helix: Gene, Organism and Environment.* Cambridge: Harvard University Press.

Limoges, Camille. 1994. "Milne-Edwards, Darwin, Durkheim and the Division of Labour: A Case Study in Reciprocal Conceptual Exchanges between the Social and the Natural Sciences." In *The Natural Sciences and the Social Sciences,* edited by I. Bernard Cohen, 317–43. Dordrecht: Kluwer Academic.

MacLeod, Roy M. 1970. "The X-Club: A Social Network of Science in Late-Victorian England." *Notes and Records of the Royal Society of London* 24 (2): 305–22.

Margulis, Lynn, and René Fester, eds. 1991. *Symbiosis as a Source of Evolutionary Innovation: Speciation and Morphogenesis.* Cambridge, MA: MIT Press.

Maynard Smith, John, and Eörs Szathmáry. 1995. *The Major Transitions in Evolution.* Oxford: Oxford University Press.

McShea, Daniel W., and Robert N. Brandon. 2010. *Biology's First Law: The Tendency for Diversity and Complexity to Increase in Evolutionary Systems.* Chicago: University of Chicago Press.

Merz, John T. 1976 [1904–12]. *A History of European Scientific Thought in the Nineteenth Century.* Gloucester, MA: P. Smith.

Michod, Richard E., and Denis Roze. 1997. "Transitions in Individuality." *Proceedings of the Royal Society B* 264: 853–57.

———. 2001. "Cooperation and Conflict in the Evolution of Multicellularity." *Heredity* 81: 1–7.

Milne Edwards, Henri. 1834. *Élémens de zoologie: ou Leçons sur l'anatomie, la physiologie, la classification et les moeurs des animaux.* Bruxelles: H. Dumont.

———. 1851. *Introduction à la zoologie générale, ou Considérations sur les tendences de la nature dans la constitution du règne animal,* premier partie. Paris: Masson.

Needham, Joseph. 1937. *Integrative Levels: A Revaluation of the Idea of Progress.* Oxford: Clarendon Press.

Nye, Mary Jo. 1993. *From Chemical Philosophy to Theoretical Chemistry.* Berkeley: University of California Press.

———. 1996. *Before Big Science: The Pursuit of Modern Chemistry and Physics, 1800–1940.* London: Prentice Hall International.

Nyhart, Lynn. 1995. *Biology Takes Form: Animal Morphology and the German Universities, 1800–1900.* Chicago: University of Chicago Press.

Nyhart, Lynn, and Scott Lidgard. 2011. "Individuals at the Center of Biology: Rudolf Leuckart's *Polymorphismus der Individuen* and the Ongoing Narrative of Parts and Wholes." *Journal of the History of Biology* 44 (3): 373–443.

Odling-Smee, F. John, Kevin N. Laland, and Marcus W. Feldman. 2003. *Niche Construction: The Neglected Process in Evolution.* Princeton: Princeton University Press.

Offer, John, ed. 2000. *Herbert Spencer: Critical Assessments.* 4 vols. London: Routledge.

———. 2010. *Herbert Spencer and Social Theory.* New York: Palgrave Macmillan.

Okasha, Samir. 2007. *Evolution and the Levels of Selection.* New York: Oxford University Press.

Oyama, Susan, Paul E. Griffiths, and Russell D. Gray. 2001. *Cycles of Contingency: Developmental Systems and Evolution.* Cambridge, MA: MIT Press.

Pearce, Trevor. 2010. "From 'Circumstances' to 'Environment': Herbert Spencer and the Origins of the Idea of Organism-Environment Interaction." *Studies in History and Philosophy of the Biological and the Biomedical Sciences* 41: 241–52.

Peel, John D. Y. 1992 [1971]. *Herbert Spencer: The Evolution of a Sociologist.* Aldershot: Gregg Revivals.

Perrin, Robert G. 1993. *Herbert Spencer: A Primary and Secondary Bibliography.* New York: Garland.

Pradeu, Thomas. 2010. "What Is an Organism? An Immunological Answer." *History and Philosophy of the Life Sciences* 32: 247–68.

Renwick, Chris. 2009. "The Practice of Spencerian Science: Patrick Geddes's Biosocial Program, 1876–1889." *Isis* 100: 36–57.

Reynolds, Andrew. 2007. "Amoebae as Exemplary Cells: The Protean Nature of an Elementary Organism." *Journal of the History of Biology* 41 (2): 307–37.

Richards, Evelleen. 1989. "The 'Moral Anatomy' of Robert Knox: The Interplay between Biological and Social Thought in Victorian Scientific Naturalism." *Journal of the History of Biology* 22 (3): 373–436.

Richards, Robert J. 1987. *Darwin and the Emergence of Evolutionary Theories of Mind and Behavior.* Chicago: University of Chicago Press.

———. 2008. *The Tragic Sense of Life: Ernst Haeckel and the Struggle over Evolutionary Thought.* Chicago: University of Chicago Press.

Rocke, Alan J. 2002. "The Theory of Chemical Structure and Its Application." In *The Cambridge History of Science, Volume 5: The Modern Physical and Mathematical Sciences,* edited by Mary Jo Nye, 255–71. Cambridge: Cambridge University Press.

Ruse, Michael. 2004. "Adaptive Landscapes and Dynamic Equilibrium: The Spencerian Contribution to Twentieth-century American Evolutionary Biology." In *Darwinian Heresies,* edited by Abigail. J. Lustig, Robert J. Richards, and Michael Ruse, 131–50. Cambridge: Cambridge University Press.

———. 2013. "Wrestling with Biological Complexity: From Darwin to Dawkins." In *Complexity and the Arrow of Time,* edited by Charles H. Lineweaver, Paul C. W. Davies, and Michael Ruse, 279–307. New York: Cambridge University Press.

Rylance, Rick. 2000. *Victorian Psychology and British Culture, 1850–1880.* Oxford: Oxford University Press.

Santelices, Bernabé. 1999. "How Many Kinds of Individuals Are There?" *Trends in Ecology & Evolution* 14 (4): 152–55.

Sapp, Jan. 1994. *Evolution by Association: A History of Symbiosis.* Oxford: Oxford University Press.

———, ed. 2005. *Microbial Phylogeny and Evolution: Concepts and Controversies.* New York: Oxford University Press.

Schütt, Hans W. 2002. "Chemical Atoms and Chemical Classification." In *The Cambridge History of Science, Volume 5: The Modern Physical and Mathematical Sciences,* edited by Mary Jo Nye, 237–54. Cambridge: Cambridge University Press.

Schweber, Silvan. S. 1980. "Darwin and the Political Economists." *Journal of the History of Biology* 13(2): 195–289.

Smith, Christopher U. M. 1982. "Evolution and the Problem of Mind. Part I: Herbert Spencer." *Journal of the History of Biology* 15 (1): 55–88.

Smith, Crosbie, and Norton Wise.1989. *Energy and Empire: A Biographical Study of Lord Kelvin.* Cambridge: Cambridge University Press.

Smith, Roger. 2010. "British Thought on the Relations between the Natural Sciences and the Humanities c. 1870–1910." In *Historical Perspectives on Erklären and Verstehen,* edited by Uljana Feest, 161–86. Dordrecht: Springer.

Spalding, Douglas. 1873. "Herbert Spencer's Psychology." *Nature* 7: 298–300.

Spencer, Herbert. 1851. *Social Statics: or, The Conditions Essential to Human Happiness Specified and the First of Them Developed.* London: J. Chapman.

———. 1852a. "A Theory of Population, Deduced From the General Law of Animal Fertility." *Westminster Review* 57: 468–501.

———. 1852b. "The Development Hypothesis." *Leader* 3: 280–81. Reprinted in *Essays Scientific, Political and Speculative,* vol. 1, 1–7.

———. 1853. "The Universal Postulate." *Westminster Review* 60: 513–50.

———. 1855. *The Principles of Psychology.* London: Longman, Brown, Green and Longman.

———. 1857a. "Progress, Its Law and Cause." *Westminster Review* 67: 445–85. Reprinted in *Essays Scientific, Political and Speculative,* vol. 1, 8–62.

———. 1857b. "The Ultimate Laws of Physiology"/ "Transcendental Physiology." *National Review* 5: 332–55. Reprinted in *Essays Scientific, Political and Speculative* , vol. 1, 63–107.

———. 1858. "Recent Astronomy and the Nebular Hypothesis"/"The Nebular Hypothesis."

Westminster Review 70: 185–225. Reprinted in *Essays Scientific, Political and Speculative,* vol. 1, 108–91.

———. 1859. "The Laws of Organic Form." *British and Foreign Medico-Chirurgical Review* 45: 189–202.

———. 1860a. "Bain on Emotions and the Will." *British and Foreign Medico-Chirurgical Review* 49 (January): 42–52. Reprinted in *Essays Scientific, Political and Speculative,* vol. 1, 241–64.

———. 1860b. "The Social Organism." *Westminster Review* 17: 90–121. Reprinted in *Essays Scientific, Political and Speculative,* vol. 1, 265–307.

———. 1862. *First Principles.* London: Williams and Norgate.

———. 1864–67. *The Principles of Biology.* London: Williams and Norgate.

———. 1870–72. *The Principles of Psychology.* 2nd ed. London: Longman, Brown, Green and Longman.

———. 1874 [1873]. *The Study of Sociology.* New York: Appleton.

———. 1874–96. *Principles of Sociology.* London: Williams and Norgate.

———. 1891. *Essays Scientific, Political and Speculative,* vol. 1. London: Williams & Norgate.

——— .1892–93. *The Principles of Ethics.* 2 vols. New York: D. Appleton.

———. 1898. *The Principles of Biology.* 2nd ed. London: Williams and Norgate.

———. 1899. "The Filiation of Ideas." In *The Life and Letters of Herbert Spencer,* edited by David Duncan, Appendix B, 533–76.

———. 1904. *An Autobiography.* 2 vols. New York: D. Appleton.

Stocking, George W., Jr. 1987. *Victorian Anthropology.* New York: Free Press.

Taylor, Michael W. 2007. *The Philosophy of Herbert Spencer.* London: Continuum.

Thomson, W. 1862. "On the Age of the Sun's Heat." *Macmillan's Magazine* 5: 288–93.

Turner, Jonathan. H. 1985. *Herbert Spencer: A Renewed Appreciation.* Beverly Hills: Sage.

Tyndall, John. 1892. "The Belfast Address" [1874] in *Fragments of Science: A Series of Detached Essays, Addresses and Reviews,* Vol. 2, 135–201. New York: Appleton.

West-Eberhard, Mary Jane. 2003. *Developmental Plasticity and Evolution.* Oxford: Oxford University Press.

Wilson, David S. 1989. "Levels of Selection: An Alternative to Individualism in Biology and the Human Sciences." *Social Networks* 11: 257–72.

Wilson, Robert A. 2007. "The Biological Notion of Individual." *Stanford Encyclopedia of Philosophy.* http://plato.stanford.edu/entries/biology-individual/.

Young, Robert M. 1967. "The Development of Herbert Spencer's Concept of Evolution." *Actes du XIe. Congrès International d'Histoire des Sciences* 2: 273–78.

———. 1970. *Mind, Brain and Adaptation in the Nineteenth Century: Cerebral Localization and Its Biological Context from Gall to Ferrier.* Oxford: Oxford University Press.

7

Biological Individuality and Enkapsis: From Martin Heidenhain's Synthesiology to the *Völkisch* National Community

OLIVIER RIEPPEL

Introduction

Individuality is generally tied to indivisibility and self-identity, and hence analyzed in terms of Leibniz's Law of the "Indiscernability of Identicals," which states that numerically identical entities share exactly the same properties. This is a strong, logical (philosophical) conception of individuality, where individuation is grounded in essential intrinsic properties. Since the time of Ancient Greek philosophers, biological individuals such as humans have accordingly, and traditionally, been individuated through their soul. More recently, in discussions of genetic engineering and gene therapy, individual crops, livestock, and humans have been individuated through their genomes, the latter construed as an essential intrinsic property of any one organism. But unlike the immaterial soul, the material basis of the genome undergoes constant change, not only in the transition from ancestor to descendant, but also during the life-cycle of a given organism. At the prokaryote level of biological organization, genomes are, or at least can be, open systems that are in constant exchange with their biological environment. Genomes are thus revealed to form eminently dynamic—that is, processual—systems individuated not by an immutable essence, but through their relational properties and propensities (Barnes and Dupré 2008).

The same is true of all biological systems. The philosophical attempt to individuate a human being on the basis of essential intrinsic properties invites insurmountable problems (Gilbert et al. 2012), which disappear if instead the human individual, as indeed any organism, gene, or species, is understood as a dynamic relational system. In his analysis of the concept of individuality, Ludwig von Bertalanffy (1932, 269) found it to be loaded with natural language connotations to such a degree that it could not perform useful theoretical work in biology. He consequently advocated abandoning the terms "individuality" and "individual" in discussions of biological complexity, and introduced the term "system" instead. A system in that sense is a spatiotemporally located complex whole, causally integrated and at the same time individuated through the relational properties prevalent amongst its parts. Significantly, Bertalanffy (1932) launched into a discussion of individuality only after having reviewed Martin Heidenhain's concept of *enkapsis*. While the fundamentals of this idea go back to German Romanticism and its proponents, such as Schelling and Goethe (Richards 2002; Rieppel 2011a), the discussion here will be based on Heidenhain's conception of the enkaptic hierarchy.

An enkaptic (nested) hierarchy is, in brief, a nested hierarchy of complex wholes with emergent properties at all levels of inclusiveness. The complex wholes that make up such a hierarchy are spatiotemporally located, functionally integrated relational systems—that is, individuals of different levels of complexity. The dynamic, indeed processual systems, or individuals, that make up the enkaptic hierarchy are subject to upwards as well as downwards causation. The enkaptic hierarchy so conceived is not only of historical importance with both biological and socio-political implications, but is implicit in many concepts of modern biology. Genes as parts of gene regulatory networks (if at least partially hierarchically structured), organs or organ-systems as parts of organisms, individual organisms as parts of a species, species as parts of a genealogical hierarchy of monophyletic taxa—all form processual systems that can be conceptualized in terms of an enkaptic hierarchy of dynamic complex wholes. Synecology also accommodates the concept of an enkaptic hierarchy, comprising organisms, populations, species, biotope, and biocoenosis. The key is that all these systems of different levels of complexity are both dynamic and historical, hence spatiotemporally located and individuated through their causal integration and interdependence, hence through relational (rather than essential intrinsic) properties. The historical sketch presented in this chapter intends to trace a long-standing concern with biological individuality and nested hierarchies in German-speaking biology; to show how these discussions have been functional in a variety of areas of comparative biology; and to demonstrate how they have even been taken beyond biology in discussions of wider philosophical and socio-political issues. The essay will trace Heidenhain's enkapsis concept from its origins in his work on the multicellular organism to its eventual extrapolation to ecosystems, to the phylogenetic system, and eventually to the *völkisch*-organicist National Community of the Third Reich. The goal here is to trace commonalities in the deployment of the concept of an enkaptic hierarchy in these different contexts that could illuminate the problem of biological identity and individuality.

The chapter concludes with a perspective of the relevance of these concepts for modern biology.

Martin Heidenhain's Synthesiology

The Tübingen anatomist Heidenhain developed his concept of an enkaptic hierarchy as a means to capture the structural and functional organization and integration of multicellular organisms. The latter he found principally hierarchically structured: cells and their organelles form complex wholes that divide and come together to form organs, themselves complex wholes of higher order that divide and combine to form the organism, again itself a complex whole of yet higher complexity. An enkaptic hierarchy then is a nested hierarchy of "individuals," "relational systems," or "complex wholes" of increasing degrees of complexity, every level of complexity instantiating emergent properties and therefore not simply reducible to the sum of its parts and their particular intrinsic properties.

Heidenhain's development of enkapsis was a response to the battles of his time over reductionism and vitalism. Born on December 7, 1864, in Breslau, Germany (now Wrocław, Poland), he was the son of the renowned physiologist Rudolf (Peter Heinrich) Heidenhain (1834–1897) (Volkmann 1935; Jacobi 1952/53; Alfert 1972). Entering the University of Breslau in 1883, he started studies in natural sciences, especially zoology, which he later continued in Würzburg under Karl Semper. Through his father's institute at Breslau University he came into frequent contact with Wilhelm Roux, professor at that university from 1879 through 1889. Roux was the founder of causal-analytic *Entwicklungsmechanik*—the term was suggested by Heidenhain's father Rudolf (Mocek 2001, 475)—which sought to reduce the organism to a mere system of forces, implicitly seeking to reduce biology to physics (Cassirer 1950, 189). Come spring 1886, Heidenhain enrolled in medical school at the University of Freiburg i. Br., where he also frequented August Weismann's institute and attended the latter's seminars. Weismann's understanding of cell theory was, at least in the eyes of some later commentators, again deeply entrenched in the reductionist-atomistic tradition and as such linked to Roux's (1881) sub-organismal struggle for existence among an organism's parts (Weidenreich 1929, 284; Heidenhain 1907, 49). After another stint in Breslau, Heidenhain earned his MD in comparative anatomy in Freiburg i. Br. under Robert Wiedersheim in 1890. In the spring of 1891 he was appointed an assistant to Albert von Kölliker in Würzburg. During the Würzburg years, Heidenhain became critical of the strictly causal-analytic, reductionistic research pursued by his contemporaries such as Roux and Weismann. Motivated by the second

German edition of Driesch's (1921) *Philosophy of the Organism* (yet rejecting Driesch's vitalism: Jacobi 1952/53, 86), Heidenhain worked out his concept of enkapsis in opposition to reductionist-atomistic interpretations of cell theory. Heidenhain (1937, xiv) found "Schwann's cell-theory [to be] purely analytic/atomistic (*zersetzend*), lacking the potential of a synthetic science which alone allows cognition of true laws of nature." In 1899 Heidenhain accepted an associate professorship in anatomy at the University of Tübingen, where he advanced to full professor and head of the anatomical institute after the death of August von Froriep in 1917. In July 1933, the government of Württemberg lowered the retirement age for university professors from 70 to 68 years, which provided an opportunity to get rid of a number of professors who remained reserved vis-à-vis the new National Socialist regime (Mörike 1988, 72). Amongst those was Heidenhain, who was a staunch republican of (remote) Jewish descent (Adam 1977, 123n18; Grün 2010, 250); he was retired as of October 1, 1933.[1] Heidenhain died in Tübingen on February 14, 1949.

Carl Gegenbaur opened his classic treatise on comparative anatomy with a conceptual and methodological separation of causal-analytic physiology from descriptive comparative anatomy (Gegenbaur 1859, 1). The same dualism is reflected in the writings of the German anatomist Wilhelm Lubosch, who distinguished the "architectural" versus the "technical" approach to comparative anatomy (Lubosch 1926, 655). In his classic review, Lubosch (1931, 60) insisted that comparative anatomy does not "explain" the human body, but instead teaches one to "comprehend it meaningfully" (*sinnvoll verstehen*). The same themes resonated in the writings of other prominent German anatomists, such as Eduard Jacobshagen (e.g., 1924, 1925) and Heidenhain (1920). Heidenhain contrasted analytic anatomy, which dissects a body into its parts, with a "synthetic anatomy" or *synthesiology* (Heidenhain 1923, 43), which seeks an understanding of the body "in its totality, in its being as an instantiation of form" (*in seiner Wesenheit als Formerscheinung*: Heidenhain 1921, 3). Heidenhain (1907, 84) opened his "introduction to a structural theory of living matter" with reference to Haeckel (1866), who had distinguished an entire hierarchy of morphological individuals (individuals of first, second, third order, etc.: Haeckel 1866, 272). Heidenhain further referred to O. Hertwig's (1906, 378) characterizations of cellular associations and colonial organisms (*Thierstöcke*) in terms of "persons" of first, second, and third order (also in Haeckel 1866), yet hastened to add: "it is possible to obtain in this way a *phylogenetic* theory of the body, not however a *histophysiological* theory." In order to obtain the latter, it is necessary to arrive at a generalization that synthesizes the histological with the physiological (Heidenhain 1907, 85).

A body, in Heidenhain's view, is constructed in terms of a nested (inclusive, i.e., enkaptic: Heidenhain 1907, 92) hierarchy: "The animal body can be decomposed into structural systems of lesser or higher order, which either effectively, or in their origin, are either divisible or have arisen through division of precursor systems of the same kind" (Heidenhain 1907, 100). The divisible systems of lesser or higher order Heidenhain called "histomeres," the complex wholes that these histomeres form at lesser or greater degrees of inclusiveness he called histosystems (Heidenhain 1907, 86). "Biosystems" that are "histosystems" are thus constituted by an association (*Vergesellschaftung*) of histomeres of different order (Heidenhain 1907, 100). Proceeding to flesh out his "synthetic theory," which was to be a "morphological systems theory," Heidenhain (1921, 5) maintained that it had to be based on "multipliable (reproducible) rudiments" (*vermehrbare Anlagekomplexe*). The proliferation of these histomeres proceeds though division, or budding and segmentation, analogous to asexual reproduction in invertebrates (Heidenhain 1923, 42). The result are histosystems of hierarchical structure—that is, an inclusive hierarchy that results from histomere-lineages splitting and splitting again. The divisible entities (histomeres) Heidenhain took to be individuals; the histosystems that result from the divisibility and consequent proliferation of the histomeres he took to be individuals of higher order—that is, complex wholes. The inclusive systems (histosystems) and the included components (histomeres) are relative concepts, however, as one moves up and down the hierarchy. If the divisible histomeres proliferate yet retain corporeal connection, there results "the formation of tissue colonies" (*gewebliche Stockbildungen*) called *histocormi* (sing. *histocormus*: Heidenhain 1921, 6; 1923, 42; *cormus* is again a term borrowed from Haeckel 1866, 285), analogous to colonies of polyps (*Polypenstock*: Heidenhain 1921, 66). This whole developmental dynamic is governed not just by the histomeres in an upward chain of causation, but more importantly so by the complex whole in downward causation. The fundamental property of living entities of all levels of inclusiveness is their potential to multiply through division (Heidenhain 1937, 6), a process governed by a similar "rhythm" (Heidenhain 1932, 2) across all domains of life, plants, animals, and humans. The histosystems thus instantiate a *Gestalt*, which is irreducible to the mere sum of its parts.

Applied to an organism, this means that the cells, indeed its organelles, are its most basic divisible parts or histomeres. From cell division results the formation of tissues, organs, and ultimately the organism itself, an hierarchically structured complex whole whose properties at any level of inclusiveness are not just the aggregate properties of its parts, but instead emergent properties of the whole that govern the synthesis of its parts into a harmonious, functionally integrated system. In this functional sense, an enkaptic system differs from an aggregate system, one characterized by a looser organization among structures that are less interdependent and integrated, less hierarchically structured in fundamental biological properties or functions. Commenting on the structure of a muscle fiber, Heidenhain (1907, 92; 1923, 22) found histosystems of varying degrees of complexity encapsulated within one another, thus forming a hierarchy that he called enkaptic (see also Hueck 1926). By analogy, the organism forms again an enkaptic hierarchy, a "socialization of individuals" (*Vergesellschaftung von Einzelpersonen*: Heidenhain 1923, 4) that represents a "living cosmos." Its development is a continuing "synthesis" that results from the division, growth, and assimilation of its parts into causally efficacious complexes (*Wirkungskreise*) of lesser and greater inclusiveness (Heidenhain 1923, 5). Transcending the old antithesis between morphology and physiology with his synthesiology, Heidenhain (1937, xv) claimed to have contributed to a "colossal change [in comparative anatomy] in favor of holism (*Ganzheitslehre*)."

Judging by the acclaim Heidenhain met from at least some of his peers, his self-congratulations were not unjustified. The anatomist Alfred Benninghoff from Kiel, later Marburg University, characterized Heidenhain's synthesiology as "the most important achievement in morphology of the last few decades" (Benninghoff 1938, 1378; on Benninghoff, see Klee 2003, 38; Aumüller and Grundmann 2002, 300). Incorporating Heidenhain's synthesiology in his *Beiträge zur Anatomie funktioneller Systeme* (1930), Benninghoff emphasized the coming together of parts in a collective functional performance according to a superordinated task that serves the complex whole (Dabelow 1953/54, 159).

In the context of 1930s Germany, enthusiasm for Heidenhain's enkaptic ideas was hardly innocent. Publishing in the *Zeitschrift für die gesamte Naturwissenschaft*, notoriously an outlet for "German Science," Gert von Natzmer argued that there was only one way to capture biological individuals: "Biological individuality is always a multiplicity in unity, and at the same time unity reigning over a multiplicity" (Natzmer 1935/36, 306). The multiplicity of parts contributes to the formation of a complex whole in an upward chain of causation, whereas the complex whole organizes the parts it comprises in a downward chain of causation. It is precisely such reciprocity of causation that characterized Heidenhain's enkaptic hierarchy, and that allowed its accommodation to the ideological matrix of the time—that is, to infuse it with the spirit of the *völkisch* National Community. The key to an understanding of the *völkisch* community propagated by Nazi ideology was the subordination of the lower-level individual to the greater good of the larger whole. Hitler wanted the German *Volk* to form "a living *organism* with the exclusive aim of serving a higher idea" (cited from Harrington 1995, 365), while a representative of the "Leipzig school of holistic psychology" recognized the holism manifest in "organic thinking" as much as in the "structure of *völkisch* life" as a weapon against "atomistic and mechanistic forms of thought" (Harrington 1995, 366). The animal psychologist Friedrich Alverdes, an adept of Nazi ideology (Klee 2003, 14; Harten et al. 2006) as well as of organicist/holistic biology (e.g., Alverdes 1935) from the University of Marburg, praised Heidenhain's synthesiology as an attempt to understand the parts of an organism out of the whole, rather than the other way around. Adopting Driesch's teleological perspective (Alverdes 1932, 91), he took Heidenhain's theory of histomeres to imply that the complex whole that is an organ system provides a preformed matrix—a *Gestalt*—into which the cells are cast as they divide and multiply (Alverdes 1932, 103). This parallels the child born into Hitler's National Community.

Another commentator, Bernhard Dürken, at the time Director of the Institute of Developmental Biology and Heredity at the University of Breslau, contrasted Heidenhain's synthesiology with classical cell theory, which he called a "bluilding block doctrine" (Dürken 1935, 63; also in Heidenhain 1907, 49). In contrast, Heidenhain is said to have shown that an "organism is not built up from cells, but instead secondarily breaks down into cells; it does not form by the aggregation of cells, but instead forms cells as it develops" (Dürken 1935, 63). Dürken (1936) went on to articulate a holistic biology (*Ganzheitsbiologie*) which pitched Hans Driesch against Wilhelm Roux, and which was rooted in the spirit of the "new times"—that is, meant as a foundational contribution to the new *Weltbild*. The concept of wholeness that forms the core of the new *Ganzgheitsbiologie*, he argued, "starts to break out from its purely scientific context and to stimulate the broad field of our intellectual culture as well as the overall attitude of our times . . . we presently experience a huge turnaround, and this turnaround relates to biology; its pivot is called wholeness" (Dürken 1936, 1). In his review of Dürken (1936), the founder of "German Biology" and at that time still head of the section for biology in the National Socialist Teachers' Association (NSLB) Ernst Lehmann (Klee 2003), a highly controversial plant geneticist at the University of Tübingen (Adam 1977; Bäumer 1990a and b; Harwood 1993; Deichmann 1996; Harrington 1996; Potthast and Hoβfeld 2010), noted that "few books have ever achieved such widespread attention [as Dürken's]" (Lehmann 1937, 396), emphasizing that contemporary holism had important scientific, ideological, as well as political implications.

Another botanist from Tübingen, however, the politically uncompromised plant phylogeneticist Walter Zimmermann (Junker 2001), turned against holistic biology in his review of Heidenhain's (1932) botanical excursions. Aligning the theoretical foundations of his research program with the *neue Sachlichkeit* (Galison 1990, 725; Friedmann 2000, 158) propagated by the Vienna Circle instead (Zimmermann 1930, 1937/1938; see also Harwood 1993, 361), he located Heidenhain in the camp of idealistic morphologists, a lifelong target of his biting critique (Rieppel 2011b): "he unquestionably is to be counted amongst the 'idealistic' morphological researchers, who are particularly influenced by Goethe's writings" (Zimmermann 1935, 26). Heidenhain himself was in no way politically incriminated, but his adoption of the Goethean (as opposed to the Newtonian) paradigm in his science predisposed his writings for adoption by German biologists who harbored sympathies for contemporary Nazi ideology and bio-politics.

Enkapsis in Ecology

In the 1930s, leading German ecologists took up ideas of hierarchy and synthesis as well, accompanied by a similar trajectory of assimilation into *völkisch* biology (Bäumer 1990a; Deichmann 1996; Potthast 2003). Haeckel (1866, 286) had introduced the term "*Oecologie*" to name the science that investigates the relations of an organism to its environment, which comprises all of its "conditions of existence": "Ecology investigates the totality of the relations of an animal to its anorganic and organic environment" (Haeckel 1870, 365). These relations were soon recognized as hierarchical, and conceptualized in terms of an enkaptic system by leading ecologists of the first half of the twentieth century.

Jax (1998, 113) identified the famous German limnologist August Thienemann (1882–1960), Director of the Hydrobiological Institute of the *Kaiser-Wilhelm-Gesellschaft* (renamed Max Planck Society after the war) in Plön and professor of hydrobiology at the University of Kiel, as amongst the first to have recognized a biotope and its biocoenosis as a tight organic unit of higher complexity (Thienemann and Kieffer 1916). In so doing, Thienemann (1918) looked back on Karl August Möbius (1886, 247n1), who had first introduced the term biocoenosis or *Lebensgemeinschaft* (Möbius 1877), as also on Friedrich Junge's influential *Der Dorfteich als Lebensgemeinschaft* (1885; see Jax 1998; Nyhart 2009). Just like Heidenhain in anatomy, Thienemann sought to replace the analytic by a synthetic approach in ecology. "I believe," Thienemann wrote, "that in contemporary zoology, genetics and physiology (including *Entwicklungsmechanik*) have passed their peak . . . Considering the contemporary intellectual life as a whole, it seems to me that a period of analysis is now again being succeeded by a period of synthesis, and that—if all the augury is not deceptive—the predominantly synthetic branches of biology will again become the focus of scientific interest."[2] Thienemann's call for synthesis was, again, embedded in "a Goethe-inspired natural philosophy" (Schwarz and Jax 2011, 239). Jax (1998; see also Potthast 1999) located Thienemann in the organicist camp within ecology (on the organicist movement in Anglo-American ecology see Potthast 1999; McIntosh 2011). Organicism attempts to transcend the old, mutually exclusive contrast between mechanism and vitalism (Ehrenberg 1929). Organicism emphasizes laws of order as expressed in systemic properties of life such as those identified by the theoretical biologist Ludwig von Bertalanffy (1932), self-proclaimed "father of organicism" (Bertalanffy 1941, 337): homeostasis and enkaptic hierarchical structure (Bertalanffy 1932). These Alverdes (1936a, 124) found to characterize not only individual organisms, but supraindividual systems as well such as ecological communities (*Lebensgemeinschaften*)—a conclusion that he found to be fully in line with Thienemann's analysis (Alverdes 1936b, 7).

In his *Lebensgemeinschaft und Lebensraum*, Thienemann (1918, 282; see also Thienemann 1935) formulated a hierarchical approach to ecology, recognizing a hierarchically structured system of biocoenoses of first, second, third, etc., order. This system of biocoenoses maintains a homeostatic equilibrium through self-regulation. Each biocoenosis at each level of inclusiveness can thus be comprehended as an organism of higher complexity (Thienemann 1918, 300)—that is, an organism of second order whose causal integration is rooted in Georges Cuvier's law of the functional correlation of parts. Subject to reciprocal causal relations, the biotope (*Lebensraum*) plus its biocoenosis (*Lebensgemeinschaft*) form a complex whole—that is, an organism of third order. Thienemann (1925, 598) articulated similar ideas in his *Der See als Lebensgemeinschaft*: "Morphology [i.e., Haeckel 1866] speaks of different levels of individuality," and the same could be done in ecology (Thienemann 1925, 594): "it is possible that Driesch's notion of 'the whole' may play a role here" (Thienemann 1925, 595). Thienemann (1935, 338; see also Thienemann 1941, 112) followed his friend and colleague Karl Friederichs' characterization of ecological individuals of greater or lesser inclusiveness as *Gestalten* (Friederichs 1927), when he invoked the concept of enkapsis to characterize the hierarchical structure of ecosystems. The anatomist Benninghoff (1935/36, 158) explicitly applauded the extrapolation of the concept of enkapsis to supraorganismal levels of complexity in ecology: "With this we have taken an essential step beyond the boundaries of the organism . . . The attempt has also been made to capture the whole living world in such an enkaptic system (FRIEDERICHS), an attempt which from the perspective of biological organicism is nothing but logical."

As sociopolitical pressure on academics increased (Potthast 2003), Friederichs (1937) succeeded in casting his science in terms of Nazi ideology (May 1937/38, 486), calling ecology the "science of *blood and soil*" (Friederichs 1937, 91; see also Deichmann 1996, 137; Stella and Kleisner 2010, 43)—a move in which he was again to be followed by Thienemann. Thienemann (1935) applied Heidenhain's concept of *Einschachtelung* [nested hierarchy] to his ecological cosmos in a talk delivered at the 37th annual meeting of the German Association for the Promotion of Teaching Mathematics and Natural Sciences (*Deutscher Verein zur Förderung des mathematischen und naturwissenschaftlichen Unterrichts*), held in Kiel from April 14 through April 18, 1935. In his talk, Thienemann acknowledged the fact that the slogan "blood and soil" had awakened political interest in ecology, and that the organicist interpretation of a *Lebensgemeinschaft* or biocoenosis as a hierarchically structured, supraindividual complex whole had invited analogies to *völkisch* society, "but I am here talking strictly as a biologist, not as a politician" (*Nationalpolitiker*: Thienemann 1935, 337). Nevertheless, he welcomed the new significance attributed to biology in the Third Reich, as it resulted in increased levels of funding for research and teaching (Thienemann 1935, 337; see also Deichmann and Müller-Hill 1994). Thienemann was, indeed, ranked fifth amongst the top funded zoologists during the National Socialist regime (Deichmann 1996, 116).

In his highly influential sketch of the foundations of ecology, Thienemann (1939; reprinted separately by Schweizerbart in Stuttgart 1939, and in Thienemann 1941) cited not only Friederich Ratzel (1901) of *Lebensraum* fame, but also the usual suspects in the camp of German holist/organicist biologists and philosophers: Friederichs of course most frequently, but also Alverdes on wholeness (*Ganzheit*), A. Meyer(-Abich) on holism and Goethe (Meyer-Abich is recognized as having "played an active role in defining the significance of holism for National Socialism": Harrington 1996, 190; see also Amidon 2009), and the philosopher Wilhelm Burkamp, whose notion of wholeness (*Ganzheit*: Burkamp 1929) Thienemann found more encompassing than that of Driesch (see also Schwarz and Jax 2011, 236). Under the section head "*Menschliches—Allzumenschliches (Angewandte Ökologie)*" [Human— Too Human (Applied Ecology)],[3] Thienemann (1939) referred to Escherich's (1934) notorious "*Termitenwahn*," ascertaining the nature of humans as part of the complex whole that is all nature, subject to the reign of universal laws of life (*Lebensgesetze*). Karl Leopold Escherich was appointed president of the University of Munich in 1933 to bring it into line with Nazi ideology. In his much acclaimed inaugural speech titled "Termite Mania—on the Education of the Political Person," he condemned both the individualism of Weimar

Republic *laissez-faire*, as well as the uniformitarianism forced by Bolshevism. Instead, he called for an education of people so they would actively embrace Nazi ideology and politics.

In 1941, Thienemann published his *Leben und Umwelt* as volume 12 of the publication series *Bios*, which he co-edited with Alverdes, Bertalanffy, Karl Beurlen (of whom more below), Driesch, and Jakob von Uexküll amongst others. The volume, inspired by Adolf Meyer(-Abich) and dedicated "in friendship" to Karl Friederichs, contains reprints of earlier publications (Thienemann 1935, 1939). In the preface, Thienemann (1941, vii) reasserted his earlier "call for synthesis," and encouraged the younger generation of academics to eschew professional specialization in favor of a holistic (*einheitlich*) understanding of the world. A reprint of his *Unser Bild der lebenden Natur* (Thienemann 1940) shows that by that time, Thienemann had dropped his earlier (Thienemann 1935) reservations with respect to analogies between enkaptic biocoenoses and the German *Volksgemeinschaft*: "The dependence of organisms on their *Lebensraum* is a truism which our *Volk* has become particularly aware of in our current times, and which is best captured by the slogan 'blood and soil'" (Thienemann 1941, 12). "The general science of the economy of nature . . . is the most important part of all natural sciences" (Thienemann 1941, 22), as it researches the eternal laws of life (*Lebensgesetze*) that govern biocoenoses as much as the German National Community. Structured in terms of an enkaptic hierarchy, complex wholes of higher order are characterized by emergent properties, the instantiation of which requires the included individuals to subject themselves to the greater good of the whole: "as part of his *Volk*, the individual will sacrifice his life if this is required for the survival of the whole . . . self-sacrifice is senseless for the hermit, but a—moral—obligation for any individual embedded in a community!" (Thienemann 1941, 12). In his review of Thienemann (1941), the zoologist turned anthropologist Gerhard Heberer (Bäumer 1990a and b; Deichmann 1996; Hoβfeld 1997; Klee 2003) praised the author as a successful, indeed leading researcher of *Biocoenotik*, a field of research that "plays a central role in biology; its significance for human racial theory is obvious" (Heberer 1942, 168).

Enkapsis, Evolution, and the Phylogenetic Hierarchy

It was the paleontologist Karl Beurlen (1901–1985; Rieppel 2012) who first explicitly portrayed the tree of life as an enkaptic hierarchy, in a theoretical edifice that fellow paleontologist Walter R. Gross (1943, 240) found to be "in every respect anchored in natural philosophy (*Naturphilosophie*), holism

(*Ganzheitslehre*), and a special theory of knowledge (*Erkenntnistheorie*)." Beurlen's (1930, 1937) phylogenetics is committed to the grotesque vision of a *völkisch* science, tainted as it was in his case by Nazi policy and ideology. He replaced Darwinian causality in phylogeny with Oswald Spengler's notion of fate (*Schicksal*), coupled with Nietzsche's notion of *Will to Power*—identified as a core concept of Nazi ideology by the philosopher Martin Heidegger (Rohkrämer 2005, 181).

The fundamental currency of Beurlen's (1930, 1937) system is the type, a phylogenetic *Gestalt*, instantiated by a well-defined body plan that characterizes a phylogenetic lineage. Analogous to an organism (and to Spengler's civilizations), such an evolutionary lineage undergoes cyclical change from youthfulness to adolescence to adulthood, followed by senescence and death (extinction). The origin of a new type from an ancestral one is discontinuous (saltational), and requires ontogenetic repatterning during early developmental stages. The fundamentals of such orthogenesis (directional and cyclical evolutionary development) and saltationism had earlier been articulated by Jaekel (1902, 1922) and others, including North American paleontologists such as Edward D. Cope and Henry F. Osborn (Levit and Olssen 2006). Beurlen's (1937) main theoretical contribution mostly consisted in glossing these fundamental ideas that were popular in early twentieth-century German paleontology with considerable philosophical (metaphysical) overburden in the spirit of a grandiose synthesis of natural sciences and humanities, infused with social and political connotations of the time.

Beurlen (1937) portrayed the evolutionary lineage that is a phylogenetic *Gestalt* instantiated by its type in its various orthogenetic manifestations as a complex whole, an individual of higher order. The phylogenetic system in its entirety thus becomes a nested hierarchy of complex wholes of increasing inclusiveness. Acknowledging the influence that the holistic schools of Hans Driesch and Jakob von Uexküll had exerted on him (Beurlen 1937, 257), Beurlen (1935/36, 1937) nonetheless chastised Driesch for not having pushed his concept of wholeness far enough, thus failing to recognize the existence of supraorganismal individuals. In characterizing this phylogenetic hierarchy as an enkaptic system, Beurlen drew on both Heidenhain's synthesiology (not cited but implied by reference to the enkapsis of histosystems: Beurlen 1937, 102), and on Thienemann's organicist approach to ecology (Beurlen 1937, 105).

Accepting Uexküll's (1926 [1920]) relativistic concept of the *Umwelt* (environment), Beurlen (1937) found the organism, as also the complex whole that is a phyletic *Gestalt*, locked in a struggle for self-differentiation and self-manifestation with respect to its abiotic and biotic environment. Rejecting the externalism and reductionism perceived to be inherent in Darwinian selection theory, Beurlen stipulated Nietzsche's *Will to Be* as an endogenous driving force of phylogenetic transformation manifest in evolving entities at all levels of complexity, from the organism to the phylum. In the struggle to persevere, the *Will to Be* manifests itself as a *Will to Power*—"as Nietzsche had very clearly recognized already" (Beurlen 1937, 223; see also Nietzsche 1968 [1901]). But the potential of individual (i.e., ontogenetic) as well as collective (i.e., phylogenetic) *Gestaltung* is not limitless nor unconstrained, "just as the egg-cell is also not omnipotent" (Beurlen 1937, 233). "Freedom does not mean randomness or chance [the Darwinian forces of evolution], but follows from the Will to Be" (Beurlen 1937, 235). The *Will to Be*, driven by the *Will to Power*, does not open the door to unconstrained individualism, but is constrained by the eternal forces of nature and history. The phylogenetic system forms an enkaptic hierarchy of complex wholes (Beurlen 1937, 131), each phylogenetic lineage (*Gestalt*) forming at its proper hierarchical level an encapsulated complex whole of lesser complexity, each level of inclusiveness characterized by emergent properties that are not just the aggregate properties of its parts. The superordinated phyletic *Gestalt* instantiates the law according to which the cycles of species (genera, families, etc.) originate, which in turn collectively constitute that phyletic *Gestalt* (Beurlen 1937, 130). Every individual (complex whole) in Beurlen's phylogenetic system is—at any level of complexity—always also part of a superordinated (encapsulating) whole: the individual parts are dependent on, and can be fully comprehended only from the perspective of, the superordinated whole: "This is exactly the conclusion which Driesch failed to draw" (Beurlen 1937, 105).

Beurlen's (1937) enkaptic hierarchy of phyletic *Gestalten*, as much as Thienemann's (1941) enkaptic hierarchy of ecological *Gestalten*, expresses the requirement of the subordinated (included) parts to be dominated by, and to submit to, the greater good of the whole, a requirement that also characterized Hitler's *Volksgemeinschaft* (Pine 2007). The very concept of wholeness was (and continues to be) built on the idea that the whole is more than the sum of its parts. Coupled with the doctrine of enkapsis in its (implicitly or explicitly) teleological interpretation (ultimately motivated by Hans Driesch's *entelechy* as in the case of Alverdes 1932; Dürken 1935; Beurlen 1937, Thienemann 1941), the whole comes to dominate its parts, the parts become subservient to the whole.

Discussion

Heidenhain originally developed the concept of enkapsis in opposition to a reductionist-atomistic interpretation of cell theory. It was soon expanded to ecology and evolutionary theory. Stipulating a nested hierarchy of biological individuals or complex wholes subject to upward and downward causation, the concept of enkapsis allowed easy accommodation to ideological pressures of the time in the context of ecology (Thienemann), and was even articulated in support of Nazi bio-politics (Weingart et al. 1992, 370; Caplan 2008, 19) in the context of evolutionary theory (Beurlen). Whatever the ultimate underlying motivations of such writing were, "every author was familiar with those political connotations, and knew what he did" (Kleβmann 1985, 366). But beyond individual motivations, such application of enkapsis shows the power this concept was believed by leading—if misled—intellectuals of the time to have in the individuation of the *Volksgemeinschaft* as a biological individual, a body or an organism. And yet the concept of the enkaptic hierarchy was able to leave such ideological connotations behind, and continued to do useful theoretical work in biology, powerfully so in Willi Hennig's phylogenetic systematics, which triggered the cladistic revolution in comparative biology (Hennig 1950, 1966; Hull 1988; Rieppel 2009a).

Are there commonalities that characterize the employment of an enkaptic hierarchy in different contexts? Benninghoff (1938, 1378) found the enkaptic hierarchy to be the quintessential expression of the part-whole relation. Zimmermann (1935, 26) identified two axioms underlying Heidenhain's synthesiology: divisibility and (nested) hierarchy. In his inaugural speech at the University of Basel, the anatomist Gerhard Wolf-Heidegger (1938, 1185) found in Goethe and Heidenhain the inspiration to proclaim the capacity for division and fusion the most fundamental properties of living substance. The holistic zoologist and idealistic morphologist Adolf Portmann, again from the University of Basel, identified the potential of living matter for self-differentiation as the most important implication of Heidenhain's synthesiology (Portmann 1960, 36). Division and association or fusion as mechanisms of proliferation and self-differentiation of hierarchically nested biological systems that are subject to the part-whole relation—these are the core concepts his readers identified in Heidenhain's synthesiology. Heidenhain's was a strictly dichotomous, divisional hierarchy, however. It is therefore interesting to see Wolf-Heidegger invoke fusion as a fundamental property of living substance, since fusion turns hierarchies into networks.

Issues concerning hierarchical structure, integration, and individuality at molecular, organismic, ecological, and social levels continue to permeate the discussion of biological systems today (e.g., Schlosser and Wagner 2004; Kolasa 2005; Klingenberg 2008; Findlay and Thagard 2012; Love 2012). On my reading, it is the mechanisms of division, differentiation, and upward as well as downward causal integration that underlie the enkaptic hierarchy, and that render the concept useful in contemporary discourse. Part and whole become relative concepts as one moves up and down the hierarchy of life. What is a part at one level of inclusiveness becomes a whole at a more inclusive level and vice versa.

Furthermore, the parts and wholes that constitute the enkaptic hierarchy are dynamic entities—that is, entities that have a discrete origin in time and space, and hence have a history. The botanist Carl W. Nägeli (1856, 1865) was perhaps the first to explicitly tie the individuality of dynamic biological entities to their origin and history. Like others before him (Rinard 1981), Nägeli recognized such biological individuals as complex entities that are composed of a multiplicity of parts (*Teilindividuen*: Nägeli 1856, 211). With respect to such composite wholes, Thienemann championed Burkamp's concept of wholeness, which presupposes unity in the togetherness of a multitude of parts (see also Sapper 1938, 111). To recognize—with Heidenhain (1907)—composite wholes as individuals that are not only themselves divisible, but also constructed from parts that are divisible individuals of lower order, was identified as a truly paradoxical comprehension of individuality by Natzmer (1935/36, 305), as it completely defies abstract logical analysis in terms of Leibniz's Law of the "Indiscernability of Identicals," for example.

Philosophy classically individuates individuals through essential intrinsic properties. The history of the concept of an enkaptic hierarchy, as well as the historical nature of biological individuals that it comprises, illustrates the insufficiency of the classical philosophical approach in the discussion of biological individuality. *Ex Ovo Omnia*—after the origin of life, living beings originate from other living beings. A biological individual has an origin, and with its origin begins its history, during which it differentiates into a hierarchically structured complex whole that changes through time as a consequence of a changing composition of parts, all the while maintaining unity, hence wholeness, through homeostasis and self-regulation (Bertalanffy 1932). Such a conception of biological individuality certainly violates Leibniz's law of the indiscernability of identicals. This latter, strong, indeed logical, conception of individuality is atomistic: the individual, a "particular," is a substantial, indivisible entity located in time and space, instantiating essential, intrinsic properties. *Logical* individuals sort into classes, subject to the membership relation.

By contrast, *biological* individuals make two out of one through fission —as in cell division, or speciation; fusion makes a whole from parts—as in genomes subject to lateral gene transfer, hybridization, or bisexual conception. Fission creates enkaptic hierarchies, fusion creates inclusive networks, both subject to the part-whole relation. Biological individuals do not sort into classes on the basis of their intrinsic essential properties, but instead sort into either a network, or a nested hierarchy of complex wholes according to their relations of descent. The dynamic biological individuals that are parts or wholes within an inclusive network or an enkaptic hierarchy are not to be individuated through intrinsic essential properties, but through the relational properties instead that maintain the causal integration of the dynamic system that is a biological individual. Such a dynamic—that is, processual and hence relational conception of biological individuality that accomodates both causally integrated hierachies as well as networks—perfectly captures modern insights in developmental genetics, comparative developmental anatomy, speciation, phylogeny, and synecology (Rieppel, 2009b; Wagner 2014; Havstad et al. 2015).

Philosophers have recognized the problems encountered when the traditional approach to individuation is applied to biological individuals such as a particular species. As a consequence they revamped traditional essentialism as an "origin essentialism," where it is the evolutionary origin—that is, a *relational* property, that becomes the *historical* essence of a species as individual (e.g., Griffiths 1999; LaPorte 2004). However, the dynamics of the complex systems that are biological individuals render "origin essentialism" too strong and hence inapplicable to those (Rieppel 2008; Pedroso 2014), especially if hierarchies are allowed to turn into networks through the fusion of subordinated parts (e.g., horizontal gene transfer in species trees). Even in the presumably most unambiguous case, its origin from a specific egg and sperm cell (Sober 1980) proves insufficient to individuate a human being (Gilbert et al. 2012). Homeostasis and self-regulation—that is, additional relational properties, are required to stabilize and maintain the complex dynamic system that is a biological individual.

Acknowledgments

I thank Lynn Nyhart and Scott Lidgard for the invitation to participate in this project, as well as for their support and input, which greatly improved the content of this chapter. The chapter derives, in part, from O. Rieppel, *Phylogenetic Systematics: Haeckel to Hennig* (2016).

Notes to Chapter Seven

1. University Archives Tübingen, UAT 117c / 16.

2. In a letter to the Dean of the Philosophical Faculty of the University of Basel, dated February 3, 1931, concerning the candidates for the succession to the professor of zoology and comparative anatomy, Friedrich Zschokke (1860–1936); Staatsarchiv Basel, StABS, ED-REG 1a 2 1402.

3. The section title borrows from the famous slogan "National Socialism is politically applied biology" by Hans Schemm, founder and head of the National Socialist Teachers' Association (Lehmann 1935).

References for Chapter Seven

Adam, U. D. 1977. *Hochschule und Nationalsozialismus: Die Universität Tübingen im Dritten Reich.* Tübingen: J. C. B. Mohr (Paul Siebeck).

Alfert, M. 1972. "Martin Heidenhain." In *Dictionary of Scientific Biography, Volume VI,* edited by C. C. Gillispie, 223–24. New York: Charles Scribner's Sons.

Alverdes, F. 1932. "Die Ganzheitsbetrachtung in der Biologie." *Sitzungsberichte der Gesellschaft zur Beförderung der gesamten Naturwissenschaften zu Marburg* 67: 89–118.

———. 1935. "Kausalität, Finalität und Gestalt." *Acta Biotheoretica* 3: 167–80.

———. 1936a. "Organizismus und Holismus: Neuere theoretische Strömungen in der Biologie." *Der Biologe* 5: 121–28.

———. 1936b. "Der Begriff des 'Ganzen' in der Biologie." *Zeitschrift für Rassenkunde und ihre Nachbargebiete* 4: 1–9.

Amidon, K. S. 2009. "Adolf Meyer-Abich, Holism, and the Negotiation of Theoretical Biology." *Biological Theory* 3: 357–70.

Aumüller, G., and K. Grundmann. 2002. "Anatomy during the Third Reich: The Institute of Anatomy at the University of Marburg, as an Example." *Annals of Anatomy* 184: 295–303.

Barnes, B., and J. Dupré. 2008. *Genomes and What to Make of Them.* Chicago: University of Chicago Press.

Bäumer, Ä. 1990a. *NS-Biologie.* Stuttgart: S. Hirzel.

———. 1990b. "Die Zeitschrift 'Der Biologe' als Organ der NS-Biologie." *Biologie in unserer Zeit* 20: 42–7.

Benninghoff, A. 1930. *Beiträge zur Anatomie funktioneller Systeme.* Frankfurt a.M.: Akademische Verlags-Gesellschaft.

———. 1935/36. "Form und Funktion. I. Teil." *Zeitschrift für die gesamte Naturwissenschaft* 1: 149–60.

———. 1938. "Über Einheiten und Systembildungen im Organismus." *Deutsche Medizinische Wochenschrift* 64: 1377–82.

Bertalanffy, L. v. 1932. *Theoretische Biologie,* vol. 1. Berlin: Bornträger.

———. 1941. "Die organismische Auffassung und ihre Auswirkungen." *Der Biologe* 10: 247–58, 337–45.

Beurlen, K. 1930. "Vergleichende Stammesgeschichte: Grundlagen, Methoden, Probleme unter besonderer Berücksichtigung der höheren Krebse." *Fortschritte der Geologie und Paläntologie* 8: 317–586.

———. 1935/36. "Das Gestaltproblem in der Natur." *Zeitschrift für die gesamte Naturwissenschaft* 1: 445–57.

———. 1937. *Die stammmesgeschichtlichen Grundlagen der Abstammungslehre.* Jena: G. Fischer.

Burkamp, W. 1929. *Die Struktur der Ganzheiten.* Berlin: Junker und Dünnhaupt.

Caplan, J. 2008. "Introduction." In *Nazi Germany,* edited by J. Caplan, 1–25. Oxford: Oxford University Press.

Cassirer, E. 1950. *The Problem of Knowledge: Philosophy, Science and History since Hegel.* New Haven: Yale University Press.

Dabelow, A. 1953/54. "Alfred Benninghoff ." *Anatomischer Anzeiger* 100: 157–65.

Deichmann, U. 1996. *Biologists under Hitler*. Translated by Thomas Dunlap. Cambridge, MA: Harvard University Press.

Deichmann, U., and B. Müller-Hill. 1994. "Biological Research at Universities and Kaiser Wilhelm Institutes in Nazi Germany." In *Science, Technology, and National Socialism*, edited by M. Rennberg, and M. Walker, 160–83. Cambridge: Cambridge University Press.

Driesch, H. 1921. *Philosophie des Organischen. Gifford-Vorlesungen gehalten an der Universität Aberdeen in den Jahren 1907–1908*. 2nd ed. Leipzig: Engelmann.

Dürken, B. 1935. "Das Verhältnis der Teile zum Ganzen im Organismus." *Zeitschrift für mathematischen und naturwissenschaftlichen Unterricht aller Schulgattungen* 66: 57–65.

———. 1936. *Entwicklungsbiologie und Ganzheit: Ein Beitrag zur Neugestaltung des Weltbildes*. Leipzig: B. G. Teubner.

Ehrenberg, R. 1929. "Über das Problem einer 'theoretischen Biologie.'" *Die Naturwissenschaften* 17: 777–81.

Escherich, K. L. 1934. *Termitenwahn: Eine Münchner Rektoratsrede über die Erziehung zum politischen Menschen*. Munich: G. Müller.

Findlay, S. D., and P. Thagard. 2012. "How Parts Make Up Wholes." *Frontiers in Physiology* 3: 1–10.

Friederichs, K. 1927. "Grundsätzliches über die Lebenseinheiten höherer Ordnung und den ökologischen Einheitsfaktor." *Die Naturwissenschaften* 15: 153–57, 182–86.

———. 1937. *Ökologie als Wissenschaft von der Natur oder biologische Raumforschung (Bios, VII)*. Leipzig: J. A. Barth.

Friedmann, M. 2000. *A Parting of the Ways: Carnap, Cassirer, and Heidegger*. Chicago: Open Court.

Galison, P. 1990. "Aufbau/Bauhaus: Logical Positivism and Architectural Modernism." *Critical Inquiry* 16: 709–52.

Gegenbaur, C. 1859. *Grundzüge der vergleichenden Anatomie*. Leipzig: Wilhelm Engelmann.

Gilbert, S. F., J. Sapp, and A. I. Tauber. 2012. "A Symbiotic View of Life: We Have Never Been Individuals." *Quarterly Review of Biology* 87: 325–41.

Griffiths, P. E. 1999. "Squaring the Circle: Natural Kinds with Historical Essences." In *Species: New Interdisciplinary Essays*, edited by R. A. Wilson, 209–28. Cambridge, MA: MIT Press.

Gross, W. R. 1943. "Paläontologische Hypothesen zur Faktorenfrage der Deszendenzlehre: Über die Typen- und Phasenlehren von Schindewolf und Beurlen." *Die Naturwissenschaften* 31: 237–45.

Grün, B. 2010. "Die Medizinische Fakultät Tübingen im Nationalsozialismus." In *Die Universität Tübingen im Nationalsozialismus*, edited by U. Wiesing, K.-R. Brintzinger, B. Grün, H. Junginger, and S. Michl, 239–77. Stuttgart: F. Steiner.

Haeckel, E. 1866. *Generelle Morphologie der Organismen, vol. 2*. Berlin: Georg Reimer.

———. 1870. "Ueber Entwicklungsgang und Aufgabe der Zoologie. Rede gehalten beim Eintritt in die philosophische Facultät zu Jena am 12. Januar 1869." *Jenaische Zeitschrift für Medicin und Naturwissenschaft* 5: 353–70.

Harrington, A. 1995. "Metaphoric Connections: Holistic Science in the Shadow of the Third Reich." *Social Research* 62: 357–85.

———. 1996. *Reenchanted Science: Holism in German Culture from Wilhelm II to Hitler*. Princeton: Princeton University Press.

Harten, H.-C., U. Neirich, and M. Schwerendt. 2006. *Rassenhygiene als Erziehungsideologie des Dritten Reichs*. Berlin: Akademie Verlag.

Harwood, J. 1993. *Styles of Scientific Thought: The German Genetics Community, 1900–1933*. Chicago: University of Chicago Press.

Havstad, J. C., L. C. S. Assis, and O. Rieppel. 2015. "The Semaphorontic View of Homology." *Journal of Experimental Zoology (Molecular and Developmental Evolution)* 9999B: 1–15.

Heberer, G. 1942. "Thienemann, A. Leben und Umwelt." *Der Biologe* 11: 168.

Heidenhain, M. 1907. *Plasma und Zelle. Erste Abteilung: Allgemeine Anatomie der lebendigen Masse*. Jena: G. Fischer.

———. 1920. "Neue Grundlegungen zur Morphologie der Speicheldrüsen." *Anatomischer Anzeiger* 52: 305–31.

———. 1921. "Über die teilungsfähigen Drüseneinheiten oder Adenomeren. Sowie über die Grundbegriffe der morphologischen Systemlehre." *Wilhelm Roux' Archiv für Entwicklungsmechanik* 49: 1–178.

———. 1923. *Formen und Kräfte in der lebendigen Natur*. Berlin: J. Springer.

———. 1932. *Die Spaltungsgesetze der Blätter. Eine Untersuchung über Teilung und Synthese der Anlagen, Organisation und Formbildung sowie über die Theorie korrelativer Systeme. Beitrag XVI zur synthetischen Morphologie*. Jena: G. Fischer.

———. 1937. *Synthetische Morphologie der Niere des Menschen. Bau und Entwicklung dargestellt auf neuer Grundlage*. Leiden: E. J. Brill.

Hennig, W. 1950. *Grundzüge einer Theorie der Phylogenetischen Systematik*. Berlin: Deutscher Zentralverlag.

———. 1966. *Phylogenetic Systematics*. Urbana: University of Illinois Press.

Hertwig, O. 1906. *Allgemeine Biologie. Zweite Auflage des Lehrbuchs "Die Zelle und die Gewebe."* Jena: G. Fischer.

Hoßfeld, U. 1997. *Gerhard Heberer (1901–1973): Sein Beitrag zur Biologie des 20. Jahrhunderts*. Berlin: VWB Verlag für Wissenschaft und Bildung.

Hueck, W. 1926. "Die Synthesiologie von Martin Heidenhain als Versuch einer allgemeinen Theorie der Organisation." *Die Naturwissenschaften* 14: 149–58.

Hull, D. L. 1988. *Science as a Process: An Evolutionary Account of the Social and Conceptual Development of Science*. Chicago: University of Chicago Press.

Jacobi, W. 1952/53. "Martin Heidenhain." *Anatomischer Anzeiger* 99: 80–94.

Jacobshagen, E. 1924. "Begriff und Formen der morphologischen Homologie." *Anatomischer Anzeiger* 58, Ergänzungsheft: 257–62.

———. 1925. *Allgemeine vergleichende Formenlehre der Tiere*. Leipzig: W. Klinkhardt.

Jaekel, O. 1902. *Über verschiedene Wege phylogenetischer Entwicklung*. Jena: G. Fischer.

———. 1922. "Funktion und Form in der organischen Entwicklung." *Palaeontologische Zeitschrift* 4: 147–66.

Jax, K. 1998. "Holocoen and Ecosystem: On the Origin and Historical Consequences of Two Concepts." *Journal of the History of Biology* 31: 113–42.

Junge, F. 1885. *Der Dorfteich als Lebensgemeinschaft*. Kiel: Lipsius & Tischer.

Junker, Th. 2001. "Walter Zimmermann (1892–1980)." In *Darwin & Co. Eine Geschichte der Biologie in Porträts*, edited by I. Jahn and M. Schmitt, 275–95. München: C. H. Beck.

Klee, E. 2003. *Das Personenlexikon zum Dritten Reich. Wer war was vor und nach 1945*, 2nd ed. Frankfurt a.M.: Fischer.

Kleßmann, Ch. 1985. "Osteuropaforschung und Lebensraumpolitik im Dritten Reich." In *Wissenschaft im Dritten Reich*, edited by P. Lundgreen, 350–83. Frankfurt a.M.: Suhrkamp.

Klingenberg, C. P. 2008. "Morphological Integration and Developmental Modularity." *Annual Review of Ecology, Evolution, and Systematics* 39: 115–32.

Kolasa, J. 2005. "Complexity, System Integration, and Susceptibility to Change: Biodiversity Connection." *Ecological Complexity* 2: 431–42.

LaPorte, J. 2004. *Natural Kinds and Conceptual Change.* Cambridge: Cambridge University Press.

Lehmann E. 1935. "Nachruf auf Hans Schemm." *Der Biologe* 4: 98.

———. 1937. "Dürken, Bernhard: Entwicklungsbiologie und Ganzheit. Ein Beitrag zur Neugestaltung des Weltbildes." *Der Biologe* 6: 396–400.

Levit, G. S., and L. Olssen. 2006. "'Evolution on Rails': Mechanisms and Levels of Orthogenesis." *Annals for the History and Philosophy of Biology* 11: 97–136.

Love, A. C. 2012. "Hierarchy, Causation and Explanation: Ubiquity, Locality and Pluralism." *Interface Focus* 2: 115–25.

Lubosch, W. 1926. "Kritische Bemerkungen über den Begriff der Biologischen Morphologie." *Gegenbaurs Morphologisches Jahrbuch* 55: 655–66.

———. 1931. "Geschichte der vergleichenden Anatomie." In *Handbuch der vergleichenden Anatomie der Wirbeltiere, Erster Band*, edited by L. Bolk, E. Göppert, E. Kallius, and W. Lubosch, 3–76. Berlin: Urban und Schwarzenberg.

May, E. 1937/38. "K. Friederichs. Ökologie als Wissenschaft von der Natur oder biologische Raumforschung 1937." *Zeitschrift für die gesamte Naturwissenschaft* 3: 486–87.

McIntosh, R. 2011. "The History of Early British and US-American Ecology to 1950." In *Ecology Revisited: Reflecting on Concepts, Advancing Science*, edited by A. Schwarz and K. Jax, 277–85. Heidelberg: Springer.

Möbius, K. 1877. *Die Auster und die Austernwirtschaft.* Berlin: Wiegand, Hempel & Parey.

———. 1886. "Die Bildung, Geltung und Bezeichnung der Artbegriffe und ihr Verhältniss zur Abstammungslehre." *Zoologische Jahrbücher. Zeitschrift für die Systematik, Geographie und Biologie der Thiere* 1: 241–74.

Mocek, R. 2001. "Wilhelm Roux (1850–1924)." In *Darwin & Co. Eine Geschichte der Biologie in Porträts*, edited by J. Jahn and M. Schmitt, 456–76. Munich: C. H. Beck.

Mörike, K. 1988. *Geschichte der Tübinger Anatomie.* Tübingen: J. C. B. Mohr (Paul Siebeck).

Nägeli, C. W. 1856. "Die Individualität in der Natur. Mit besonderer Berücksichtigung des Pflanzenreiches." *Monatsschrift des wissenschaftlichen Vereins in Zürich* 1: 171–212.

———. 1865. *Entstehung und Begriff der naturhistorischen Art.* Munich: Verlag der königlichen Akademie der Wissenschaften.

Natzmer, G. v. 1935/36. "Individualität und Individualitätsstufen im Organismenreich." *Zeitschrift für die Gesamte Naturwissenschaft* 1: 305–16.

Nietzsche, F. 1968 [1901]. *The Will to Power.* Translated by Walter Kaufmann and R. J. Hollingdale. Edited by Walter Kaufmann. New York: Vintage.

Nyhart, L. K. 2009. *Modern Nature: The Rise of the Biological Perspective in Germany.* Chicago: University of Chicago Press.

Pedroso, M. 2014. "Origin Essentialism in Biology." *Philosophical Quarterly* 64: 60–81.

Pine, L. 2007. *Hitler's "National Community": Society and Culture in Nazi Germany.* London: Hodder Arnold.

Portmann, A. 1960. *Neue Wege der Biologie.* Munich: Piper.

Potthast, Th. 1999. "Theorien, Organismen, Synthesen: Evolutionsbiologie und Ökologie im angloamerikanischen und deutschsprachigen Raum von 1920–1960." In *Die Entstehung der Synthetischen Theorie. Beiträge zur Geschichte der Evolutionsbiologie in Deutschland. 1930–*

1950, edited by T. Junker, and E.-M. Engels, 259–92. Berlin: VWB Verlag für Wissenschaft und Bildung.

———. 2003. "Wissenschaftliche Ökologie und Naturschutz: Szenen einer Annäherung." In *Naturschutz und Nationalsozialismus*, edited by J. Radkau, and F. Uekötter, 225–54. Frankfurt a.M.: Campus.

Potthast, Th., and U. Hoβfeld. 2010. "Vererbungs- und Entwicklungslehren in Zoologie, Botanik und Rassenkunde / Rassenbiologie an der Universität Tübingen im Nationalsozialismus." In *Die Universität Tübingen im Nationalsozialismus*, edited by U. Wiesing, K.-R. Brintzinger, B. Grün, H. Junginger, and S. Michl, 435–82. Stuttgart: F. Steiner.

Ratzel, F. 1901. *Der Lebensraum. Eine biogeographische Studie.* Tübingen: Laupp.

Richards, R. J. 2002. *The Romantic Conception of Life: Science and Philosophy in the Age of Goethe.* Chicago: University of Chicago Press.

Rieppel, O. 2008. "Origins, Taxa, Names and Meanings." *Cladistics* 24: 598–610.

———. 2009a. "Hennig's Enkaptic System." *Cladistics* 25: 311–17.

———. 2009b. "Species as a Process." *Acta Biotheoretica* 57: 33–49.

———. 2011a. "Species Are Individuals—The German Tradition." *Cladistics* 27: 629–45.

———. 2011b. "Wilhelm Troll, Physics, and Phylogenetics." *History and Philosophy of Life Sciences* 33: 321–42.

———. 2012. "Karl Beurlen (1901–1985), Nature Mysticism, and Aryan Paleontology." *Journal of the History of Biology* 45: 253–99.

———. 2016. *Phylogenetic Systematics: Haeckel to Hennig.* Boca Raton: CRC.

Rinard, R. G. 1981. "The Problem of the Organic Individual: Ernst Haeckel and the Development of the Biogenetic Law." *Journal of the History of Biology* 14: 249–75.

Rohkrämer, Th. 2005. "Martin Heidegger, National Socialism, and Environmentalism." In *How Green Were the Nazis? Nature, Environment, and Nation in the Third Reich*, edited by F.-J. Brüggemeier, M. Cioc, and T. Zeller, 171–203. Athens: Ohio University Press.

Roux, W. 1881. *Der Kampf der Teile im Organismus. Ein Beitrag zur Vervollständingung der mechanischen Zweckmässigkeitslehre.* Leipzig: W. Engelmann.

Sapper, K. 1938. "Zur Kritik der Ganzheitsbiologie." *Acta Biotheoretica* 4 (2): 111–18.

Schlosser, G., and G. P. Wagner. 2004. *Modularity in Development and Evolution.* Chicago: University of Chicago Press.

Schwarz, A., and K. Jax. 2011. "Early Ecology in the German-Speaking World through WWII." In *Ecology Revisited: Reflecting on Concepts, Advancing Science*, edited by A. Schwarz and K. Jax, 231–75. Heidelberg: Springer.

Sober, E. 1980. "Evolution, Population Thinking, and Essentialism." *Philosophy of Science* 47: 350–83.

Stella, M., and K. Kleisner. 2010. "Uexküllian *Umwelt* as Science and as Ideology: The Light and the Dark Side of a Concept." *Theory in Bioscience* 129: 39–51.

Thienemann, A. 1918. "Lebensgemeinschaft und Lebensraum." *Naturwissenschaftliche Wochenschrift* N. F. 17 (20): 281–90; (21): 297–303.

———. 1925. "Der See als Lebensgemeinschaft." *Die Naturwissenschaften* 13: 598–600.

———. 1935. "Lebensgemeinschaft und Lebensraum." *Unterrichtsblätter für Mathematik und Naturwissenschaften* 41: 337–50.

———. 1939. "Grundzüge einer allgemeinen Ökologie." *Archiv für Hydrobiologie* 35: 267–85.

———. 1940. "Unser Bild der lebenden Natur." *90. u. 91. Jahresbericht der Naturhistorischen Gesellschaft zu Hannover 1940*: 27–51.

———. 1941. *Leben und Umwelt.* Leipzig: J. A. Barth.

Thienemann, A., and J.-J. Kieffer. 1916. "Schwedische Chironomiden." *Archiv für Hydrobiologie* (Suppl. 2): 483–554.

Uexküll, J. v. 1926 [1920]. *Theoretical Biology.* Translated by D. L. Mackinnon. New York: Hartcourt, Brace.

Volkmann, R. v. 1935. "Martin Heidenhain und die mikroskopische Technik (Zu seinem 70. Geburtstag)." *Zeitschrift für Mikroskopie* 51: 309–15.

Wagner, G. P. 2014. *Homology, Genes, and Evolutionary Innovation.* Princeton: Princeton University Press.

Weidenreich, F. 1929. "Vererbungsexperiment und vergleichende Morphologie." *Palaeontologische Zeitschrift* 11: 275–86.

Weingart, P., J. Kroll, and K. Bayertz. 1992. *Rasse, Blut und Gene. Geschichte der Eugenik und Rassenhygiene in Deutschland.* Frankfurt a.M.: Suhrkamp.

Wolf-Heidegger, G. 1938. "Vereinigung und Teilung als Gestaltungsvorgänge im Entwicklungsgeschehen." *Schweizerische Medizinische Wochenschrift* 19: 1181–89.

Zimmermann, W. 1930. *Die Phylogenie der Pflanzen. Ein Überblick über Tatsachen und Probleme.* Jena: G. Fischer.

———. 1935. "Heidenhain, M.: Die Spaltungsgesetze der Blätter. Beitrag XVI zur synthetischen Morphologie." *Der Biologe* 4: 26.

———. 1937/38. "Strenge Objekt/Subjekt-Scheidung als Voraussetzung wissenschaftlicher Biologie." *Erkenntnis* 7: 1–44.

8

Parasitology, Zoology, and Society in France, ca. 1880–1920

MICHAEL A. OSBORNE

Introduction

Parasites complicate our notions of individuality, autonomy, and dependence. Their problematic nature commands attention from biologists, philosophers, and historians. For biologists, parasitism and related phenomena (such as commensalism and mutualism) may be thought of as interactive relationships that exemplify various continua of symbiosis (Leung and Poulin 2008). The boundaries delimiting these various forms of symbiosis are vague, with complex dynamics of benefit, cost, and sometimes pathogen-mediated termination (Sapp 2004; Pérez-Brocal et al. 2013). Moreover, parasitism and other forms of symbiosis provide eukaryotes with opportunities for horizontal gene transfer and may represent important evolutionary mechanisms (Nardon and Grenier 1993; Sapp 1994; Wijayawardena et al. 2013). Some contemporary biologists see the host-parasite relationship itself not only in terms of gain or harm between separate entities, but also as a feature of a larger co-evolving adaptive complex system forged by interdependencies (Dujardin and Dei-Cas 1999; Horwitz and Wilcox 2005; Zook 2015).

For philosophers and historians, parasites are interesting for their scientific and metaphorical uses. In late nineteenth-century France, parasites and the condition of parasitism found diverse and pluralistic uses by biologists and social theorists. Study of parasites also intersected with considerations of the specialization of biological and social labor. Physicians, naturalists, and sociologists pondered the status of parasites as they considered the relatedness of individuals and groups of organisms to each other, posited concepts of social wholes, and contemplated the moral valences of dependence.

Instances of the application of the term *parasite* to the biological world rose substantially in the two decades prior to World War I. There are a number of reasons for this, including Europe's engagement with the colonial world, fundamental research done on malaria, and Ronald Ross's 1898 determination of its mosquito-borne nature. The era also saw the emergence of a medical specialty of parasitology and the related founding of schools of tropical medicine in London, Liverpool, and Paris. The French academic physician and zoologist Raphaël Blanchard (1857–1919) coined the term *parasitology*. He viewed parasitism as a generalized phenomenon and argued that control of parasites and their vectors were key elements in France's effort to colonize Africa and to bring colonies into the socioeconomic body of a Greater France (Blanchard 1912).

Parasitism in this colonial era was hardly a topic for parasitological research alone. It spoke to issues already of interest among scientists engaged in the study of biological individuality, association, and part-whole relations in organisms. Beginning as early as 1828, the zoologist and marine biologist Henri Milne-Edwards had promoted the concept of a division of physiological labor (d'Hombres 2012; Limoges 1994). In the 1870s a new body of biological research considered part-whole relationships. Claude Bernard's *Introduction à l'étude de la médecine expérimentale* (1865; translated in 1957 as *An Introduction to the Study of Experimental Medicine*) constituted a primer on part-whole relationships and considered the lives of individual cells and how they related to the whole organism. The volume addressed the dual lives of cells—that is, the relationship where cells have one life of their own and another in their functional relations with other cells and with the integrated whole of the organism. Bernard noted, "It is doubtless correct to say that the constituent parts of an organism are physiologically inseparable one from another, and that they all contribute to a common vital result" (Bernard 1957 [1865]). Bernard's words resonated with those of cell theory co-founders Schleiden and Schwann, who had also viewed cells as having a double life (Sapp 2003, 76). Thus there is a kind of division of labor at work in Bernard's conceptualization of the relationship between cell and organism. Bernard's use of animal vivisection, and his conviction that experimentation could render clear the law-bound nature of biological processes, energized physiological investigation.

Bernard's program gained adherents as the nineteenth century wore on, including his successor Paul Bert and through him, Blanchard (who also collaborated with Milne-Edwards's son, Alphonse). Blanchard's contemporary, the zoologist Edmond Perrier, was a leading theorist of biological part-whole relations who also drew both on Bernard and on the division of labor theories of the elder Milne-Edwards. For all of these men, the study of biological part-whole relations raised questions about dependence and autonomy similar to those raised by the phenomena of parasitism.

Parasites stimulated reflection on individuality, association, and dependence not only in biology but also in the social thought of an era that drew heavily on zoology and was replete with organic analogies and metaphors (Bernardini 1997; Clark 1984; Schneider 2002). The spheres of sociology and biology, however, converged, merged, and diverged in manifold ways. Although this study focuses on part-whole relationships in biology and society from the vantage point of French parasitology, scientists in other national contexts also considered parasites and the problems of defining biological and social wholes and individuals. For example, the pathologist, anthropologist, and cell theorist Rudolf Virchow employed research on roundworms accomplished by the zoologist Rudolf Leuckart to argue against the theory of spontaneous generation and consider the parasite's role in disease and infection (Virchow 1958 [1895], 199–200). Virchow also wrote on diverse aspects of part-whole relationships, and while he was most interested in cells, he referred to them in terms of atoms and individuals and concluded that biological individuals were really an "array of systems" constituting a "*unified commonwealth*" (Virchow 1958 [1859], 124, 138). These pronouncements and others, like "*The organism is not a single unit, but a social system,*" provided robust inspiration for social theory (Virchow 1958 [1898], 223). The political trajectories of France and Germany could hardly have been more divergent, but the use of biological analogy for social purposes was broadly similar.

Late nineteenth-century social and political tensions forced consideration of the nature of the parts and wholeness of France itself. The country endured decades of strikes, the loss to Germany of Alsace and part of neighboring Lorraine, the French civil war of the Commune, and international communist and socialist movements. By century's end the Dreyfus affair, in which a young Jewish military officer of Alsatian descent was unjustly convicted of providing information to the German embassy in Paris, revealed additional fissures in French society. Dreyfus was pardoned in 1906, but the episode unleashed anti-Semitic fervor and stimulated heated discussion of the nature of the social body and its functions with regard to individual rights. All these events forced consideration and debate on the proper configuration of society and the right mode or modes of governance, and French biologists engaged in these debates.

The zoologist and administrator Edmond Perrier was the best-known naturalist of his era to engage the question of the "social whole." Perrier's zoological work investigated the nature of complex wholes, and he applied it to the social sphere. He explicitly considered the evolutionary transition of single-celled organisms to multicellular ones, pondered the status of individual cells in multicellular organisms, and investigated the biological and societal implications of mutualism, commensalism, and parasitism. Perrier's ideas provided biological foundation for various metaphors of the body and nation, finding their way into political economy and the organicist socio-logy of René Worms (1869–1926), among others.[1] Woven throughout these biological and social discussions are themes that question individuality in terms of varying degrees of functional autonomy and dependence, often in language indicating hierarchical relationships of individuals within a greater organic or societal whole.

Defining the Parasite as a Biological Entity

The many discussions of parasitism in the largest of all French medical dic-tionaries, the 100 volume *Dictionnaire encyclopédique des sciences médicales* (1864–89), are revealing of the era's understandings of association among organisms. A lengthy article on parasites from the 1880s provides context (Davaine and Laboulbène 1885). Its primary author, Casimir Davaine, was a physician and medical microbiologist credited with demonstrating the pathogenic role of bacteria in human and animal diseases. He also made fun-damental studies on anthrax later used by Louis Pasteur, and investigated phagocytosis. An associate of veterinarians and agriculturalists, he was drawn to the study of nematodes and plant diseases caused by worms (Théodoridès 1968). His article enumerates the many classes of parasites and damages caused by them. Most interesting, however, is Davaine's struggle to define parasitism (Davaine and Laboulbène 1885, 67), a task not easily accomplished in categorical fashion even today (Zook 2015). His starting point was that parasitism was a condition of dependence, "*An association between two indi-viduals of different species such that one of the two cannot live except with the assistance of the other*" (Davaine and Laboulbène 1885, 68, italics in original). In this definition, then, one member of the pair was dependent on the other but no harm to the host was implied.

We might ask, as biologists then did, about the functional autonomy of organisms. Just how dependent was the parasite on the host, and how did metabolism and space bind the two together? In the late 1870s the German botanist Anton de Bary had argued that parasitism was merely a special case of symbiosis, where organisms of two different species shared common phe-nomena of life (Perru 2002). Yet the definition of a parasite and specification of its relationship to its host remained contested, and was not soon resolved. Decades later the biologist Maurice Caullery could still write that it was im-possible to distinguish between commensalism, parasitism, and symbiosis. These relationships, he held, were not separated in nature by any discontinuity, and the phenomena merely displayed aspects of the same general laws (Caullery 1922; Perru 2002). By the interwar period, the student laboratory manual for the Paris Faculty of Medicine's course on medical parasitology echoed Caullery's skepticism. First published in 1928, the manual noted that an insect would be termed a free predator if it killed an organism to eat, but this same insect would be labeled a parasite if it sucked blood from the host organism and went on its way. The manual continues that the definition of parasitism is "vague enough and it is often very difficult to establish a clear distinction between parasitism, mutualism and commensalism" (Brumpt and Neveu-Lemaire 1958, 2). Similar definitional issues over parasitism and symbiosis persist (Leung and Poulin 2008; Martin and Schwab 2013; Zook 2015).

The term parasite had been used at least since Greek antiquity where it signified an assistant to a priest who took part in common meals. While various meanings of the term evolved in theater and other cultural activities, it often connoted people who dined next to one another. The term was not always used in a biological sense and was employed in seventeenth century mineralogy to signify stones formed from other rocks. By 1798, the zoologist and comparative anatomist Georges Cuvier referred to lice as parasites, and in the subsequent century the term incorporated the biological notion of the parasite harming its host ("Parasite" 1971–94). Yet even in late nineteenth-century parlance, the term parasite might signify simply an organism on the surface of another with no physiological relationship. Davaine distinguished true parasites from commensal organisms, a group recently defined by the Belgian zoologist Pierre-Joseph Van Beneden (1809–94) as organisms that dined together but not at the expense of one other.

Davaine responded to Van Beneden by asserting that neither habitat nor mode of nutrition were sufficient to qualify an organism as a parasite. The term was not, he stressed, a natural category on par with that of species but more of an umbrella term describing a range of specific organisms after the fashion of the terms carnivore or herbivore. For Davaine, and for many biologists and ecologists today, proximity, duration, and the "intimacy" of the parasite-host relationship constitute important factors for defining parasitic and symbiotic relationships (Martin and Schwab 2013). Thus Davaine excluded mosquitoes, bedbugs, and leeches from the class of parasites as they fed on their host but then went elsewhere (Brumpt and Neveu-Lemaire 1958, 68). By necessity, then, his definition inclined toward plant and animal sociology, and social theorists instrumentalized the multiple valences of the parasite and host relationship in their considerations of the social whole and its constituent parts.

Davaine defined parasitism as a state of enduring physiological relationships between organisms of different species. This may have been a response to descriptive teratological studies of the 1830s by Étienne and Isidore Geoffroy Saint-Hilaire, who had labeled the smaller of two members of a doubled fetus as parasitic upon the larger form (Geoffroy Saint-Hilaire 1832–36). The different species criterion is important too, for it differs from the way the term worked out in late-century social thought. In addition to being members of species different from their hosts, true parasites were less complete in function than most members of the animal kingdom, although degrees of autonomy and subordination varied. Davaine viewed parasites that spent a portion of their lives away from their hosts as more perfected, or we might say as possessing autonomy in a broader range of physiological functions, than those dwelling within hosts for most of their life cycle. Other zoologists were also studying parasites and reflecting on the nature of their degrees of perfection, dependence, and autonomy. The Dane Japetus Steenstrup, for example, whose work on alternation of generations is considered by Lynn Nyhart and Scott Lidgard in this volume, was thinking along similar lines as he investigated parasitic worms in the 1840s. Davaine, working some four decades later, hazarded the generalization that all parasites were degraded or imperfect in organization with reference to their host species. In short there was a kind of subordination of the parasite's individuality in relation to the host. This was particularly so for tapeworms and other internal parasites, which seemed to contract out many of their anatomical structures and physiological activities to their hosts.

Other physicians and zoologists offered divergent definitions of the parasite-host relationship. In the 1840s Van Beneden had focused on the life cycle of cestode flatworms, especially *Taenia* infections in dogs, and was able to show that the cysticercus, formerly classified as a separate organism, was actually the larval stage of an adult worm. A convinced biological "fixist," foe of spontaneous generation, and a believer in perfect adaptation, Van Beneden marveled at the intricacies of God's creation and authored a popular book in 1875 entitled *Les commensaux et les parasites dans le règne animal* (translated into English in 1883 as *Animal Parasites and Messmates*), which reviewed parasitism, mutualism, and commensalism (Van Beneden 1875, 1883; d'Hombres and Mehdaoui 2012; Hamoir 2002). Unlike Davaine, Van Beneden included organisms that visited one another in the class of parasites, including those that "fed at the expense of a neighbour, either establishing themselves voluntarily in his organs, or quitting him after each meal, like the leech or the flea" (Van Beneden 1883, xix). He also believed that every class of the animal kingdom included some parasitic forms, which he noted could have widely divergent life paths, with some living "at first like true Bohemians . . . certain of getting invalided at last in some well-arranged asylum"

(Van Beneden 1883, xix). Organisms like Van Beneden's Bohemians might thus occupy different places in the social fabric as they aged, commencing as innocuous members of life's fabric but finishing as a burden. Generally, true parasites shared both food and abode with their hosts, but Van Beneden classified parasites by their relative dependence on or autonomy from their hosts.

For Davaine, genuine parasitism required sustained host-parasite relations. The parasite could not immediately kill the host as it was dependent on it and needed to develop simultaneously. Hosts also accommodated parasites in various ways by forming a cyst or envelope around the invader. Davaine described this relationship as a dependent one similar to the biological ties between a mother and her fetus or child. In this, his ideas on parasitology and maternity resonated with those of the physician Patrick Manson (Li 2004). Still, Davaine believed that in animals, parasite and host retained a measure of their individuality—defined largely by autonomy. In the plant kingdom, however, individuality might be lost; some forms of plant-upon-plant parasitism seemed to produce an "intimate and insoluble" relationship, with parasite and host seemingly forming "a single body" (Davaine and Laboulbène 1885, 14). Davaine made no comment on the evolutionary significance of parasitism, but perhaps a medical dictionary was not the place for that.

Davaine's review article, like earlier work by de Bary and Van Beneden, failed to produce an agreed-upon definition of parasitism. A range of positions persisted on the topic of the individuality of parasites. In some relationships parasite and host retained a measure of respective autonomy; in other more intimate relationships one member of the relationship was more dependent. Davaine, who envisioned parasitism as the subordination of one individual to another from a different biological family, wrote of lichens and especially plant-upon-plant parasitism where two organisms formed an intimate relationship and seemingly fused into a single body (Davaine and Laboulbène 1885, 68, 74). The study of parasitism expanded as physicians, particularly those engaged in colonial medicine or the study of tropical diseases, and metropolitan and colonial governments, found utility in the study of parasites. Simultaneously, social theorists pondered the status of colonies within the body of a Greater France (meaning both France proper and the colonies), and the relationship of colonized people to metropolitan French citizens.

The Development of Parasitology as a "Colonial Science" in the Work of Raphaël Blanchard

In the late nineteenth century the medical study of parasites and negatively valued metaphors of dependence co-mingled in debates over the development of the French colonial empire. I have not found the term parasite explicitly used to describe relations between the modern French colonies and France proper. Nonetheless, arguments over the nature of Greater France and the cost and value of the colonial venture framed the colony as a dependent part of the social whole. An older theory of colonial relations, mercantilism, had regarded colonies as existing solely for the good of the governing nation. Mercantilism became less viable after the French Revolution, which enshrined citizenship and proclaimed the universal rights of man. Yet the French still regarded colonized peoples as subjects, not as citizens, and colonies and colonial subjects persisted in subordinate political and economic relationships to the citizens of France (Conklin 1998). By the 1890s, the French pondered how to develop their investments in empire, and politicians and businessmen thought long and hard about how the colonies could be made less dependent on the French nation and ideally fund themselves rather than drain French money and resources (Betts 1961).

Here is where the coiner of the term parasitology, the physician-zoologist Raphaël Blanchard, inserted himself into the debate as a builder of networks to support medical parasitology and keep the colonies part of a Greater France. Blanchard arrived in Paris in 1874, where he sampled courses at the Paris Faculty of Medicine and devoted himself to microscopy after hearing lectures by the naturalist and histologist Charles Robin. Blanchard was a founder of the Société zoologique de France and became its general secretary. There he cultivated relationships with scientific mandarins including Perrier and the Paris Muséum of Natural History's Professor of Mammals and Birds, Alphonse Milne-Edwards. Society duties also kept him in touch with the problems of zoological nomenclature, a pressing issue in medical parasitology. Blanchard had also worked as a *préparateur* for the physiologist and colonial functionary Paul Bert, who succeeded his mentor Claude Bernard in the Sorbonne's Chair of Physiology in 1869. Blanchard claimed his medical degree in 1880, about the time Davaine was composing his medical dictionary article on parasites.

In 1883 Blanchard began teaching zoology to first-year medical students at the Paris Faculty of Medicine, where he focused on medical zoology, especially the life cycles of parasites. He later gained a professorship as the Faculty's Chair of Parasitology and Medical Natural History (Osborne 2008). Parasitology developed not only in Paris but also at other French institutions. In 1894, at the Faculty of Medicine at Lille, Alfred Giard preceded Blanchard by becoming France's first incumbent of a university chair in parasitology. In 1902 the malariologist Alexander Le Dantec took up a chair of colonial medicine focused on parasitology at the Faculty of Medicine at Bordeaux (Barrière

1982). The Paris Faculty in this period was not an engine of research and publication. But Blanchard, who authored more than 600 articles and books over his career, was an exception. He was also not a clinician and took a dim view of clinical medicine if it was unsupported by laboratory results. This likely challenged perspectives held by older colleagues.

Blanchard joined parasitology and colonial activities in several venues. For example, in 1902 he founded an Institute of Colonial Medicine at the Paris Faculty of Medicine that trained about thirty postdoctoral students per year. By 1909 the Faculty required all third-year students to take a laboratory course on medical parasitology and be examined on the subject (Blanchard 1908–9). Clinicians may have disagreed with Blanchard's hierarchy of knowledge that championed natural history over medicine. But most colonial diseases and many others, he argued, were parasitical in origin and could be addressed only through extra-clinical measures and study. In ferreting out life cycles and taxonomy, and experimenting with disease-causing organisms, the naturalist was the "advisor and authorized guide of the hygienist and physician" (Blanchard 1905b, 132). Colonization itself re-valorized natural history, and this was due to more than just an increase in specimens for study. Medical parasitology, maintained Blanchard, was important for metropolitan medicine and would only become more so as colonial commerce grew. Parasites found in the colonies, those far-distant parts of Greater France, also posed dangers for France proper, and he raised the possibility of malaria victims from Senegal or Indochina transmitting the disease to *Anopheles* mosquitoes in the Paris region (Blanchard 1905a, 107).

Blanchard admired Pasteur and the bacteriological sciences, but he argued that parasitology now constituted the research front of the healing arts. It was formerly said, he wrote, "that bacteriology was going to be the last word of medicine; it is now outstripped and considerably so by animal parasitology and medical entomology, the sphere[s] of which are truly without limits." We can predict, he continued, a new golden age for humanity where henceforth "immense territories open up to the white race where formerly, until now, it was stalked by a hundred unknown enemies now revealed. These discoveries, and those of tomorrow, will change the face of the world" (Blanchard 1912, 122–23). His vision was of a smoothly functioning Greater France with parasitology aiding imperial and economic goals by assuring the health of colonized peoples and enabling them to labor under the direction of a few French supervisors (Osborne 2014, 200–205).

Blanchard's activities bring us back to the political dimensions of parasitology and the nature of the body in the complex whole of the extended nation-state. The acquisition of the French colony of Madagascar occasioned a public lecture course in 1901 that drew an audience of 500 to the Paris Muséum of Natural History. Blanchard's mixing of scientific parasitology with colonialism made it seem as if the many problems of the French colonial mission would soon be solved. It was formerly said, he wrote, "that Madagascar was the tomb of Europeans: until quite recently, it was thought that climatic conditions were unique in supporting this uncleanliness" (Blanchard 1902, 398). Now, however, parasitology had revealed the weaknesses of neo-Hippocratic climatic doctrines and associated place-based ideas of illness and health. The true source of malaria, a disease that had accounted for some 3,000 deaths in the 1895 French expeditionary force to Madagascar, had now been solved, he claimed. Its cause was neither freshly tilled soil nor swamps, but "always and uniquely, an insect bite" (Blanchard 1902, 423). Thus medical parasitology enabled a new style of French colonial activity. Provided the transmission of parasitic diseases could be controlled, the colonies would then have sufficient and healthy labor to produce products and reduce their dependency on metropolitan budgets, becoming functioning—if still lesser—parts of a Greater France.

Edmond Perrier, Zoology, and Sociology

The marine zoologist and administrator Edmond Perrier, a longtime director of the Paris Muséum of Natural History (1900–1919), provides a zoological counterpart to Blanchard. Like Blanchard, Perrier too was a member of the Académie de Médecine and the Société zoologique de France and was deeply involved in France's colonial turn. Perrier, who co-edited a six-volume study on the useful plants of French West Africa and also wrote on the birds of the region, is best known for his work on the origins and habits of colonial animals (by which he meant animals living in groups) rather than the French colonies per se. His major achievement in this regard was *Les Colonies animales et la formation des organisms* (Animal colonies and the formation of organisms), which first appeared in 1881 and was followed by a second edition in 1898. After the fashion of Bernard, Perrier, who was principally interested in intraspecific rather than interspecific relationships, argued that all multicellular animals were collectivities (Perrier 1898, 741). The volume treated Darwin's ideas, but Perrier elaborated a theory of neo-Lamarckian transformism by cooperation and association as an alternative to competition and natural selection. Like Blanchard, Perrier admired the social philosophy of Solidarism embraced by many politicians in the 1890s and first decade of the twentieth century. He signaled the lessons to be learned from biological association, the division of labor, and parasitism, and how those lessons might be applied to human society via Solidarism and an organicist conception of society.

Historians now interpret Solidarism as an outcry against *laissez-faire* and corporatist economic and social policies and as an attempt to balance individual rights with collective responsibilities (Dutton 2002; Jackson 2009; Sheradin 2000). It also incorporated many aspects of zoological thought, particularly neo-Lamarckian views of nature that many theorists found in Perrier's writings. Situated on the political continuum between liberalism and socialism, advocates of Solidarism abhorred the struggle for existence. In contrast to Darwinian and Spencerian selectionism, advocates of Solidarism promoted mutual assistance between individuals and sought security against poverty, illness, and unemployment. The majority of them examined society and individualism from organicist viewpoints and charted a research program reminiscent of Claude Bernard's views of the dual lives of cells in higher organisms. High-profile architects and promoters of Solidarism, frequently in disagreement, included Charles Gide, Léon Bourgeois, and René Worms, whom I discuss below. Gide's definition of Solidarism nicely encapsulates the philosophy's organicism and concern for whole and part relationships. "The living being," he wrote, "the individual, can scarcely be defined except as the solidarity of function which unites distinct parts" (Bourgeois 1912, 20). The sentiment was much the same at the level of society, which Solidarists saw as a collectivity of individuals differentiated though the division of labor and progressing through growing interdependence (d'Hombres 2010).

Solidarism, sometimes termed the official French government social philosophy around 1900, squarely confronted how individuals ought to relate to the whole of society and also how societies might progress and prosper (Hayward 1961). It emphasized cooperation rather than conflict in human affairs, and looked to the state to balance interventions for social justice with maintenance of some—but not absolute—individual freedoms. No citizen had complete autonomy and agency, and the entire social organism prospered though cooperation. Solidarism was explicitly and deeply rooted in biological thought, particularly Perrier's writings, and had much to say about autonomy, individuality, and wholes. Within this framework, social theorists selectively adopted biological concepts and adapted ideas of what constituted an individual or a parasite to their own ends.

The metaphor of the parasite permeated Solidarist thought. For example, the political economist and economic historian Charles Gide advocated a cooperativist model for society. In an 1893 speech in Montpellier entitled "The enemies of cooperation," Gide attacked the town's small and independent shopkeepers by calling them "parasites" and telling them that they should prefer to be eaten by the consumer cooperative movement rather than by the large private department stores that had sprung up in Paris and other European cities. Although ingestion is one way host organisms encounter parasites, Gide's use of the term diverged from contemporary biological understandings of parasite-host dynamics. Ingestion by either private enterprise or the cooperative movement spelled "death," at least commercial death, although the latter case implied a continued if ill-defined commensal relationship. The shopkeepers were unimpressed and blocked legislation, blacklisted workers favorable to cooperation, and burned down a warehouse known to sell to cooperatives. Gide was undeterred and continued to appeal to laws of the natural sciences and try to fashion a society in conformity with them (Sheradin 2000, 147–55). Significantly, he used the term parasite to indicate harm to the social whole and also applied it to members of the same species. In this way too he diverged from contemporary biological meanings of the term, which did not view parasite-host relations as universally harmful.

Perrier's book on animal colonies was a tour de force tracing an arc from protoplasm to humans. It narrated the story of how primordial single-celled simple organisms, or *plastides*, had first reproduced their like as a repeating unit but then diverged to become complex organisms where associations of *plastides* developed specialized functions while still retaining their association with others in the colony. These first simple multicellular animals, or *merides*, were quite similar to one another and differed no more than hydras differed from rotifers (Perrier 1898, ix–x). Perrier even wrote of *plastides* as sharing "solidarity" with one another. "All higher organisms," wrote Perrier, "were first *associations* of *colonies* of individuals similar to one another" (Perrier 1898, 144). The elementary *plastides* changed in response to environmental conditions in neo-Lamarckian fashion and also in response to their associations with other *plastides*. Bernard's concept of the organism and his ideas of how parts relate to the whole thus shine through Perrier's prose. Perrier distinguished two laws. The first was that of association and concerned limitations in the size of the tiny *plastides* that, because they were composed of protoplasm and had soft cell walls, could not themselves expand indefinitely. Even the higher-level multicellular *merides* were limited in scale. A second law, the law of the independence of anatomical elements, dictated that the *plastides* conserved some of their primordial independence (Perrier 1898, 702, 707). The natural sciences, he continued, could guide social thought as they

teach us that in everything, association prospers. Associated elements, while keeping a freedom from one another which is the necessary condition of progress, remain united by constant compromises, and confirm the place always held higher among the social virtues of the practice of *solidarity.* (Perrier 1898, xxxii)

Perrier's philosophy thus combined perspectives on individuality and accommodation and linked the interior dynamics of the parts and wholes of a single organism with a holistic perspective on the social organism.

Perrier assigned parasites a general role in the animal and human economy. His concept of parasitism recalled the broad sweep of de Bary's notion of symbiosis. Perrier even wrote of nonharmful domesticated animals as a kind of parasite. A discussion of parasitism appeared in reference to his law of reciprocal adaptation, which stated that once two or more organisms entered into enduring relations with one another, the couplet of organisms were altered by the relationship. Like Davaine, he too cited Van Beneden on mutualism and commensalism and wrote of how fungi and algae had co-adapted to form lichens and thus combined their respective individualities. He also commented on the enhanced reproductive capacity of many parasites and how at times they forced early reproduction by their hosts (Perrier 1898, 713). But the parasite changed too, and Perrier wrote of parasites such as nematodes as having evolved by regression from an extinct line of arthropods through the Lamarckian mechanism of disuse (Perrier 1898, xxi–xxiii). Despite these apparent adaptive benefits, Perrier viewed the parasitic condition as somewhat degraded: it is for this reason that he equated specialized non-parasitic reproductive individuals within a hydrozoan colony as "veritable parasites" (Perrier 1898, 712).

He also speculated on how individual members of a colony might confer an advantage to the colony in terms of its survival, yet at the cost of each individual's range of functionality—again reducing the individual's autonomy and increasing its dependence, much like parasites. Drawing an analogy from the social sphere and applying it to lower organisms, Perrier likened members of a colony to French civil servants or *fonctionnaires* with specialized duties yet allegiance to the whole. The same facts, he wrote, appeared in all animal associations,

> whether they were composed of individuals of a different species, or individuals of the same species, or whether these individuals are physiologically separated from one another as they are in simple *societies,* or whether they are united among themselves as we see in certain cases of parasitism or in *colonies.* (Perrier 1898, 713)

Perrier's perspective on parasitism and animal associations is significant because he goes beyond Davaine to envision the parasite to host relationship as equivalent or nearly equivalent to relationships between individuals of the same species. This conceptual move may have energized the application of the term to human-to-human associations by Charles Gide and two additional social theorists considered below, Léon Bourgeois and René Worms.

Solidarism and Social Theory

Perrier's comprehensive theory covered primordial cells, groups of cells and embryological matters, and scales of complexity. A number of social theorists embraced his version of the division of labor theory and association; here I examine just two of them, Léon Bourgeois and René Worms. Bourgeois did not write directly of parasites, but addressed the proper functioning of the social organism and drew inspiration from biology. A moderate Republican politician and civil servant who popularized Solidarism in a book of the same title in 1896, he was prominent in reformist circles and was honorary president of the Musée Social, an institution founded in 1894 for the scientific study of working-class issues (Bourgeois 1896). A winner of the Nobel Peace Prize in 1920, he twice refused offers of the presidency of the French Republic but served as prime minister. William Schneider portrays Bourgeois as following a middle path between the harsh individualism of the Social Darwinists and programmatic versions of communism purporting to subsume individualism to a communal will, a stance encapsulated in the phrase "union for life" (Schneider 2002, 29–30). In this, his conception of the social organism mirrored Claude Bernard's ideas on how cells related to one another in the internal environment of the organism, which was itself influenced by the forces of the external environment.

Organicist imagery and scientism permeated Bourgeois's version of Solidarity. His writings commented on the social debt of individuals and argued against selfish behaviors (Bourgeois 1912). Although his prose was replete with references to biological theory, he was careful to note that human society was not rigorously similar to an organism because it did not comprise "a living being where the parts are, as in the biological aggregate, materially unified to one another" (Bourgeois 1912, 27). Others too grounded their vision of society in biology but also found limitations in it.

René Worms, a jurist, statistician, and social theorist, addressed the social organism in his 1896 book *Organisme et Société* (*Organism and Society*), and another of 1910, *Les principes biologiques de l'évolution sociale* (*The Biological Principles of Social Evolution*), where he embraced the notion of biological wholes and of society as a kind of organism. Like Bourgeois he too claimed that society was more complex than a biological individual, although the main features of individual organisms were met again in the study of societies. Thus the anatomy, physiology, and pathology of societies were in essence, although not in all particulars, the same sort of phenomena encountered in the study of single organisms (Worms 1896, 7). He peppered his book with citations to Perrier's work on animal colonies, Claude Bernard's writings, and the sociologist Auguste Comte's notion of the social organism. The influential Comte had conceptualized societies as progressing through primitive theological and metaphysical stages to arrive at the positive or scientific stage that adhered to observation and experiment as sources of knowledge. Worms too wanted a science of society and maintained that the laws governing the members of the social body were, "at least in part, those governing the cells of an organism." Society itself then was analogous to a single organism, but its complexity made it a kind of "*supra-organisme*" (superorganism) (Worms 1896, 8–9). In short, society was analogous to an organism whose parts, tissues, and internal workings were analogous to the inter-relationships of cells in multicellular organisms. Worms held a more robust organicist vision than Bourgeois, and while Worms conceded that sociologists were not biologists, he argued that sociology had a discrete and dynamic object of study, and thus like biologists sociologists too merited respect and a place in the pantheon of the sciences. Like Émile Durkheim, the most famous of all French sociologists, Worms also adopted scientific methods in pursuit of social facts.[2]

Worms's scholarship on the social body manifests a fundamental transition in the use of the term parasite. While Perrier had muddied the waters by comparing degraded reproductive individuals within a colony to parasites, in biology, most often, the term had been applied to a relationship between two different species. Worms, like the political economist Gide, was explicit in applying the term to human-to-human relationships within the context of the national body. The contours of that body itself, and the status of colonies vis-à-vis France proper, were in dispute during Perrier's era. Worms wrote of the "parasitic" pirates of Algiers who had troubled the European powers during the nineteenth century (Worms 1896, 316). His use of the term violated Davaine's interspecific criterion for parasitism, since the Algerians were also members of the human species. These pirates, presumably Arabs, caused "lesions" in the host, by which Worms meant Western European society. In his telling of the story, the pirates preferred to hold prisoners for ransom rather than kill them immediately, so in this they weakly mimicked parasite-to-host accommodation. Yet Worms made little room for this accommodation of parasite to host, and vice versa, and conceptualized the relationship in stark terms. Sometimes the lesions inflicted on the host were mortal—in this case the populations around the Mediterranean who were not engaged in piracy. At other times they provoked a reaction that did away with the parasite much as the French had done in 1830 when they deposed the Bey of Algiers and began forming their second colonial empire (Worms 1896, 316n1). This contrasts with Perrier's treatment of parasites, which in many ways normalized them and gave them a kind of value in his neo-Lamarckian view of transformism.

Conclusion

In reviewing late nineteenth-century perspectives on the role and status of parasites, I have considered how we might use the idea of parasites to look at biological and sociological conceptions of parts, the ways in which those parts associate with each other, and how parts relate to biological and social wholes. Davaine and Blanchard declared biological definitions of parasitism to be a relationship between two separate species, but as we have seen, the lines between parasitism and commensalism remained indistinct. Moreover, whatever the relationship of parasite to host, it was dynamic. While it is true that turn of the century social philosophies of cooperation like Solidarism, and Perrier's versions of them in particular, drew inspiration from neo-Lamarckian ideas, they were also naturalized with reference to concepts such as the division of labor stemming from the work of Perrier's Muséum colleagues, Henri and Alphonse Milne-Edwards. Biological analogies were plentiful in late-century philosophy. They were also supple and easily modified.

In a France beset by anarchist activities and syndicalism, Solidarism seemed a quiescent balm to the pressing social questions of the day. Citizenship in the social body was a coveted prize for the French, and citizens constituted the essential parts of the nation. But full membership in that social body, itself an entity in dispute, rarely extended to the colonized peoples of Greater France. In 1894, just prior to publication of Worm's *Organism and Society*, France entered the trauma of the Dreyfus Affair, which unleashed waves of anti-Semitism and other practices of marginalization.

Jews like Worms who lived through the Dreyfus Affair and the overt anti-Semitism of the era felt strongly the currents of marginalization in French society. Worms's comments on the parasitical pirates of Algiers, who presumably were Arab and Muslim, were inflected by the social forces of the era and the move to achieve full citizenship and inclusion in the social body for the Jewish population in North Africa—though not for all French North Africans. Worms himself was a defender of the Decree of Crémeux, a statute of 1870 that had given both Algerian Sephardic Jews and European settlers in French Algeria (*pieds noirs*) French citizenship while simultaneously erecting barriers to Arab Muslim and Berber populations. His focus on the activities of Algerian pirates connotes an image of the parasite as marginal and as an exception in violation of the laws of social physiology. Perrier, Davaine, and Van Beneden, on the other hand, recognized the ubiquity of biological parasitism and were reluctant to define parasites as universally harmful. As I have tried to show, their biological notions of parasitism as an interspecific phenomenon did not translate well into social thought even as Gide, Bourgeois, and others tried to adapt those ideas to render the study of society scientific.

Acknowledgments

The author wishes to thank the volume editors and Elizabeth M. Nielsen for careful reading and comments on this paper.

Notes to Chapter Eight

1. Perrier's neo-Lamarckian ideas of cooperation influenced the sociologist Émile Durkheim (1857–1917), although the latter was not an organicist and distanced himself from Perrier's brand of biology after the Dreyfus affair (Gissis 2011).

2. For Durkheim, who turned away from Perrier's neo-Lamarckian biology, organic solidarity was more typical of simple societies rather than the complex whole encountered in industrialized Europe.

References for Chapter Eight

Barrière, Georges. 1982. "Raphaël Blanchard (1857–1919), Sa Vie—Son Oeuvre." PhD diss., Université de Aix-Marseille II.

Bernard, Claude. 1957 [1865]. *An Introduction to the Study of Experimental Medicine*. Translated by Henry C. Green. New York: Dover.

Bernardini, Jean-Marc. 1997. *Le darwinisme social en France (1859–1918). Fascination et rejet d'une idéologie*. Paris: CNRS Éditions.

Betts, Raymond F. 1961. *Assimilation and Association in French Colonial Theory, 1890–1914*. New York: Columbia University Press.

Blanchard, Raphaël. 1902. "Climat, hygiene et maladies." In *Madagascar au début du XXème siècle*, edited by Raphaël Blanchard, 397–452. Paris: Société d'éditions scientifiques et littéraires.

———. 1905a. "La médecine colonial." *Archives de Parasitologie* 9: 95–121.

———. 1905b. "Zoologie et medicine." *Archives de Parasitologie* 9: 129–44.

———. 1908–9. "Projet de réorganization du Service de la Parasitologie." *Archives de Parasitologie* 13: 311–42.

———. 1912. "[résumé] L'Entomologie et la Médecine." *Premier Congrès International d'Entomologie, Bruxelles, 1–6 August 1910.* 1: 122–23.

Bourgeois, Léon. 1896. *Solidarité*. Paris: Armand Colin.

———. 1912. *Solidarité*. 7th ed. Paris: Armand Colin.

Brumpt, Émile, and Maurice Neveu-Lemaire. 1958. *Travaux pratiques de parasitologie [revue et completé par Lucien Brumpt]*. Paris: Masson et Cie.

Caullery, Maurice. 1922. *Le parasitisme et la symbiose*. Paris: Octave Doin.

Clark, Linda L. 1984. *Social Darwinism in France*. Tuscaloosa: University of Alabama Press.

Conklin, Alice L. 1998. "Colonialism and Human Rights, A Contradiction in Terms? The Case of France and West Africa, 1895–1914." *American Historical Review* 103: 419–42.

Davaine, Casimir-Joseph, and Joseph Alexandre Laboulbène. 1885. "Parasites, parasitisme." In *Dictionnaire encyclopédique des sciences médicales*, edited by Amédée Dechambre and Léon Lereboullet, Series 2, vol. 21: 66–116. Paris: Asselin.

d'Hombres, Emmanuel. 2010. "Le solidarisme. De la théorie scientifique au programme de gouvernement." *Revue d'éthique et de théologie morale*, no. 260: 81–107.

———. 2012. "The 'Division of Physiological Labour': The Birth, Life and Death of a Concept." *Journal of the History of Biology* 45: 3–31.

d'Hombres, Emmanuel, and Soraya Mehdaoui. 2012. "'On What Condition Is the Equation Organism-Society Valid?' Cell Theory and Organicist Sociology in the Works of Alfred Espinas (1870s–80s)." *History of the Human Sciences* 25 (1): 32–51.

Dujardin, L., and E. Dei-Cas. 1999. "Towards a Model of Host-Parasite Relationships." *Acta Biotheoretica* 47 (3–4): 253–66.

Dutton, Paul V. 2002. *Origins of the French Welfare State: The Struggle for Social Reform in France, 1914–1947*. Cambridge: Cambridge University Press.

Geoffroy Saint-Hilaire, Isidore. 1832–36. *Histoire générale et particulière des anomalies de l'organisation chez l'homme et les animaux . . . ou, Traité de tératologie*. 3 vols. Paris: Baillière.

Gissis, Snait B. 2011. "Lamarckism and the Constitution of Sociology." In *Transformations of Lamarckism: From Subtle Fluids to Molecular Biology*, edited by Snait B. Gissis and Eva Jablonka, 89–100. Cambridge, MA: MIT Press.

Hamoir, Gabriel. 2002. "La révolution évolutionniste en Belgique. Du fixiste Pierre-Joseph Van Beneden à son fils darwiniste Edouard." *Annales de Médecine Vétérinaire* [Liège] 46: 43–48.

Hayward, J. E. S. 1961. "The Official Philosophy of the French Third Republic: Léon Bourgeois and Solidarism." *International Review of Social History* 6 (1): 19–48.

Horwitz, Pierre, and Bruce A. Wilcox. 2005. "Parasites, Ecosystems and Sustainability: An Ecological and Complex Systems Perspective." *International Journal of Parasitology* 35: 25–32.

Jackson, Jeffrey H. 2009. "Solidarism in the City Streets." *French Cultural Studies* 20 (3): 237–256.

Leung, T. L. F., and R. Poulin. 2008. "Parasitism, Commensalism, and Mutualism: Exploring the Many Shades of Symbioses." *Vie et Milieu* 58 (2): 107–15.

Li, Shang-Jen. 2004. "The Nurse of Parasites: Gender Concepts in Patrick Manson's Parasitological Research." *Journal of the History of Biology* 37: 103–30.

Limoges, Camille. 1994. "Milne-Edwards, Darwin, Durkheim and the Division of Labour: A Case Study in Reciprocal Conceptual Exchanges between the Social and Natural Sciences." In *The Natural Sciences and the Social Sciences*, edited by I. Bernard Cohen, 317–43. Dordrecht: Kluwer Academic.

Martin, Bradford D., and Ernest Schwab. 2013. "Current Usage of Symbiosis and Associated Terminology." *International Journal of Biology* 5 (1): 32–45.

Nardon, Paul, and Anne-Marie Grenier. 1993. "Symbiose et évolution." *Annales de la Société Entomologique de France*, New Series, 29 (2): 113–40.

Osborne, Michael A. 2008. "Raphaël Blanchard, Parasitology, and the Positioning of Medical Entomology in Paris." *Parassitologia* 50: 213–20.

————. 2014. *The Emergence of Tropical Medicine in France.* Chicago: University of Chicago Press.

"Parasite." 1971–94. In *Trésor de la langue française; dictionnaire de la langue du XIXe et du XXe siècle (1789–1960),* edited by Paul Imbs and Bernard Quemada. Paris: Éditions du Centre national de la recherche scientifique. http://atilf.atilf.fr/tlf.htm.

Pérez-Brocal, Vicente, Amparo Latorre, and Andrés Moya. 2013. "Symbionts and Pathogens: What Is the Difference?" In *Between Pathogenicity and Commensalism,* edited by Ulrich Dobrindt, Jörg H. Hacker, and Catharina Svanborg, 215–43. Current Topics in Microbiology and Immunology 358. Berlin: Springer.

Perrier, Edmond. 1898. *Les colonies animales et la formation des organismes.* 2nd ed. Paris: Maisson et Cie.

Perru, Olivier. 2002. "Du parasitisme à la symbiose dans le règne animal: 1875–1883." *Revue des Questions Scientifiques* 173 (1): 19–38.

Sapp, Jan. 1994. *Evolution by Association: A History of Symbiosis.* New York: Oxford University Press.

————. 2003. *Genesis: The Evolution of Biology.* New York and Oxford: Oxford University Press.

————. 2004. "The Dynamics of Symbiosis: An Historical Overview." *Canadian Journal of Botany* 82: 1046–56.

Schneider, William H. 2002. *Quality and Quantity: The Quest for Biological Regeneration in Twentieth-Century France.* Cambridge: Cambridge University Press.

Sheradin, Kristin A. 2000. "Reforming the Republic: Solidarism and the Making of the French Welfare State, 1871–1914." PhD diss., University of Rochester.

Théodoridès, Jean. 1968. *Un grand médecin et biologiste, Casimir-Joseph Davaine, 1812–1882.* Oxford and New York: Pergamon.

Van Beneden, Pierre-Joseph. 1875. *Les commensaux et les parasites dans le règne animal.* Paris: Germer Baillière.

————. 1883. *Animal Parasites and Messmates.* New York: D. Appleton.

Virchow, Rudolf. 1958 [1859]. "Atoms and Individuals." In *Disease, Life, and Man: Selected Essays by Rudolf Virchow,* edited by Lelland J. Rather, 120–41. Stanford, CA: Stanford University Press.

————. 1958 [1895]. "One Hundred Years of General Pathology." In *Disease, Life, and Man: Selected Essays by Rudolf Virchow,* edited by Lelland J. Rather, 170–215. Stanford, CA: Stanford University Press.

————. 1958 [1898]. "Recent Progress in Science and Its Influence on Medicine and Surgery." In *Disease, Life, and Man: Selected Essays by Rudolf Virchow,* edited by Lelland J. Rather, 216–45. Stanford, CA: Stanford University Press.

Wijayawardena, Bhagya K., Dennis J. Minchella, and J. Andrew DeWoody. 2013. "Hosts, Parasites, and Horizontal Gene Transfer." *Trends in Parasitology* 29 (7): 329–38.

Worms, René. 1896. *Organisme et Société.* Paris: V. Giard et E. Brière.

————. 1910. *Les principes biologiques de l'évolution sociale.* Paris: V. Giard et E. Brière.

Zook, Douglas. 2015. "Symbiosis—Evolution's Co-Author." In *Reticulate Evolution: Symbiogenesis, Lateral Gene Transfer, Hybridization and Infectious Heredity,* edited by Nathalie Gontier, 41–80. Cham: Springer.

9

Metabolism, Autonomy, and Individuality

HANNAH LANDECKER

Introduction

In the work of the twentieth-century philosopher Hans Jonas, metabolism plays a central role. Metabolism, he observed, is often considered no more than the elementary level underlying higher functions such as perception, locomotion, and desire. Jonas, by contrast, opined that not enough thought had been given to metabolism as the *basis* for those functions: "It presages them by enacting within itself the cardinal polarities which those functions will expand and span with their more determinate relationships: the polarity of being and not-being, of self and world, or freedom and necessity" (1974, 196). In his view, things need to eat and drink of other life and of the external world, but they do not thereby lose distinctiveness. Instead, metabolism and its constant dynamic inter-conversions simultaneously bind organisms to the environment and free them from it—this he called the dialectic of "needful freedom."

Metabolism is the antinomy of freedom at the roots of life, "its liberty itself is its peculiar necessity." It constitutes the organization of organisms for "inwardness, for internal identity, for individuality," but it simultaneously turns the organism outward, "toward the world in a peculiar relatedness of dependence and possibility" (Jonas 1966, 84). In this analysis, metabolism's function is not to be a boundary between organism and environment, but to produce that distinction in the first place. It produces "inwardness" at the basis of individuation and selfhood, inwardness that is present in all life from the very beginning.

Jonas, who began his career in philosophy as a student of Heidegger writing a two-volume work on ancient Gnosticism, is perhaps an unexpected observer of the mid-twentieth-century biochemistry of intermediary metabolism. The work provides a unique diagnostic reading of the existential implications of modern concepts of metabolism, which by this time had become enshrined as a set of maps of circular interlinked enzymatic reactions of the synthesis and breakdown of molecules and the generation of energy. Whatever one thinks of the philosophy he thereby developed, his diagnosis stands: metabolism is intrinsic to concepts of individuality. To be more precise, metabolism has been intrinsic to conceptualizing the process of individuation: how an organism becomes an individual, set off from the world, that becomes the unit of analysis for those who study locomotion, perception, and desire—physiology, neurology, psychology, etc.

In this essay, I am concerned not with the philosophy of metabolism, but in the historical development of this particular metabolism of philosophy.[1] How did we get in the mid-to-late twentieth century to this metabolism, the understanding of which underlies an image of the modern individual organism: the spatially bounded, immunologically defended, developmentally produced, and genetically coherent body that persists in time and space? (Gilbert et al. 2012). Such autonomous bodies depend on maintenance of the daily, hourly energy and substance stability of the bounded body even in the face of constant necessity to ingest other organisms in order to stay individuated, a self. To this question—how does the organism-as-individual persist over time as the self-same entity?—has come this answer, so unquestioned that it becomes almost invisible: by converting the substance of others into the self. The other ceases to be itself, because it is both broken down into pieces for reassembly as needed by the eater, and transformed in the processes of energy production. This answer may be understood as a fundamental *logic of conversion*—food is converted to constitute the eating body's substance, motion, and heat in metabolism. This logic of conversion also implied a particular distribution of agency: there is the *converter* and the *converted*, the eater and the eaten.

I argue below that a very particular logic of conversion was established and consolidated in the nineteenth century, first in physiology and animal chemistry, and then in biochemistry, and has remained relatively stable from the mid-nineteenth century to the opening of the twenty-first. This historical excavation is selective, focusing on early articulations of the intermediary tissue processes of nutrition and metamorphosis thought to happen after digestion and before excretion. This demarcation of metabolism as distinct from other parts of the animal economy opened out the "epistemic space" of the concept (Müller-Wille and Rheinberger 2012, xi). This history of establishment of the idea that food is completely converted into the eater gives insight into when and why it became "common sense" to think of the consumed as inert raw material whose only role in life was to provide energy and building blocks to the consumer. The term *metabolism* emerged to capture the space and time of inter-conversion, that ambiguous zone between eater and eaten in which two become one.

The history of nutrition science, animal chemistry, and the biochemistry of intermediary metabolism is already well-described in the historical literature. I reread this literature to draw out French physiologist Claude Bernard's theory of "indirect nutrition" in relation to his German contemporaries' work on *Stoffwechsel*—the metamorphosis of matter in living tissues— through input-output experiments with animals. My aim is to characterize the emergence of a common logic of conversion, even in ostensibly opposed frameworks for understanding what happened to food inside animal bodies. Subsequently, the development of the science of *intermediary* metabolism in the twentieth century, the pursuit of which was an organizing force for the discipline of biochemistry, did not so much break the explanatory patterns established in the nineteenth century so much as filigree them in extraordinary molecular detail.

In short, I maintain that a historically peculiar metabolism is constitutive to the emergence of the modern notion of organisms as individuals. This is of course not a one-way process in which ideas and practices of metabolism contour possibilities for understanding individuality—what counted as an individual also contoured, very materially, the possibilities for the emergence of metabolism as a scientific concept. The nineteenth-century science of nutrition and assimilation from which the twentieth-century biochemically elaborated notion of metabolism comes is itself built with particular assumptions about individuals, in the sense that metabolism is worked out in some kinds of bodies and not others in experimentation.

Concepts don't travel alone. Understanding the historical specificity of the logic of metabolism will help articulate its role in building commonly held notions of individual organisms as alone in the world, held apart from the world and others by the very need to constantly ingest the outside in order to continue being an autonomous entity. Jonas, as with most philosophers, was trying to generate a universally applicable analysis of what *is*; we are situated in a very different historical moment in which the science of metabolism is becoming radically destabilized, as is the notion of stable discrete biological individual organisms. From the current vantage point, then, his becomes a very clear statement of what *was*. The essay ends with a reflection on how contemporary developments in biology can reorient our sense of metabolism not as a universal, but as a historical configuration, one that has played a central role in the biological and philosophical question of what is, or makes, an individual.

The Logic of Conversion: Claude Bernard and the Nutritive Reserve

A central finding on which physiologist Claude Bernard's illustrious career was built was the demonstration that the animal body did not just break down substances such as sugar, but also built them up. His 1848 work, "On the origin of sugar in the animal economy," departed from existing assumptions that animals only decompose the complex substances received from plants (Bernard 1987 [1848]). Animals, Bernard argued, both create and destroy sugar, evidenced by the fact that dogs fed sugar, starch, meat, or nothing at all for two days, all have sugar in their blood. This work was followed by a series of experimental efforts to show that the liver could "produce sugar from a substance within itself," a substance which could be extracted from the tissue, and then fermented to yield sugar (Young 1957, 1432).

The demonstration of synthetic powers of liver tissue was done by excising a liver from a living dog, washing it of all blood, proving the absence of detectable sugar, waiting, and then later showing the presence of sugar. This "vital activity" was detected when the animal was dead but the tissue was not, enhancing its perception as a power of the tissue itself. Rhetorically, such disembodied action on the part of the organ was further underlined by Bernard's statement that one could get a piece of liver from a butcher's shop in order to demonstrate this tissue's ability to produce sugar. The procedure of washing and waiting could be repeated several times, and it was not the death of the tissue that brought the production of sugar to an end, but the exhaustion of the substance that the liver was making sugar out of.

In 1857 Bernard named that substance "glycogenic matter," a starch-like material whose existence proved, he argued, that sugar is made from something within the tissues themselves. The liver generated both "external secretions" such as the bile that went into the intestine, and "internal secretions," such as glycogenic matter (Olmsted and Olmsted 1961, 88). This work led to Bernard's famous elaboration of animals having and continuously making for themselves a an internal environment, a *milieu intérieur* that was always being adjusted in relation to an external environment, but created a stability and continuity that enabled a life independent from the fluctuations of temperature, light, and food of the outside.[2] Glycogen was a "nutritive reserve," which the animal could make at one time and draw upon at another; it was a cushion against the world.[3]

The context for this work in the 1840s was what historian F. L. Holmes has described as the confrontation of physiology and chemistry on questions of animal respiration, digestion, and nutrition (Holmes 1974). Lavoisier's late eighteenth-century theory that combustion of carbon and hydrogen to carbonic acid and water, either in the lungs or the blood, was the source of animal heat, was both contested and experimentally recalcitrant, as were questions about the chemical processes that go on inside animals between eating and excretion (Mendelsohn 1964; Fruton 1999). Food was redefined in terms of its chemical constitution, via practices of extraction, titration, distillation, crystallization, and combustion analysis performed on the flesh and fluids of plants and animals, including the elementary analysis of common foodstuffs such as milk, vinegar, and sauerkraut (Moulton 1942; Klein 2003; Orland 2010). Vital action gained new chemical terminologies, such as Berzelius' newly minted term *catalysis*, to refer to the property of some organic substances to accelerate chemical changes in other substances while themselves remaining unchanged. A much-debated question at the center of chemical investigation was the relation of the nitrogen in plants to the nitrogen in animals—since animal bodies had a higher concentration of this element, it was widely held that plant matter was "animalized" on consumption, but in the main provided all the elements, pre-formed, that the animal needed to live and grow. In many ways, empirical investigations in organic chemistry centered on questions of continuity—elementary analysis focused on the steady presence of the elements of hydrogen, carbon, oxygen, and nitrogen even as they were recombined through reaction into new products.

In France, the chemist Jean-Baptiste Dumas together with the agronomist Jean-Baptiste Boussingault generated a model of life on earth as a general system of exchanges (Goodman 1972; Simmons 2006). Dumas argued that each of the major classes of animal substances was represented in plant matter by chemically equivalent substances. Because animals do not create organic materials but only destroy what plants provide for them, one could see the vegetable kingdom as "an immense reduction apparatus," and animals as, "from the chemical point of view, combustion apparatuses" (quoted in Holmes 1974, 24). Dumas became engaged in a long-standing personal and scientific battle with chemist Justus Liebig, who took a very similar stance on the matter of nitrogenous substances, writing that, "animals are a higher form of vegetable" (1964 [1842], 48). Dumas and Liebig disagreed violently, however, on the agriculturally important issue of the source of animal fat (whether animals make fat out of starch as Liebig held, or get it only from plants, as Dumas thought), as well as on the issue of who should have priority over the various ideas they were expounding (Simmons 2015).

Bernard, trained in the laboratory of physiologist François Magendie, wrote in his famous *Introduction to the Study of Experimental Medicine* that he was pushed by these debates between the chemists to investigate the role of sugar in the animal economy for himself (Olmsted and Olmsted 1961,

66). Bernard was influenced also by Magendie's criticism that neither side was well supported by experimental evidence in actual animal bodies. Quite the reverse situation prevailed in Bernard's ensuing experimental work. An American visitor observing Bernard's lectures commented on how "it was curious to see walking about the amphitheater of the College of France dogs and rabbits, unconscious contributors to science, with five or six orifices in their bodies from which at a moment's warning, there could be produced any secretion of the body, including that of the several salivary glands, the stomach, the liver, and the pancreas" (Donaldson, quoted in Olmsted and Olmsted 1961, 73). The dog was the preferred subject of experimentation; it was by opening up this interior world of large animals that Bernard came to his theory of nutrition.

This experimental work assertively separated nutrition from digestion, and gave it a space and a time of its own, anchored in the inner reserve. Where digestion was the process by which food came into the body, nutrition named the processes of change occurring in the internal fluids, tissues, and organs. In the 1878 *Lectures on the Phenomena of Life Common to Animals and Plants,* Bernard presented nutrition as one of the general characteristics of living beings.

> Nutrition is the continuous mutation of the particles which constitute the living being. The organic edifice is the site of a perpetual nutritive movement which leaves no part at rest; each one, without cease or respite, takes its food from the medium that surrounds it and into it rejects its wastes and its products. This molecular renovation is imperceptible to the sight, but as we see its beginning and its end, the intake and output of substances, we conceive of its intermediate phases, and we represent to ourselves a flow of material which passes incessantly through the organism and renovates it in its substance and maintains it in its form. The universality of such a phenomenon in the plant, in the animal, and in all their parts, and its constancy which suffers no interruption, make it a general sign of life. (1974, 26)

This depiction of nutrition as perpetual movement was not particular to Bernard, as the notion of the inside of the animal as a kind of place of constant inner motion or "*Wechsel der Materie*" is visible in French and German physiological writing from about 1800 onward (Parnes 2000, 74).

What *was* particular to Bernard was the idea that this constancy served *indirect nutrition.* Food is not converted directly into body and action, he argued, but organisms can and must maintain nutritive reserves. To have a reserve means, by definition, that the chemical reaction from input to output can not be instantaneous, and suggests an intermediate form—remember the lapse of time between looking for sugar and looking for sugar again in the excised liver. Food does not pass directly from plants to animals; rather,

> nutrition is indirect. The food first disappears, as a definite chemical material, and it is only after extensive organic work, after a complex vital elaboration, that the food comes to constitute the reserves, always identical, that serve for the nutrition of the organism. *Nutrition and digestion are completely separate;* the nature of the food is essentially variable, and never has any effect in the normal state on the formation of the reserves, which remain constant, like the composition of the fluids and the organic tissues. In a word, the body never nourishes itself directly from the various foods, but always by means of identical reserves, prepared by a sort of work of secretion. (1974, 103 emphasis added)

By this logic, food loses autonomy, and ceases to be itself. It disappears, through a vital elaboration. Even if the food has fats in it, Bernard observed, the animal "creates fat instead of finding it fully formed" (1974, 103). The dog, for instance, "does not get fat on mutton fat, it makes dog fat" (1974, 104).

Indirect nutrition, and the necessity for every body to make its own nutrition within itself, was related very directly by Bernard to degrees of autonomy that an organism could have from its physical and nutritional environment. Because the diet of an animal is highly varied, Bernard explains, animals must have "within themselves mechanisms that derive similar materials from these varied diets and regulate the proportion of them that must enter the blood" (1974, 90). I want to draw attention to this articulation of the animal's *ability to turn the environment into itself.* The result is a vision of self-enclosed organisms, plant or animal, each with the whole of life within it:

> The living organism is made for itself, and it has its own intrinsic laws. It works for itself and not for others. There is nothing in the law of evolution of grass that implied that it should be cropped by a herbivor; nothing in the law of evolution of the herbivor that indicates that it must be devoured by a carnivor; nothing in the law of growth of cane that announces its sugar must sweeten man's coffee. The sugar formed in the beet is not destined, either, to maintain the respiratory combustion of the animals that feed upon it; it is reserved for consumption by the beet itself in the second year of its growth, when it flowers and fructifies. (1974, 107)

Here we see the autonomy of beings that live for and in themselves. Once dead, by virtue of being digested, matter ceases to have creative agency of its own, because it is subsumed to the being of the eater. "Living beings cannot exist except with materials from other beings that have died before them or were destroyed by them" (1974, 108). It is not problematic to be composed of dead others, because they have ceased to be themselves in any meaningful way: they "disappear as a definite chemical material" and become the reserves, a kind of matter always identical with itself rather than remaining distinct in terms of its source. Being eaten meant having the means of one's autonomy subsumed into someone else's freedom—becoming someone else's reserves. The space and time of the nutritive reserve thus simultaneously allowed each organism to live in and from itself, and to simultaneously share, from the lowliest tardigrade to the sessile plant to the complex animal, a "common" phenomenon of life. The nutritive reserve, always identical with itself, freed the animal from any debt or relation to what it ate.

It is not so much that Bernard articulated something that no one else did, and then everyone followed his lead. Rather he distilled and then to a certain extent made idiosyncratic enough to be distinctive the reigning logics of the debates that he was embedded in. The logic of conversion was not just the problem of how one thing became another; it was how stability and continuity was maintained even in the face of constant "molecular renovation"; how something that was constantly transforming could remain itself, free in the world by constituting its own world through internal secretion.

From Nutrition to the Metabolic Individual of Homeostasis

Claude Bernard never used the word metabolism in referring to glycogen's chemical transformations. *Stoffwechsel*, metamorphosis, and metabolism were all terms used by others working on animal chemistry. Bernard studiously ignored them in favor of other words for change of form—renovation, mutation, oxidation, fermentation, putrefaction, synthesis, assimilation, dissimilation. He intentionally avoided the language of his German counterparts such as the physiologist Jacob Moleschott, who described the very dualist "cycle of life" that Bernard was decrying: that plants used the energy of the sun to construct what animals destroyed, and that animals after death decompose only to be reconstructed into useable form by plants (Kamminga 1995). Or perhaps, as Holmes has observed, the early uses in physiology of the word metabolism were associated with the cell, and in particular the cell theory, and was therefore defined "at a level of organization remote from the level at which Bernard, Helmholtz and Dumas attacked the problem chemically" (Holmes 1992, 11). However, as is so often the case, historical actors who saw themselves as violently opposed to one another were, even in fairly immediate retrospect, together responsible for the emergence of new scientific concepts.

Justus Liebig and his students, for example, insisted on chemical *Stoff-wechsel* in animal tissues. Bernard mocked them for their study of animal ingestions and excreta without any idea of what happened in between: "I am the first who has studied the intermediate," he asserted in his notebook (quoted in Fruton 1999, 341). Yet Liebig—as much as Bernard—was responsible for the delineation of the space and time after digestion and before secretion and excretion as a distinct object of scientific inquiry, extracting it and setting it apart from a more general natural historical understanding of eating, digesting, breathing, vegetative growth, fermentation, and putrefaction as continuous components of the "animal economy" (Klein 2003, 44). Here, in this inside space, both Liebig and Bernard located the conversion of matter from other (foodstuff) to self (to tissue, or nutritive reserve). Digestion served to break down food into smaller parts and make it soluble in the blood so it could get to the tissues, and excretion was the return of matter to the outside world; by contrast, nutrition or *Stoffwechsel* was the crucial conversion process *inside* by which the animal turned food into itself. Where digestion made food portable or smaller or made its form amenable to chemical conversion, nutrition and *Stoffwechsel* were conceptualized as terms specific to complete chemical conversion: a metamorphosis that saw the recombination of elements to form new chemical compounds (in the chemical lexicon) or the synthetic construction and organic secretion of new substances (in the physiological).

The logic of nutrition as conversion in the nineteenth century also bears close relation to the "conversion processes" between electricity, magnetism, and heat explored in laboratories in the wake of the invention of the voltaic pile (Kuhn 1962, 323). In Germany, the students of Liebig pursued intake-output experiments on animals in search of an exact accounting of the "the continuous exchanges of matter with their surroundings" performed by living organisms, without reference to Bernard's theories of nutrition—they focused on what went in and what went out, not on the physiological process in between (Holmes 1987, 269). However, in many ways it didn't matter whether an investigator believed that plants lived to feed the fire of human action or not. The minute study of food in relation to excrement, inhalation in relation to exhalation, imbibition in relation to perspiration was equally driven by a logic of conversion, asking what was turned into what; how did some matter get turned into different forms of matter; where did movement and heat come from? These experimental and theoretical investigations were also focused on conversions, mostly of protein: the question of how nitrogen remained constant in quantity even as the elements were recombined and transformed into different overall forms.

The notions of the nutritive reserve, the *milieu intérieur*, and Liebig's ideas about the chemical conversions of *Stoffwechsel* were all ultimately subsumed to the term *metabolism*, which is now taken to be the translation of *Stoffwechsel* in English. Understanding how these ostensibly opposed visions were synthesized as metabolism would require too long a detour through cell and protoplasm theory in the nineteenth century; suffice it to say that by the last third of the century, these different concepts came together via the stuff of the cell—specifically, its proteinaceous protoplasmic matter (Brain 2015). In a set of famous passages on the "catholicity of assimilation" Thomas H. Huxley said that when he ate a piece of mutton—once the living protoplasm of a sheep—"a singular inward laboratory, which I possess, will dissolve a certain portion of the modified protoplasm, the solution so formed will pass into my veins, and the subtle influences to which it will then be subjected will convert the dead protoplasm into living protoplasm, and transubstantiate sheep into man" (1869, 21). An individual, within the protoplasmic continuum, was one possessed of a "singular inward laboratory," what would very shortly come to be termed a *metabolism*.

It lay to Michael Foster, British physiologist and a biographer of Claude Bernard, to gather those "subtle influences" of transubstantiation under the banner of metabolism. In a synthesis of cell theory, protoplasm theory, and nutrition, his influential *Textbook of Physiology* (first issued in 1877 and in multiple editions thereafter) placed the chemical work of *Stoffwechsel* within the cytological space of the cell, where the protoplasm was the space—and the agent—of inner nutrition identified by Bernard. It was not much of a leap, and it was not commented upon as a novel formulation; metabolism was simply offered as the right word for these phenomena. The space, and to a certain extent the time, in between digestion and excretion had been established in animal chemistry and physiology as situated in the tissues. With the rise of cell theory, the tissues were now understood to be *cellular*, in particular the muscle tissues, the much-investigated site of bodily work and motion (Geison 1969). The image of *Stoffwechsel* as the process of replacement of tissues worn down by use, together with the depiction, via chemistry, of protoplasm as "living proteoid," made the stuff of cells the logical place to situate this biology and chemistry of the in-between where food ultimately ceased to be itself and was converted into the tissue of the eater's body (Foster and Sheridan Lea 1897, 3).

By the end of the nineteenth century, the interest in material transformations from food to body was to a certain extent overshadowed by the discourse of energy—not the conversion of matter to different matter, but matter to energy, within the new and exciting logics of the second law of thermodynamics (Rabinbach 1992). There is not space here to go into the rise of the concept of the food calorie, nor the sequence of investigations and arguments that led to the location of the phenomenon of catalysis in the physical objects that came to be called enzymes, in cells—making the work of conversion more specifically attached to objects that could be investigated experimentally in and of themselves, without having to rely on input and output alone to fathom the intermediate (Kohler 1973; Holmes 1992). It has been my purpose to map out the establishment of the space of metabolism in the nineteenth century, which, once established, became a capacious housing for the discipline of biochemistry in the twentieth, as it sought to trace every molecule and transformation involved in the "intermediate" zone identified but not pinned down by nineteenth-century science.

The protagonists in the story of metabolism seem always to have been moving the bounds of continuity and discontinuity as they fought for the proper description of the living. Huxley, for example, claimed a fundamental dichotomy of plant and animal—"plants are the accumulators of the power which animals distribute and disperse"—while at the very same time asserting "no logical halting place" between the lowly fungus and foraminifer and the matter that gives rise to thought in the human (1869, 26). It was the power to interconvert, thanks to chemical action, one and the other, the dead and the living protoplasm, that allowed Huxley to settle, finally, on the unity of life where he might just as well (as one of his more hysterical critics put it) have located fundamental difference: "if you *identify* all life in protoplasm you must equally *differentiate* all life in protoplasms; for of no one living thing, and of the organs of no one living thing, is the protoplasm interchangeable with that of another; and this involves, instead of Mr. Huxley's universal *identity* in power, in form, and in substance, infinite *difference* in all these respects" (Stirling 1872, 4, original emphasis).

While completely ineffectual, this Hegelian protest pinpoints the role played by the space of metabolism: as the process mediating difference and sameness, where one being ceases to be in the process of maintaining another. The work of narrating the history of metabolism shares this oscillation between continuity and discontinuity, but in this case I have chosen to emphasize what I see as characteristics of the concept that actually possess remarkable historical stability. Even as speculative theories of enormous vital protein molecules in the protoplasm were discarded, and early twentieth-century scientists noted "radical changes" in "our notions regarding the relation of the food proteins to tissue proteins," the conviction deepened that food did not change the character of the tissues of the eater: "whatever may be the source, or chemical make-up of the [food] previous to its involvement in the nutritive processes, the resulting tissue cells and fluids remain characteristic and specific for the species" (Osborne and Mendel 1912, 473). From this early twentieth-century period comes the idea that food is broken down to provide the *bausteine*, or building blocks, from which the complex molecules of the protoplasm are built (Kossel 1912).

The study of *intermediary metabolism*, around which the science of biochemistry was built, focused on pinning down the exact nature of the constant chemical conversions that maintain organisms in a steady homeostatic autonomy from the fluctuating world. In the 1930s, British biochemists took up Bernard's maxim about the constancy of the internal medium being the condition of free life as a kind of slogan (Cooper 2008). It became widely read and quoted after Joseph Barcroft used it to structure his 1934 work *Features in the Architecture of Physiological Function* (Needham 1943, 213). In biochemistry, the nineteenth-century industrial image of the body as combustion engine and food as fuel was gradually displaced. As Rudolf Schoenheimer put it in *The Dynamic State of Body Constituents*, "new results imply that not only the fuel but the structural materials are in a steady state of flux," and there were no known machines that continually renewed themselves even as matter flowed through them and was transformed (1942, 65). But even as the concept of what was involved in the chemical reactions of metabolism became much more detailed, the image of complete conversion was accentuated, with the total breakdown and re-synthesis of things that came into the body only underlined by the mutual constant breakdown and buildup of the body itself.

With the rise of the information paradigm in genetics and the widespread theoretical influence of cybernetics after the mid-twentieth century, this focus on the flow of energy and matter was demoted to a supporting rather than a determining role in the understanding of life. Hans-Jörg Rheinberger reflects that the notion of information flow "created a completely new register in which to speak about living systems, thereby distinguishing them from purely physical and chemical systems characterized by flows of matter and energy alone. I personally remember this as being something like a liberation" (2013, 485). Metabolic inter-conversion provided energy and stuff, while the gene provided form. Norbert Wiener, for example, waxing lyrical in *The Human Use of Human Beings: Cybernetics and Society*, explained to readers that "life is an island here and now in a dying world," and that humans are able to carry with them—for a short period of time—the special environment that they make for themselves through the homeostatic mechanisms of temperature control, salt balance, and most importantly, metabolism:

It is the pattern maintained by this homeostasis, which is the touchstone of our personal identity. Our tissues change as we live: the food we eat and the air we breathe become flesh of our flesh and bone of our bone, and the momentary elements of our flesh and bone pass out of our body every day with our excreta. (1950, 96)

The pattern—the "personal identity" of the individual—was maintained by metabolism, not determined by it. "In these discussions" of the determinants of form and pattern, said W. Ross Ashby, "questions of energy play almost no part—energy is taken for granted" (1956, 5). Ashby used intermediary metabolism as the departure point for cybernetic questions, explicitly contrasting the two:

> The older point of view saw, say, an ovum grow into a rabbit and asked "why does it do this? Why does it not just stay an ovum?" The attempts to answer this question led to the study of energetics and to the discovery of many reasons why the ovum should change—it can oxidise its fat, and fat provided free energy; it has phosphorylating enzymes, and can pass its metabolites around a Krebs cycle; and so on. In these studies the concept of energy was fundamental. Quite different, though equally valid, is the point of view of cybernetics. It takes for granted that the ovum has abundant free energy, and that it is so delicately poised metabolically as to be, in a sense, explosive. Growth of some form there will be; cybernetics asks "why should the changes be to the rabbit-form, and not to a dog-form, a fish-form, or even to a teratoma-form?" (1956, 5)

Even as cybernetics relegated matter and energy to a maintenance role, it accepted the logic of conversion carried forward from Bernard, via the science of intermediary metabolism and the concept of homeostasis. Information systems controlled energy systems to generate stable individual forms over time.

Metabolisms and Individuals Today

When we ask, today—"what is metabolism?"—the entity depicted by textbooks and dictionaries is the one built from these nineteenth- and twentieth-century sciences: the sum of the physical and chemical processes in the organism by which its material substance is produced, maintained, and destroyed, and by which energy is made available. Some definitions talk about cells, some don't, but a basic characteristic of the term is that it stands for a set, or a sum of life processes, some of them synthetic, some of them destructive, concerning the transformation of matter and energy in the process of sustaining life. Metabolism is not exactly a thing, but it is a set. None of the things in the set—the oxidation of this molecule or breakdown of that one—by themselves constitute metabolism, because it is the character of metabolism that individual reactions are interlinked in cyclical sets of reactions, one producing the conditions for the other, and it is only as a sum that metabolism happens.

In these definitions, metabolism is always thing-like, and also something that happens; it is a word that contains a particular temporal character. There is no metabolism that does not happen in time, that is not a process of one thing becoming another. This cumulus of transformative processes in time, these "interconnected sequences of mostly enzyme-catalyzed chemical reactions by which a cell, tissue, organ, etc. sustains energy production, and synthesizes and breaks down complex molecules," to quote the *Oxford English Dictionary*, is furthermore understood to describe the way living things interact with the world through food. The need for matter and energy in the outside world to be transformed internally into matter and energy in the body is met by metabolism. Without metabolism, there is not life. Without metabolism, the bounded body does not continue, but dissolves into the environment.

Metabolism has thus, historically, come to us as "thing" that sustains the bounded body through time. This metabolism is both historically and culturally particular. In making a map of all the chemical and energetic transformations undergone by food in the body, metabolism was thought to be solved. The intense focus on transformation, of the ability of this set of processes to convert matter into other forms, has left little room for considering the ways in which foodstuffs—that is to say, other organisms—might persist across bodies. Or, very practically speaking, it has left little room for thinking about the ways in which the specific nature of foods might matter, since they become interchangeable when viewed as a set of building blocks. When the variability of the environment is a problem that has been solved by its destruction and re-synthesis into more convenient form, the specific nature of food matters very little. If the logic of total conversion has been central to the theorization of individuation, individuality, and organisms' autonomy from the environments they depend on, then what consequences might realizations of less total conversions hold?

Interesting changes are afoot in the life and ecological sciences with profound implications for the modern ontologies of energy and information, metabolism and genetics, and eating and autonomy that I have just described. These changes are fracturing traditional demarcations between the converter and the converted, the eater and the eaten at many levels of investigation. For clarity, I will characterize these challenges as taking three different forms. First is the idea that genetic material, either DNA or RNA, comes to persist in the eater without losing its distinctive regulatory or genetic function. DNA or RNA that is ingested is not necessarily broken down into constitutive nucleotides or smaller molecular components, but maintains some integrity and is carried forward in the body and life of the eater over time. Second is the ingestion of other whole organisms that are not broken down but come into some symbiotic or commensal relation with the eater, an arrangement that comes to persist over time, even intergenerational time. Third is the notion that ingested food functions as a representation of the external world in the body of the eater, causing changes in bodily regulation and function *not* by being the substrate for energy production or growth, but by acting as a signal or cue to change or grow; in short, food *as* information.

First, eating DNA. The ability to rapidly sequence large amounts of genetic material, or metagenomics, has led to a heightened appreciation of the role of horizontal gene transfer between organisms and their food. Horizontal gene transfer refers to the transmission of genetic material between "distinct, reproductively isolated genomes" (Richards et al. 2011, 98). Unicellular eukaryotes that engulf what they eat have obtained genes from their prey (Loftus et al. 2005). Even fungi, which do their digesting outside of their cells by exuding enzymes into their surroundings and then ingesting the resulting stuff, seem to have acquired significant portions of their genomes— particularly clusters of genes involved in metabolism—through horizontal transfer from other fungal species, from bacteria, and from plants (Richards 2011; Slot and Rokas 2011). These studies begin to reveal just how much this kind of transmission of genetic material between species, rather than vertical inheritance and mutation between generations of the same species, is responsible for genome changes over time (Dupré 2012).

Eating RNA can also lead to an interaction between the genome of the eaten and the genome of the eater. Genetic material that is eaten may play a more transient regulatory role in the eating body even if it is not incorporated into the eater's genome. The model organism *Caenorhabditis elegans* is a nematode worm that feeds on bacteria such as *Escherichia coli*; while the bacteria themselves are "mechanically broken down before entering the gut and do not live in the animal," the ingested bacteria nonetheless secrete diffusible molecules into the worm's body that function as an "interspecies signal" that impacts the rate of the worm's aging via gene regulatory changes (Heintz and Mair 2014, 408). Humans and other animals are of course not fungi or nematodes. However, it is far from clear that everything that comes into the animal body is dissolved away and converted, either.

For example, human and bovine milk contains microRNAs, encased in lipid envelopes called exosomes, that affect gene regulation in the drinker's cells (Melnik et al. 2014). Although it is not entirely clear just how the RNA gets into which cells, experiments with cultured human cells in vitro show bovine microRNA-containing exosomes enter cells; and there is a great deal of trafficking of exosomes containing microRNAs *within* the cells in a body, which suggests that it can also happen between the cells of different bodies (Lässer et al. 2013). It appears that drinking raw milk (either human or cow) can prevent atopic diseases such as eczema in human babies, and that the microRNAs are playing a regulatory role in gene expression pathways in the developing thymus. MicroRNAs are very short strands of RNA with important, but still quite unclear roles in gene regulation; that milk is the vehicle for inter-individual transmission of gene-regulating genetic material has led some scientists to describe milk as an "exosomal microRNA transmitter" and an "endocrine signaling system" (Melnik et al. 2014). Milk, with a long history of scientific investigation as the quintessential food, has in addition to its nutritional or energetic content also become a medium of communication (Orland 2010).

Of course, the area of microRNA biology is extremely new and fraught with the technical detail of studying such tiny molecules; it changes from day to day at a rate that hardly can be captured by ponderous studies of the long history of metabolism such as this one. Nonetheless, for the purposes of this paper, these recent findings are of interest because they explicitly break with the classic model of metabolism elaborated above, and thus in many ways serve to further underline the assumptions on which that model was based: in particular, the assumption that living things that become food for others cease to exercise any agency in the conduct of life as they become completely subsumed to the agency of the autonomous eater (Wilson 2005).

A second instance of a breakdown in the logic of conversion in contemporary life science is in the study of ingested living matter that maintains a degree of cellular integrity after being eaten. The explosion of research around the microbiomes and mycobiomes of insects and other animals indicates that one role of ingested food is to constitute and shape the populations of commensal bacteria and fungi that live in and on other organisms, either by providing the microbes themselves, or providing things for the microbes to live on. For example, oligosaccharides in human breast milk are "minimally digested by the infant" but are nonetheless necessary to support its newly established microbiome, which it also gets in part from breast milk (Sela and Mills 2010). Microbes that live in guts of various kinds are not digested but instead actively participate in the nutritive process going on in their hosts, in a kind of metabolic community. The microbiomes of humans and other animals are shaped by what they eat and how they behave; equally, commensal microbes shape their hosts' metabolic function and their behavior (thus making the host/guest binary distinction a bit arbitrary). Fruit flies show mating preferences for other flies that eat the same diets; hyenas have distinctive clan-specific body odors determined by the bacteria resident in their scent glands; humans' different microbiomes are differentially attractive to mosquitoes; and laboratory mice can have their anxiety levels manipulated through the administration of probiotics (Ezenwa et al. 2012).

In light of these and other examples of metabolism that happens "collaboratively" across many organisms of many kinds, philosophers John Dupré and Maureen O'Malley argue that evolutionary forces act on reproductive-metabolic systems, not on "traditional" individual organisms. If microbes participate in speciation, mating patterns, social behavior, and disease transmission in wasps, fruit flies, hyenas, mice, and humans, while at the same time such organisms selectively enable and suppress microbes through their eating behaviors, then their metabolic and reproductive fates are entwined (Brucker and Bordenstein 2013).

> Standard discussions of characteristics of life . . . tend to prioritize one or other of two fundamental but very different features of living things: the capacity to form lineages by replication and the capacity to exist as metabolically self-sustaining wholes. We suggest that this tension can best be resolved by seeing life as something that arises only at the intersection of these two features: matter is living when lineages are involved—directly or indirectly—in metabolic processes. (Dupré and O'Malley 2009)

The very fact that metabolism can be seen as something that is not a foundational generator of boundaries between inside and outside, between one organism and another, demonstrates the historical shift from the metabolism of the twentieth century, whose philosophy we began with at the outset of this article. Shared metabolism does not seem to found individuation, but rather, to show differing degrees of intercalation of "animals in a bacterial world," or any other combination of things that live in such close proximity it is difficult to say where one begins and another ends: plants in a fungal world, fungi in an insect world, all of the above in a viral world (McFall-Ngai et al. 2013; Hughes et al. 2012). Inwardness is not a hallmark of the concept of metabolism today either, and with its diminishing, the scaffolding for the architecture of the autonomous, free-living, bounded individual fractures.

At the same time, Dupré and O'Malley employ the classic definition of metabolism generated in the history recounted above: "the transformative biochemical reactions that sustain life processes" (2009, 2). It is just spread out more, between more transformative agents. Their analysis follows the logic of twentieth-century uses of metabolism as an analytic tool for the investigation of ecology, in which a community of living things could be regarded collectively as an organism, and it should "therefore be possible to study the metabolism of that organism," as G. Evelyn Hutchinson put it in 1940 (quoted in Mitman 1992, 140). Hutchinson too defined metabolism as the processes of transfer of energy and matter from one form to another. This is a seemingly basic definition, and has been powerful for a century or more. Living things eat and metabolize in order to continue living—organisms eat (alone or commensally) in order to move, sleep, reproduce, and so on. Living things subsume other living things to them in order to do these things. It seems to me, however, that something more profound is happening than the reordering of metabolism as a boundary-denying communal event rather than a table for one.

A third way in which the logic of total conversion is being disrupted can be seen in the changing agency of the metabolite—a small molecule that is an intermediary or product of metabolism. Even if ingested organic matter does not maintain some kind of bodily or cellular integrity, but becomes disarticulated into metabolites, even for the metabolite a category of action is emerging that is not fuel, not building block, not indigestible poison, not drug. Witness, for example, this discussion of the relationship between metabolites and cell signaling, by cancer biologists Kathryn Wellen and Craig Thompson. They begin by recognizing the traditional hierarchy by which the organism/cell runs metabolism in order to sustain ongoing life: "Metabolic pathways and signal transduction pathways have often been viewed as separate entities—each crucial to the life and activity of the cell, but performing different functions—with signal transduction executing functions within the cell and metabolism maintaining cellular integrity and function." They note that "it is now well accepted that signal transduction regulates metabolism," such as the example of insulin binding to its receptor to stimulate the uptake of glucose by fat and muscle cells (Wellen and Thompson 2012, 270).

The opposite is less well accepted: "Reciprocally, signaling pathways have been shown to be regulated by metabolism through intracellular nutrient-sensing signaling molecules." As we saw with the treatment of metabolism by cybernetics, matters of energy and matter were considered to be downstream, processes executed by other kinds of information entities such as genes in the service of the maintenance and replication of form. This hierarchy is being questioned. In light of the increasing appreciation of epigenetic modes of control of differentiation and cell function, the production of metabolic intermediaries can become a mode of control in itself: "Modifications such as acetylation, methylation, glycosylation and phosphorylation are all generated from metabolites," protein modifications that in turn drive signal transduction and gene expression, "as well as cellular metabolism itself" (Wellen and Thompson 2012, 270). Cell proliferation and cell death can by modulated by protein acetylation, for example, which itself occurs in response to "changes in extracellular nutrient supply" (Johnson and Kornbluth 2012, 1).

At the level of the organism, changes in the food environment such as the amount and kind of food available to an organism are now being theorized as being transduced, through this upstream role of metabolites, as a signal that controls development and reproduction. Extensive study of the stem cell and cell progenitor systems in fruit flies, for example, shows that germ-line, neural, and blood stem cell progenitors all respond rapidly and markedly to nutritional signals such as amino acids (Shim et. al. 2013). It is important to note that these cells do not respond because of a direct lack of substrate or energy—they are not directly starved. Rather, they respond to food-derived signals such as the concentration of amino acids, and these signals are theorized to act as a representation of the external world, warning of imminent starvation or abundance and the need to shift physiology to live in a changing environment. Similarly, food ingestion can act as a temporal signal. Circadian rhythms, run by molecular clocks, shift in relation to the periodicity and frequency of food ingestion (Rutter et al. 2002); eating likewise is rhythmic in relation to sleeping and light-dark cycles (Sassone-Corsi 2013).

This is not an image of life extracted from the environment and subdued to the internal world of the eating body. Classic biochemistry, organized around the task of the "complete description of intermediary metabolism, that is to say, of the transformations undergone in matter by passing through organisms," seems stood on its head (Haldane 1937, 1). We might as well say that contemporary metabolic science is directed at the question of what happens to organisms as they pass through matter (Landecker 2013b). Thus, the character of metabolism today is not only a question of whether it is distributed through organisms that always exist as congeries or holobionts, but a rethinking, across all kinds of fields of life and biomedical science, from fungal biology to human oncology, from microbiology to circadian biology, of what metabolism is in the first place. This has generated a diverse set of new descriptions of ingested matter—whether it be genetic, cellular, molecular—as much more than fuel for the fire. Ceasing to be depends, radically, on one's perspective.

Conclusion

It is not generally the role of historians of science to comment on scientific papers published yesterday or even just last year. Historical excavation is often pursued in order to understand how present concepts and ways of doing things came about. Michel Foucault called this the "history of the present"—understanding how the contingencies and possibilities of the past shape modern individuality (Foucault 1977). The act of identifying these contingencies and "conditions of possibility" is a way of destabilizing the given nature of present concepts and practices, by showing that they could have been otherwise. However, today, "to destabilize our present does not seem such a radical move," as Nicolas Rose has observed in his proposal for a "modest cartography" of the present that "would not so much seek to destabilize the present by pointing to its contingency, but to destabilize the future by recognizing its openness" (2007, 5). This analysis has sought to destabilize neither the present nor the future, but to look back. The instability and rupture in the present moment—the glimpses of an "otherwise" around ideas of metabolism, conversion, and individuality—provide an opportunity to revisit and rethink our history of the past—to see the same history of biology that we have always had, but with renewed curiosity for the canonical stories of modern physiology and protoplasm, cell theory and biochemistry.

It has been my aim to excavate the ways in which the nineteenth-century logic of conversion provided important resources for ideas about individuality as a form of autonomy from the world, enabled by the physiological and chemical powers of the organism to turn the world into itself. This was a logic of fairly total conversion even as it supported a materialist argument for the unity of the world in physical substance: metabolism was the place where difference was literally taken apart or burned, where chemical continuity could be detected even as the excreted reaction products recombined the elements into new forms. I am able to see and articulate this logic of conversion because of my observations of a rupture in this logic underway today in a diverse array of life and biomedical sciences. Ideas of metabolism and individuality have been co-constitutive in the past; as their historical footings become unmoored and these categories undergo significant change in the present, their interrelation becomes all the more interesting. What will the philosophy of *this* metabolism be?

Notes to Chapter Nine

1. For a more extended analysis of the metabolism of Jonas' philosophy, see Landecker 2013a.

2. It is always a bit treacherous to wade into a historiography as deep as that which concerns Bernard. There are many things to say about this work. For F. L. Holmes, for example, the interest of Bernard was manifold: the face-off between physiology and chemistry in the nineteenth century, the watching of experimental practice unfold in notebooks and how that differed from the published works, the significance for all who followed of separating digestion from nutrition, and of adding fermentation to oxidation as part of what had to be considered in animal chemistry (Holmes 1974, 1992). For others, including philosopher Georges Canguilhem, Bernard is a figure central to the history and philosophy of vitalism, the idea of the animal as a machine, and to the history of the concept of regulation (1988). The concept of the *milieu intérieur* too has generated much historical writing, in part through historians' interest in the social metaphors of the division of labor used to explain the individual and differentiated work of cells to build and maintain the internal environment shared by all of them as a body; in part through the argument that this articulation of a internally regulated constancy lays the ground for twentieth-century ideas of homeostasis in the work of Walter Cannon, and negative feedback control in cybernetics (d'Hombres 2012).

3. See Landecker 2013a for a further reading of the role of glycogen as a mediating third entity used by Bernard to counter the binary of organism and environment.

References for Chapter Nine

Ashby, W. Ross. 1956. *An Introduction to Cybernetics.* New York: Science Editions.

Barcroft, J. 1934. *Features in the Architecture of Physiological Function.* New York: Macmillan.

Bernard, Claude. 1974. *Lectures on the Phenomena Common to Animals and Plants.* Translated by Hebbel E. Hoff, Roger Guillemin, and Lucienne Guillemin. Springfield, IL: Charles Thomas.

―――. 1987 [1848]. "The Formation of Sugar in the Animal Body." In *Intermediary Metabolism,* edited by Otto Hoffmann-Ostenhof, 32–42. New York: Van Nostrand Reinhold. Originally published as "De l'origine du sucre dans l'économie animale," in *Archives Générales de Médécine* 18 (1848): 303–19.

Brain, Robert. 2015. *The Pulse of Modernism: Physiological Aesthetics in Fin-de-Siècle Europe.* Seattle: University of Washington Press.

Brucker, Robert M., and Seth R. Bordenstein. 2013. "The Hologenomic Basis of Speciation: Gut Bacteria Cause Hybrid Lethality in the Genus *Nasonia.*" *Science* 341 (6146): 667–69.

Canguilhem, Georges. 1988. *Ideology and Rationality in the History of the Life Sciences.* Cambridge, MA: MIT Press.

Cooper, Steven J. 2008. "From Claude Bernard to Walter Cannon: Emergence of the Concept of Homeostasis." *Appetite* 51: 419–27.

d'Hombres, Emmanuel. 2012. "The 'Division of Physiological Labour': The Birth, Life and Death of a Concept." *Journal of the History of Biology* 45 (1): 3–31.

Dupré, John. 2012. *Processes of Life: Essays in the Philosophy of Biology.* Oxford: Oxford University Press.

Dupré, John, and Maureen A. O'Malley. 2009. "Varieties of Living Things: Life at the Intersection of Lineage and Metabolism." *Philosophy and Theory in Biology* 1:e003.

Ezenwa, Vanessa O., Nicole M. Gerardo, David W. Inouye, Monica Medina, and Joao B. Xavier. 2012. "Animal Behavior and the Microbiome." *Science* 338 (6104): 198–99.

Foster, Michael, and Arthur Sheridan-Lea. 1897. *A Textbook of Physiology, with an Appendix on the Chemical Basis of the Animal Body*. London: Macmillian.

Foucault, Michel. 1977. *Discipline and Punish: The Birth of the Prison*. Translated by Alan Sheridan. New York: Pantheon.

Fruton, Joseph. 1999. *Proteins, Enzymes, Genes: The Interplay of Chemistry and Biology*. New Haven: Yale University Press.

Geison, Gerald L. 1969. "The Protoplasmic Theory of Life and the Vitalist-Mechanist Debate." *Isis* 60 (3): 272–92.

Gilbert, Scott F., Jan Sapp, and Alfred I. Tauber. 2012. "A Symbiotic View of Life: We Have Never Been Individuals." *Quarterly Review of Biology* 87 (4): 325–41.

Goodman, David C. 1972. "Chemistry and the Two Organic Kingdoms of Nature in the Nineteenth Century." *Medical History* 16 (2): 113–30.

Haldane, John B. S. 1937. "The Biochemistry of the Individual." In *Perspectives in Biochemistry*, edited by Joseph Needham and David E. Green, 1–10. Cambridge: Cambridge University Press.

Heintz, Caroline, and William Mair. 2014. "You Are What You Host: Microbiome Modulation of the Aging Process." *Cell* 156 (3): 408–11.

Holmes, Frederic L. 1974. *Claude Bernard and Animal Chemistry*. Cambridge, MA: Harvard University Press.

———. 1987. "The Intake-Output Method of Quantification in Physiology." *Historical Studies in the Physical and Biological Sciences* 17 (2): 235–70.

———. 1992. *Between Biology and Medicine: The Formation of Intermediary Metabolism*. Berkeley: University of California Press.

Hughes, David, Jacques Brodeur, and Fréderic Thomas. 2012. *Host Manipulation by Parasites*. Oxford: Oxford University Press.

Huxley, Thomas H. 1869. "The Physical Basis of Life." *Fortnightly Review* 5: 129–45.

Johnson, Erika, and Sally Kornbluth. 2012. "Life, Death, and the Metabolically Controlled Protein Acetylome." *Current Opinion in Cell Biology* 24 (6): 876–80.

Jonas, Hans. 1966. *The Phenomenon of Life: Toward a Philosophical Biology*. New York: Harper & Row.

———. 1974. *Philosophical Essays: From Ancient Creed to Technological Man*. Englewood Cliffs, NJ: Prentice Hall.

Kamminga, Harmke. 1995. "Nutrition for the People, or the Fate of Jacob Moleschott's Contest for a Humanist Science." *Clio Medica* 32: 15–47.

Klein, Ursula. 2003. *Experiments, Models, Paper Tools: Cultures of Organic Chemistry in the Nineteenth Century*. Stanford: Stanford University Press.

Kohler, Robert E. 1973. "The Enzyme Theory and the Origin of Biochemistry." *Isis* 64 (2): 181–96.

Kossel, Albrecht. 1912. "The Chemical Constitution of the Cell." Harvey Lectures, Series 7, 33–51. Philadelphia: J. B. Lippincott.

Kuhn, Thomas. 1962. "Energy Conservation as an Example of Simultaneous Discovery." In *Critical Problems in the History of Science*, edited by Marshall Claggett, 321–83. Madison: University of Wisconsin Press.

Landecker, Hannah. 2013a. "The Metabolism of Philosophy, in Three Parts." In *Dialectic and*

Paradox: Configurations of the Third in Modernity, edited by Bernhardt Malkmus and Ian Cooper, 193–224. Bern: Peter Lang.

———. 2013b. "Post-industrial Metabolism: Fat Knowledge." *Public Culture* 25 (3): 495–522.

Lässer, Cecilia, Maria Eldh, and Jan Lötvall. 2013. "The Role of Exosomal Shuttle RNA (esRNA) in Cell-to-Cell Communication." In *Emerging Concepts of Tumor Exosome-Mediated Cell-Cell Communication*, edited by Huang-Ge Zhang, 33–45. New York: Springer.

Liebig, Justus. 1964 [1842]. *Animal Chemistry, or Organic Chemistry in Its Application to Physiology and Pathology*. Translated by William Gregory. New York: Johnson Reprint Corporation

Loftus, Brendan, Iain Anderson, Rob Davies, U. Cecilia M. Alsmark, John Samuelson, Paolo Amedeo, Paola Roncaglia, et al. 2005. "The Genome of the Protist Parasite *Entamoeba histolytica.*" *Nature* 433 (7028): 865–68.

McFall-Ngai, Maureen, Michael G. Hadfield, Thomas C. G. Bosch, Hannah V. Carey, Tomislav Domazet-Lošo, Angela E. Douglas, Nicole Dubilier, et al. 2013. "Animals in a Bacterial World, a New Imperative for the Life Sciences." *Proceedings of the National Academy of Sciences USA* 110 (9): 3229–36.

Melnik, Bodo C., Swen Malte John, and Gerd Schmitz. 2014. "Milk: An Exosomal MicroRNA Transmitter Promoting Thymic Regulatory T Cell Maturation Preventing the Development of Atopy?" *Journal of Translational Medicine* 12: 43.

Mendelsohn, Everett. 1964. *Heat and Life: The Development of the Theory of Animal Heat*. Cambridge, MA: Harvard University Press.

Mitman, Gregg. 1992. *The State of Nature: Ecology, Community, and American Social Thought, 1900–1950*. Chicago: University of Chicago Press.

Moulton, Forest Ray, ed. 1942. *Liebig and after Liebig: A Century of Progress in Agricultural Chemistry*. Washington, DC: American Association for the Advancement of Science.

Müller-Wille, Staffan, and Hans-Jörg Rheinberger. 2012. *A Cultural History of Heredity*. Chicago: Chicago University Press.

Needham, Joseph. 1943. *Time: The Refreshing River*. London: George Allen & Unwin.

Olmsted, James M. D., and E. Harris Olmsted. 1961. *Claude Bernard and the Experimental Method in Medicine*. New York: Collier.

Orland, Barbara. 2010. "The Invention of Nutrients: William Prout, Digestion and Alimentary Substances in the 1820s." *Food and History* 8 (1): 149–68.

Osborne, Thomas B., and Lafayette B. Mendel. 1912. "The Role of Gliadin in Nutrition." *Journal of Biological Chemistry* 12 (3): 473–510.

Parnes, Ohad. 2000. "The Envisioning of Cells." *Science in Context* 13 (1): 71–92.

Rabinbach, Anson. 1992. *The Human Motor: Energy, Fatigue, and the Origins of Modernity*. Berkeley: University of California Press.

Rheinberger, Hans-Jörg. 2013. "Heredity in the Twentieth Century: Some Epistemological Considerations." *Public Culture* 25 (3): 477–93.

Richards, Thomas A. 2011. "Genome Evolution: Horizontal Movements in the Fungi." *Current Biology* 21 (4): R166–68.

Richards, Thomas A., Guy Leonard, Darren M. Soanes, and Nicholas J. Talbot. 2011. "Gene Transfer into the Fungi." *Fungal Biology Reviews* 25: 98–110.

Rose, Nicolas. 2007. *The Politics of Life Itself*. Princeton: Princeton University Press.

Rutter, Jared, Martin Reick, and Steven L. McKnight. 2002. "Metabolism and the Control of Circadian Rhythms." *Annual Review of Biochemistry* 71 (1): 307–31.

Sassone-Corsi, Paolo. 2013. "When Metabolism and Epigenetics Converge." *Science* 339 (6116): 148–50.

Schoenheimer, Rudolf. 1942. *The Dynamic State of Body Constituents.* Cambridge, MA: Harvard University Press.

Sela, David A., and Mills, David A. 2010. "Nursing Our Microbiota: Molecular Linkages between Bifidobacteria and Milk Oligosaccharides." *Trends in Microbiology* 18 (7): 298–307.

Shim, Jiwon, Shubha Gururaja-Rao, and Utpal Banerjee. 2013. "Nutritional Regulation of Stem and Progenitor Cells in *Drosophila.*" *Development* 140: 4647–56.

Simmons, Dana. 2006. "Waste Not, Want Not: Excrement and Economy in Nineteenth-century France." *Representations* 96: 73–98.

———. 2015. *Vital Minimum: Need, Science and Politics in Modern France.* Chicago: University of Chicago Press.

Slot, J. C., and A. Rokas. 2011. "Horizontal Transfer of a Large and Highly Toxic Secondary Metabolic Gene Cluster between Fungi." *Current Biology* 21: 134–39.

Stirling, John H. 1872. *As Regards Protoplasm.* London: Longman's, Green & Co.

Wellen, Kathryn E., and Thompson, Craig B. 2012. "A Two-Way Street: Reciprocal Regulation of Metabolism and Signalling." *Nature Reviews Molecular Cell Biology* 13: 270–76.

Wiener, Norbert. 1950. *The Human Use of Human Beings: Cybernetics and Society.* New York: Houghton Mifflin.

Wilson, R. A. 2005. *Genes and the Agents of Life: The Individual in the Fragile Sciences.* Cambridge: Cambridge University Press.

Young, Frank G. 1957. "Claude Bernard and the Discovery of Glycogen." *British Medical Journal* 1 (5033): 1431–37.

10

Bodily Parts in the Structure-Function Dialectic

INGO BRIGANDT

Introduction

This essay discusses bodily parts and their epistemological role in biological theorizing. Many contributions to this volume discuss what constitutes an individual, also addressing the relation between unicellular individuals and colonies of cells (Herron, this volume) or between cells and a multicellular individual (Reynolds, this volume). But cells are not the only relevant parts of individuals (Love and Brigandt, this volume), and other kinds of bodily parts also persist as particular parts across developmental time and as homologues across evolutionary time, so that they have some integrity and biological identity. The various parts of an organism stand in part-whole relations, raising the theoretical issue of hierarchical organization, which can also be found in discussions of biological individuality (Nyhart and Lidgard, this volume; Rieppel, this volume). Moreover, epistemic issues about how to draw boundaries arise not only in the case of complicated cases of individuality (e.g., symbioses and evolutionary transitions in individuality), but also in the delineation of the parts that come to form and sustain the individual. Practice in many biological disciplines concerns the identification and individuation of bodily parts and how their relations and interactions explain complex organismal phenomena, where what qualifies as a meaningful part depends on the epistemic aims that biologists pursue in a particular context.

Throughout history, *structure and function* have been major organizing principles for the study of organisms and their organization, with implications for the individuation of bodily parts and the characterization of their nature. However, often structure and function have been seen as antagonistic (Russell 1982 [1916]; Gould 2002; Amundson 2005; Love 2013a).[1] In the first half of the nineteenth century, the Geoffroy-Cuvier debate in France was driven by the divergent viewpoints that structure determines function or that

function determines structure (Appel 1987; Hall 1998). Structural considerations were closer to those who like Geoffroy favored an individuation of
bodily parts in terms of homology; and in Great Britain Owen (2007 [1849])
terminologically contrasted homology with analogy and used homologies as
a phenomenon that the functionalist natural theologians of the Bridgewater
Treatises could not explain. While Darwin introduced with the idea of natural selection a convincing functionalist component to biological theorizing,
in the remainder of the nineteenth century many practitioners of the tradition of evolutionary morphology focused on homology and the phylogenetic
transformation of structures without regard to functional considerations
(Coleman 1967, 1971; Nyhart 1995). With the rise of the Modern Synthesis,
natural selection and function became central pillars of evolutionary theorizing, leading to morphology gaining relevance again in the form of functional
morphology (Wake 1982; Gans 1985). Yet the notion of developmental constraint (sometimes termed architectural constraint) introduced a structural
component often seen in opposition to natural selection (Brigandt 2015b).
And recent approaches such as phylogenetic systematics and evolutionary
developmental biology have reinvigorated the notion of homology, and made
the explanation of the evolutionary origin of novel structure—often regardless of considerations about adaptation—a focal concern.

My aim is to lay out a perspective on bodily parts that does not view
structure and function as antagonistic notions. The starting point is that
while there are different kinds of function (and different legitimate notions
of function), many of the uses of "function" in the generation of biological
knowledge can be encapsulated in the idea of a function as an activity taking place internal to an organism, so that such a function is a bodily part.
After pointing out that in some cases even structures are individuated using functional considerations, I argue that functions (as activities) are bodily
parts just like structures are. While functions are usually seen as a mere attributes of structures, from my perspective structures and functions are on a
par, so that function is an indispensable and fundamental ingredient of any
biological ontology.[2] I detail how this perspective clarifies some current debates and in several contexts permits fruitful biological investigation beyond
the structure-function dichotomy. One reason is that function in my sense is
compatible with but conceptually independent of natural selection. Another
reason is that a body composed of functions is at the same time composed of
structures, where the structures and functions can stand in hierarchical part-
whole relations among each other, enabling the multilevel explanation of
organismal organization and its developmental formation and evolutionary

modification. I conclude with a pluralistic approach to individuation, which assumes that different individuation schemes for bodily parts are to be used in different biological contexts, relative to the epistemic aims in operation (see also Lidgard and Nyhart, this volume; Love and Brigandt, this volume; Sterner 2015).

Individuating a Structure Can Take Us to Its Functional Context

Apart from traditional accounts of biological individuality that address physiological functioning and autonomy, nowadays evolutionary approaches are prominent that focus on individuals as units that are subject to selection and figure in transgenerational evolutionary change (Godfrey-Smith 2013; Love and Brigandt, this volume). Especially on the latter approach, an important characteristic of biological individuality is the ability to reproduce. A bodily part—a part of an organism—usually does not have this ability (with the exception of the cells of a multicellular individual, which retain a significant degree of individuality). Bodily parts may still be said to form cross-generational lineages, as in cases of homology the same kind of part reappears across generations (the bodily part is replicated), though the bodily part alone does not have the causal means of replicating itself.[3] But despite their lack of full biological individuality, it is of central epistemic importance to *individuate* bodily parts, by recognizing parts and deciding on their boundaries. In taxonomic and evolutionary contexts, comparison among different species requires the use of meaningful traits, and a properly identified homologue's transformation in evolution can be tracked and explained. In many different areas of biology, from molecular biology to physiology and functional anatomy, a whole is understood in terms of its parts, often by explaining the properties or functioning of the whole in terms of the organization and interaction among its parts, an issue that in the last decade has become the subject of extensive philosophical investigation under the label of "mechanistic explanation" (Bechtel and Richardson 1993; Bechtel and Abrahamsen 2005). Generating such biological understandings presupposes that a whole is first decomposed into meaningful parts (Craver 2007; Winther 2011). Rasmus Winther (2006) distinguished between compositional biology (which offers part-whole explanations) and formal biology (which explains using mathematical models) as distinct styles of biological reasoning. But even in contexts where formal-mathematical models are prevalent, finding the right units matters (as Winther also recognizes in his 2011 piece). Systems biologists, for instance, have to break down an overall system's complex functioning into components, and they have to decide which among the plethora of cellular entities and molecular pathways to include in their mathematical models (Brigandt 2013b).

Bodily parts can be individuated by different kinds of criteria, which can be used in combination. A part can be delineated by spatial boundaries, and in the case of developmental traits, temporal boundaries such as a particular developmental stage often bears on the trait's identity (see also Nyhart and Lidgard, this volume). Alternatively, the part's internal structure or internal interactions can be deemed to be more relevant than its boundaries. Molecular systems as studied by systems biologists have blurry boundaries at best (something similar holds for developmental processes), and entities are considered to be parts of a system not because they are within some visible spatial boundaries, but because they have such properties or engage in such interactions that these entities contribute to the system property of interest. In fact, many of the molecular entities within a cellular region do not contribute to the system function under consideration and for this reason are not considered parts of the system studied. Sometimes a bodily part is individuated as being an entity of a certain *type*—that is, belonging to the same kind as other such entities. For instance, a gene is not spatially disjoint but part of a continuous DNA molecule. A particular gene such as fushi tarazu may indeed be spatially delineated by a start and a stop codon (where the start codon is preceded by a promoter), but what makes such structural features significant is that they make fushi tarazu a gene, and are features shared by any other gene. Something similar applies to homologues as bodily parts, as in this context a spatial part of one particular individual is a meaningful biological part only if it is of the same type as parts of other organisms.

Biologists generally emphasize that organismal parts cannot be studied in isolation, as they interact with other parts in an organized fashion. Moreover, one part is often transformed by the impact of others.[4] But in addition to being *causally* affected by its context, I want to take this issue further as there are cases where the ontological *identity* conditions of a bodily part include biological factors outside of this part. Consider a stem cell, which has two basic characteristics. It can divide and self-renew for an extended period of time (by dividing such that at least one of the two daughter cells is a stem cell), and it can give rise to various differentiated cell types. When the latter happens and what type of differentiated cell is produced are influenced by the stem cell's niche—that is, its molecular and cellular microenvironment, which includes other cells, their surface structures, factors secreted by them, and the extracellular matrix. A stem cell maintaining its capacity to self-renew (and thus remaining a stem cell) is likewise influenced by its microenvironment (Li and Xie 2005; Moore and Lemischka 2006; Morrison and Spradling 2008). If so, a cell being a stem cell is partially determined by factors external to this cell. Even a gene as a specific DNA segment is a gene in virtue of its larger context. It is well-known that due to such processes as alternative splicing, the particular product resulting from a gene strongly depends on the cellular context (Griffiths and Stotz 2013), but this carries over to a given DNA segment qualifying as a gene at all. Being a gene is being able to produce a molecular product—a protein or at least an RNA—and a DNA segment may lose this ability in a different regulatory context or when it is part of a different genome that renders the DNA segment nonfunctional because of modified enhancers or changes to other genes that used to regulate it. A gene is not a cellular *process* (that would include many non-DNA entities and their activities), it just is a particular DNA segment as a *structure*. But its identity as a gene is still constituted by conditions and entities outside of this DNA segment. The relevance of such an ontological situation is that even when the focus is on a structural part alone, it is *epistemically* necessary to take the relevant outside factors into account.

Thus, ontological identity conditions can take us beyond a bodily part. Moreover, this context to be epistemically taken into account often includes *functional* aspects; some biological entities are defined in terms of having certain causal and functional capacities, which are context-dependent (Brigandt 2009, 2011b). The above example of individuating genes foreshadowed this. Consider pseudogenes, DNA sequences similar to genes, but no longer coding for proteins or other material products or processes in a cell. A pseudogene may have many structural hallmarks of a gene, but it is not a gene precisely because it is nonfunctional. A structure is a gene by virtue of a function—the ability to produce a molecular product. Whether a DNA segment is involved in the formation of gene products and what the products are depends on various transcriptional and post-transcriptional processes, all of which implicate molecular and cellular functions (Griffiths and Stotz 2013). From the perspective of functional genomics, Finta and Zaphiropoulos (2001) suggest that genes are "statistical peaks within a genome-wide pattern of expression of the genetic information" (160). On this approach, the extent to which a genomic region (a segment of DNA) qualifies as a gene depends on how the degree of its activity *compares to the activity* of other genes. In fact, a larger functional context—genome-wide expression due to a cell's activities—matters.

Another example is homologues as seen from the perspective of evolutionary developmental biologists, who are interested in the developmental basis of homology (Brigandt 2003). A homologue has a certain type of individuality as a morphological unit of evolutionary transformation, given that it is able to change across generations while still being the same character. The homologue is also quasi-autonomous from other homologues (Laubichler 2000) in the sense that across generations it can vary relatively independently from the organism's other bodily parts, making it a part that is meaningfully individuated from an evolutionary perspective (Wagner and Stadler 2003; Brigandt 2007; G. Wagner 2014). These variational abilities—in particular how one homologue can be modified independently of others—are due to the complex mechanisms that underlie the development of these homologues, which are addressed by evo-devo in line with its agenda of accounting for the developmental basis of morphological evolvability (homology is explicitly tied to evolvability in Brigandt 2007).[5] Thus, even if the focus is on an anatomical structure, this structure is a bodily part in the sense of being a homologue (capable of undergoing morphological change) in virtue of *function* features, to wit, the causal, functional, and dynamical aspects of an organism's development, which can generate morphological variation across generations. Contrary to the temptation to assume that a structure is defined by structural-spatial features alone, an organism's structural organization into different homologues and their boundaries are ontologically constituted by the organism's functional-developmental dynamics, which bolsters epistemic agendas that pay attention to organismal context and developmental processes.

While homology pertains to a bodily part's identity across evolutionary time, an issue germane to a part's development and its identity across developmental time is robustness. Robustness is the ability of an organismal system to produce or maintain a certain trait despite modifications internal and external to the system (Kitano 2004; Masel and Siegal 2009). Often robustness is an evolved ability where the organismal system actively responds to disturbances, so as during embryonic development to permit the reliable generation of the adult phenotype in the face of perturbations to development, and subsequently to maintain functioning and an adaptive phenotype in a changing environment. There is also robustness to genetic modifications. While this is most commonly known from experimental knockout studies, where the deactivation of a gene implicated in a molecular pathway surprisingly does not yield a changed phenotype, robustness to genetic change is also of evolutionary significance (discussed in Brigandt 2015a). Traits on various levels of organization can exhibit robustness, from the structure of RNAs and proteins, up to metabolic networks and organismal traits (A. Wagner 2005b). Different aspects of the early segment formation in *Drosophila* exhibit robustness (Umulis et al. 2008), including the formation of the expression patterns of gap genes (Manu et al. 2009) and segment polarity genes (von Dassow and Odell 2002). In the nematode *Caenorhabditis elegans*, the vulva of hermaphrodites develops such that the final pattern is reached in spite of cellular disturbances (Braendle and Félix 2008; Milloz et al. 2008). In such a complex anatomical structure as the tetrapod limb, muscles are reliably innervated and supplied with blood because in development a surplus of nerves and blood vessels grow from the body core toward their target, guided by chemical signals (with those nerves degenerating that do not find a target), so that the target muscle becomes innervated and connected to blood vessels even if this process is disturbed (Kirschner and Gerhart 2005).

Occasionally robustness can be due to structural redundancy (the situation where an additional copy of a structure is present so that the loss of one does not have any negative impact), but often robustness is a *distributed process* in that the overall system undergoes various functional changes to compensate for the loss of one component (A. Wagner 2005a and b; Ihmels et al. 2007). In the case of a gene regulatory network, the experimental deactivation of a gene may lead to compensatory changes in the activities of the other genes of the network. The above example of the development of a functioning limb is an instance of exploratory behavior, which is likewise a distributed process. As a result, while it is a particular structure (e.g., a spatial pattern of gene expression or an anatomical structure) that is robust to certain perturbations, its robustness lies in the underlying developmental and physiological processes, including how their functioning adapts to disturbances, implicating again a bodily part's functional context.

Functions as Bodily Parts

So far I have pointed out that even if a structure is at stake, in many cases its individuation as a structure makes reference to functions involving the structure's surrounding context. Now I want to make a more thought-provoking point about function by arguing that functions are bodily parts just like structures are. There are different legitimate notions of function, and the type of function relevant for my purposes is already implicitly used in different biological fields, but the task is to articulate this notion and to show how function on this construal is on par with structure in scientifically important respects. The latter involves first that structures and functions are bodily parts and stand in part-whole relations, so as to enable the study of organismal organization (discussed in this section). It also includes that structures as well as functions can be homologized across species and their change in the course of evolution can be studied (addressed in the next section).

The term "function" can mean different things in different biological contexts (Wouters 2003). Sometimes it refers to what a structure has been selected for—a structure is an evolutionary adaptation for a certain function. Another notion of function, which like the sense of function of concern to me does not invoke selection history, pertains to how a structure currently contributes to the bodily system of which it is a part or to the whole organism's survival and reproduction. Sometimes called a causal-role function, an example would be to say that the heart's function is the circulation of blood, in that given how it is situated within the circulatory system, the heart contributes to blood circulation. On this use of "function," a function is an attribute of a structure, yet the notion of function I am after is to make structures and functions bodily parts with equal standing. This is the notion of *function as activity*, by which I mean the activity performed by a structure or an organized collection of structures. The heart muscle's activity is its rhythmic contraction in a specific pattern across time; and the heart's activity may include blood being pushed out of it. An organismal part's activity is what takes place internally to this part (contracting) or includes its most proximal effects on adjacent parts (pushing blood into the aorta), but its activity-function does not involve how this part contributes to the surrounding organismal system, on which other notions of function focus. The heart contributes to the circulation of blood, but only because of how this part is related to other organismal parts; and a structure contributes to an organism's survival and fitness in a certain fashion, given the organism's relation to its environment. Causal-role functions (and other notions of function not of concern to me) capture such relations by addressing what a bodily part is *for*, but an activity-function focuses on what a bodily part *does* (Love 2007, 2013a).

A structure (e.g., the heart) as well as an activity-function (e.g., the contracting of the heart) is a physical part of an organism. Consisting in quite specific changes across time, an activity always takes place during a certain period of time, but given that I conceive of an organism as a living being across time, an activity is still a physical part of the organism. (And a structure such as a mature bone likewise exists during a longer period of time, even when not undergoing any particular change.) The fact that both structures and activity-functions are *bodily parts* (being present or taking place within an organism) is a crucial reason for why on my approach structure and function are on a par. A bodily part, regardless of whether it is a structure or an activity-function, is characterized by features *internal* to it, even though it does contribute to the surrounding system (which is captured by the notion of function as causal role). Indeed, although in standard examples causal-role functions are attributed to a structure, also an activity-function has causal-role functions—for example, the contracting of the heart contributes to the circulation of the blood. This reinforces the equal standing of structure and function as activity. Unlike a causal-role function, an activity-function is not a mere attribute of some bodily part, it is a bodily part.

Functions in my sense are studied in a variety of biological disciplines. (This is the same sense of "function" that was used in the previous section on the relevance of functional context.) Anatomical functions—as this term is often used—are functions as bodily activities. Examples are limb movement and mastication in tetrapods, both of which consist in the coordinated activities of skeletal elements, among other components. Functional anatomists study how such functions contribute to survival and have been shaped by natural selection, but important research questions pertain to a particular anatomical function as such (regardless of fitness contribution), in particular the analysis of its internal biomechanical operation. One obvious example concerns how articulated bones, given the contraction of attached muscles, result in the bones' coordinated movement so as to enable the mastication of food, which is the anatomical function investigated. Behavioral patterns are likewise activity-functions, as they involve the movement of some parts of the body or the whole body (Ereshefsky 2007).

There are of course kinds of biological activities other than relative movement and contraction. In neurobiology, while the structure of individual neurons and the connectivity structure of many neurons is investigated, so is neuron function (the electrophysiological activity at the membrane of an individual neuron) and neural network function (the electrophysiological activity within the network). At the molecular level, a gene's function is its transcriptional activity, which may include the cell types and spatial regions of the organism in which the gene is expressed. A developmental process is a complex activity-function in that a developmental process consists in the activities and interactions of several entities across time, though when speaking about "developmental process" one naturally focuses on the process's many ingredients rather than the overall activity taking place. A similar situation holds for "system" as used by systems biologists, as this term foregrounds the system's various active components rather than the resulting system functioning. Many such molecular-cellular activities have neither clear spatial nor clear temporal boundaries to other activities. But this underscores the need to pay attention to the particular epistemic considerations underlying a contextual individuation decision (and even in the case of structures I have pointed to different possible individuation criteria).

Overall, there is quite a variety of activities within an organism, but the notion of an activity-function highlights common themes across different domains and disciplines. One issue already illustrated by the above examples is that an activity-function is a part of an organism and thus is just as much a bodily part as a structure is. A living organism (extended through time) is composed of many activity-functions from gene functions to complex anatomical functions; however, this is not to deny that an organism is also composed of structures. Structure and function are not notions that are in competition, nor does the biological study of one exhaust the other.[6] In fact, there are important connections between structures and functions, to wit, *part-whole relations.* I have mentioned that structures and activity-functions contribute to the larger system (of which they are a part). This is more precisely contributing to the system's *activity,* which shows that this activity-function has as its components structures and lower-level activities (e.g., activities performed by these individual component structures). Just as one chimpanzee's hand as a concrete structure does not exist without its lower-level component structures (e.g., bones, muscles, nerves), so does this chimpanzee's grasping movement as a concrete activity not exist without its component structures performing characteristic activities—for example, the contraction of the opponens pollicis muscle—which, as a *component part* of the hand grasping activity, is an activity-function on a lower level (Fig. 10.1). Importantly, part-whole relations also obtain *across* structures and functions. A function such as a grasping-activity performed by the hand has not only other functions, but also structures (e.g., the opponens polis muscle) among its component parts. And the contraction of the opponens pollicis is an activity-function that is a physical part of the chimpanzee's hand, persist-

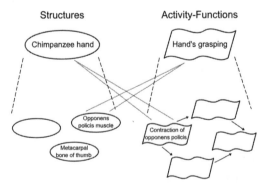

FIGURE 10.1. Part-whole relations among structures and activity-functions. Just as a structure consists of several lower-level structures (left side), so can a function be decomposed into lower-level functions (right side, the arrows indicate temporal-procedural relations in addition to the spatial relations of structures). Moreover, a structure (e.g., the opponens pollicis muscle) can be a component part of a function (e.g., a grasping activity of the hand), and an activity can be a physical part of a structure.

ing as a structure across time. Even though it can be epistemically legitimate to consider an organism's structure in abstraction from function, in reality a concrete structure (existing across time) performs activities or in any case is maintained by (component) activities.

While so far I have emphasized part-whole relations among structures and activity-function to underscore the parity of structure and function, in the next section part-whole relations will play a much more important role by showing what benefits my framework has for scientific theorizing. But beforehand, I have to address the related topic of hierarchical organization.[7] Organismal organization is a well-known principle; however, my focus on function introduces a special issue. Structures are hierarchically organized in that one structure can be a spatial part of another one. But a purely spatial containment cannot hold for functions, which are intrinsically temporal, and sometimes quite short-lived. Hierarchy and part-whole relations still obtain, provided that *procedural organization* is taken into account in biological theorizing. Procedural organization involves the temporal order and causal relations among different activities (Fig. 10.1, right side; see also Love 2007). In a developmental process there are various structures that exhibit specific activities (e.g., the binding of a transcription factor to a DNA segment), where the temporal organization of these activities—together with the spatial locations of structures being present and activities occurring at certain points of time—yields the overall operation of the developmental process. The binding of a certain transcription factor at specific genomic region as a particular step of the developmental process is one molecular function that is a lower-level part of the overall functional process. One component of such a complex anatomical function as the closing movement of the mammalian jaw is the contraction of the medial pterygoid, which itself is an activity-function, so that one function can indeed be a part of another function (even though it is not a purely spatial part, but includes containment among regions of time), yielding a hierarchical, spatial-procedural organization.

Fruitful Investigation beyond the Structure-Function Dichotomy

I now show how the notion of a function as an activity and a body having activity-functions (as well as structures) as its parts alleviates some longstanding tensions between structure and function in biology. These nowadays arise in different contexts related to evolutionary issues, often because some epistemic principles or explanatory agendas rightly distinguish between structure and natural selection, but turn this into an opposition of structure and any kind of function.

One issue is the important practice of homologizing bodily parts, which recognizes that such parts have some identity across evolutionary time. Homology has traditionally been distinguished from analogy—that is, structures that are similar because they have the same function. Even if nowadays homology is contrasted with homoplasy (similar traits not derived from common ancestry), it is still not to be conflated with analogy. Well before the advent of Darwin's evolutionary theory, the contrast between homology and shared function was clearly made, and it was structures but not functions that were seen as homologous across species (Brigandt 2011a). The historical origin of the homology concept was tied to disputes about whether form or function was prior (Russell 1982 [1916]; Appel 1987; Nyhart 1995). But one can fruitfully *homologize functions* in different species, provided that these are activity-functions and thus bodily parts. For in this case, the "function" pertains to the things that are homologized, but not to the reason why the relation of homology obtains (which still is common descent of bodily parts), so that there is no illicit conflation with analogy.[8]

Alan Love (2007) has made this point in the context of comparative developmental genetics, where talk about functional homology is quite common. He argues that the phrase "functional homology"—which sounds like a contradiction in terms—can be made coherent if it is understood as "homology of function." In such cases, gene activities are the traits homologized, solely because of their common descent and regardless of their fitness-contribution. Both a gene as a structure and a gene's expression activity as a function can be present within a taxon and thus be homologous. In case of an evolutionary transformation in gene activity, it may well be that homology of the gene obtains but not of the gene's function. For instance, while *distal-less* is a highly conserved gene, found in one copy in nearly all invertebrates and in multiple copies in vertebrates, its expression patterns differ substantially across these taxa, sometimes being expressed in different tissue types or in a structure not possessed in some taxa. Within echinoderms alone, Lowe et al. (2002) point to autapomorphic *distal-less* expression regions (where they associate this novel gene activity with the origin of a new anatomical structure or a change in the species' life history). Even when *distal-less* is expressed in similar bodily regions—consider vertebrate and arthropod appendages— this expression pattern did not exist in the appendage-less common ancestor, entailing non-homology of gene expression activity. While such examples have been used to argue that similarity of gene expression does not entail homology of the developed structures (Nielsen and Martinez 2003), the point I make here is that a particular gene (a structure) and the gene activity (an activity-function) are non-identical bodily parts and thus can operate as different evolutionary characters.

I have stated that a gene's function is its (changing) expression pattern across developmental time, in line with my above articulation of the notion of an activity-function, which includes the internal activity of a bodily part or at most its most proximal effect on adjacent parts. To be sure, some uses of "gene function" refer to which other genes are regulated by the gene, and thus more distal effects of the gene. But this can be captured by my approach as well, and highlights again the relevance of hierarchy and part-whole relations. Uses of "gene function" that include regulatory relations among different genes are best seen as the activity of a gene regulatory network, and thus as an activity-function on a higher level than the activity of a single gene (which is a *part* of and thus on a lower level than the regulatory network's activity).

Other domains where functions are homologized include functional anatomy. Anatomical functions are features that are conceived to be shared across taxa in a cladistic context (Amundson and Lauder 1994; Lauder 1994). As mentioned, anatomical functions are functions in the sense of activity-functions, and so can stand in relations of homology without running afoul of the homology-analogy distinction. The same holds for behavioral homology, which more precisely means homology of behavioral patterns (Ereshefsky 2007). Under the label of "process homology," a few have suggested that developmental processes can be homologized across species (Gilbert et al. 1996; Gilbert and Bolker 2001; Minelli 2003). While many shy away from this idea on the grounds that it conflates homology and function,[9] in my view it is a perfectly coherent notion provided one views a developmental process as a complex activity-function. In this case the focus is on the *internal* operation of a developmental process (and how the internal operation compares to other species), regardless of the fitness-impact and adaptive purpose of this developmental process.

Another domain that is fundamentally concerned with function is neuroscience and cognitive science. While comparative accounts in these areas have for the most part homologized the brain structures of different species, which is conceptually uncontroversial, there is an emerging literature on cognitive homology that addresses brain activities and cognitive activities (Platt and Spelke 2009; García 2010; Murphy 2012). Here again, clarity is served by determining what kind of biological traits are individuated (as activity-functions, cognitive activities belong to the same basic category as many other bodily functions), how they are related to other traits (e.g., to cognitive structures of the same organism), and why they can be meaningfully homologized. Moreover, the need to relate different cognitive functions of an individual raises the issue of *serial homology* in the context of functions. A single performance of a neuronal or cognitive activity is a concrete bodily activity, and several such performances within an individual are different particular bodily activities of the same type, which are thereby serially homologous in a trivial fashion. More interesting is the situation where one cognitive activity is a real modification of another activity within the same individual—for example, when an activity is redeployed for a quite different purpose, as in the common phenomenon of neural reuse (Anderson 2007, 2010). If one is a modification derived from the other they are serially homologous, in spite of differing somewhat in their internal operation and fulfilling substantially different cognitive tasks—the latter as the activity's impact on the larger organismal system is irrelevant to activity-function and homology. For instance, Jason Clark (2010a and b) argues that some higher cognitive emotions are serially homologous to basic emotions, and Taylor Murphy (2012) suggests that mentalization (our ascribing mental states to ourselves and others) is serially homologous to the activity of the default network (a network of brain regions active even when the brain is at wakeful rest).

Apart from merely homologizing bodily parts across species, their evolutionary transformation has to be tracked and explained. Here the presence of part-whole relations is important as they are the precondition for structures and functions to be able to evolve. A structure can reappear in different generations as the "same structure" while undergoing "structural change." If this sounds paradoxical, it can be clarified by distinguishing between characters and character states (Brigandt 2007; G. Wagner 2007, 2014). Bodily parts in different organisms that are homologous are the same character, so that a character can form an historical lineage, while the character undergoes evolutionary modification in the sense that its character state changes. In the case of a structure, this means that across generations some of its parts (component structures) or their spatial relations change. An activity-function can likewise undergo evolutionary modification because it has both structures and lower-level activity-functions as its parts, some of which can change or the spatial-procedural organization of which can be subject to modification.

Another reason for recognizing functions as an important type of bodily part is the origin of *evolutionary novelty*. An issue more general than transitions in evolutionary individuality, accounting for the evolution of novelty is one of the major challenges for contemporary evolutionary biology, with scientific efforts being devoted to many individual cases (Brigandt and Love 2010, 2012). But whereas this research often has a bias towards novelties that are structures, functions on different levels of organization can likewise be interesting evolutionary novelties in need of explanation. Rather than appealing to structural features (e.g., complex cell-cell junctions as seen in derived taxa), Sally Leys and Ana Riesgo (2012) consider an aggregate of cells a true epithelium if it has the *capacity* to seal and control the ionic composition of the internal milieu, so that an epithelium is considered a function feature. Based on this they argue that the epithelium is also present in sponges and thus a novelty of metazoans. On a higher level of morphological organization, the fin movement underlying locomotion in fish is a novel anatomical function of vertebrates.

There are at least two reasons why so many studies of evolutionary novelty focus on the origin of novel structures. One is pragmatic: it is easier to explain the development and evolutionary origination of a few structures (e.g., the skeletal elements of a fin; Hall 2007), compared to an anatomical function that consists in different types of structures interacting and changing in time so as to perform a complex activity (e.g., how muscles attach to skeletal elements of the fin, how they are supplied with blood, are innervated, and contract so as to create characteristic fin movements; Brigandt 2010). However, the difficulty of explaining the evolutionary origin of complex anatomical functions should not make us overlook their biological importance (Love 2003, 2006, 2013a). In fact, since individual structures usually do not originate in isolation but together with the evolutionary advent or modification of other structures as part of a structure-function-complex performing new bodily activities, developmental and evolutionary studies scrutinizing a structure in isolation may in fact offer a false account of its origination. Recognizing this entails the need for interdisciplinary research and integrative explanations that capture complex, multilevel characters (Love 2013a and b).

My account of part-whole relations and functions as activities is helpful in this context, as it clarifies that a complex bodily part is composed of both structures and functions as its parts (arranged in a spatial-procedural organization), which themselves have components parts on yet a lower level. There is notorious disagreement about how to define evolutionary novelty and which traits count as novelties, given the difficulty of distinguishing between a mere quantitative variant (which does not qualify as a novelty) and a qualitatively different trait (Brigandt and Love 2010, 2012; Palmer 2012). No matter how novel a structure may look, there are always some evolutionary precursors, at least on lower levels. For example, in the transition from fins in fish to limbs in tetrapods, the tetrapod autopodium (including the digits) as the most distal element has traditionally been deemed a novelty (Hall 2007). But more recent evidence has shown that fin rays and limb digits share gene expression patterns—a precursor on a lower level—which together with paleontological evidence has led some to conclude that the tetrapod autopodium and not even the digits are novel (Johanson et al. 2007; Boisvert et al. 2008; Hall and Kerney 2012). Yet from the perspective of my hierarchical framework, the important task is not to decide whether or not an evolutionary change counts as "novel," but to empirically investigate which of a derived trait's component parts on several levels (structures as well as functions) were already present in the ancestor and which were not. Subsequently, an evolutionary origin explanation can fruitfully be put forward by detailing how precursor structures and activity-functions on different levels change during developmental time, and how such developmental processes and their spatial-procedural organization were modified across evolutionary time so as to result in the novelty (Love 2006; Brigandt and Love 2012).

Tetrapod limbs and digits are structures, in line with the bias of studies of novelty toward structure, but even here (lower-level) functions are implicated in the overall explanation—for example, conserved and modified gene expression activities. A case where the very explanatory target—the evolutionary novelty—can be seen to be a function is neural crest cells. Neural crest cells have various features, including the capacity for long-range migration and for forming various cell types (so as to ultimately give rise to diverse cranial and other anatomical structures). Such capacities and in particular the performance thereof (i.e., migrating, differentiating) are not structures, but activity-functions, so that I would argue that to be a neural crest cell is best defined functionally. Neural crest cells have long been known to be a novelty of vertebrates, but also in this case ancestral precursors have been discovered, in the form of neural crest-like cells in tunicates (which most likely are the sister-group of vertebrates). Neural crest-like cells have some of the function features of neural crest cells, most notably the capacity to migrate, yet they lack other capacities—for example, for long-term migration and for forming ectomesenchyme derivatives (Jeffery 2007; Abitua et al. 2012). Here again, my approach is not to ponder the applicability of the label "novelty," but to track and explain evolutionary change in terms of the modification and acquisition of activity-functions and structures relative to the ancestral condition (as seen in neural crest-like cells), from higher levels (e.g., change in cell migration activity) to the molecular level (e.g., changes in gene expression activities).

A second reason for some approaches to evolutionary novelty to be centered on structure is that they stem from an agenda that focuses on the developmental generation of novelty while setting aside consideration about the fitness contribution of structures. Though research on so-called key innovations centers on how the fitness contribution and ecological impact of a new trait furthers subsequent adaptive radiation, more structuralist approaches focus on the very origination of the structure by means of development regardless of the influence of natural selection (Müller 1990; Müller and Newman 2005). The latter approach may even motivate researchers to endorse a definition of novelty that stipulates a novelty to be a new structure, thereby excluding new functions (Müller and Wagner 1991; see also Peterson and Müller 2013). However, developmental-morphological origination explanation vs. adaptation explanation (the kind of explanation sought) is an issue orthogonal to whether the focus is on bodily structures as opposed to bodily functions (the kind of bodily part subject to explanation). The two issues are (loosely speaking) "structure vs. function" pairs that should not and—using my perspective—cannot be conflated.

Even in the case of a novel *function* such as fin movement we need to ask how such a complex bodily part could have been brought about by changes in ancestral *developmental* mechanisms, or what changes in cellular-developmental mechanisms led to the evolution of the epithelium as the functional ability of tissues to seal. Thus, also those with a more structuralist outlook on evolution should recognize functions as genuine bodily parts that can be homologous across species or be evolutionary novelties (not restricting definitions of novelty to structure) and view novel functions as in need of explanation.

A Pluralistic Coda

I have addressed considerations about function, not to prioritize it over structure, but to stress the need to conceptualize both structure and function, by pointing to instances where considerations about structure implicate function as well, and by arguing that both structures and functions are genuine bodily parts. My account of structures and activity-functions as bodily parts has articulated their part-whole relations within an organism's spatial-procedural organization, so as to highlight multilevel investigation and explanation. While most of my discussion has centered on two generic types of bodily parts—structures and functions—this is not to say that bodily parts are individuated in a rather uniform fashion. Quite on the contrary, as mentioned earlier, there are many types of individuation considerations, which can be used in combination. A bodily part can be delineated by spatial or temporal boundaries. Alternatively, the part's internal structure can determine its identity, or some of its internal activity can be decisive. Some bodily parts are individuated in a relational fashion, where its intrinsic features alone are insufficient, while its impact on or its difference from other bodily parts is crucial (e.g., one cell ceasing to exist upon dividing into two cells). Sometimes what matters is that a particular bodily part belongs to the same class (is of the same type) as many other such parts, including parts of other organisms—for instance, a particular segment of DNA is considered as a relevant part because it is a gene. Given the variety of distinct individuation considerations, there are different kinds of bodily parts, which differ in the nature of their identity and integrity.

Overall, this yields a *pluralism*, which is ontologically due to the complexity of nature that cannot be captured by a single individuation scheme and epistemologically made salient by there being different legitimate scientific aims. Different classificatory or explanatory aims (pursued in different research contexts) often require different individuation considerations and representation schemes. Breaking a model organism's development into normal stages is fruitful for the purposes of developmental biology, yet explaining the evolution of development and morphology requires a different individuation, as normal stages obscure natural variation in development, phenotypic plasticity, and how developmental stages are created and transformed in evolution (Love 2009, 2010, 2013a). There are different ways of abstracting an organismal system into parts (called partitioning frames by Winther 2011), and such partitioning frames are often cross-cutting and have different advantages and disadvantages. Scientific representations are tools used for particular epistemic purposes. Taking such epistemic aspects as methodological, classificatory, and explanatory aims into consideration enables one to reflect on why a certain individuation scheme is used in a certain context, what makes it appropriate given its purpose, and what its limitations are relative to the intended purpose or relative to other possible epistemic aims (Brigandt 2013a; Love and Brigandt, this volume). Paying attention to epistemic aims matters beyond the individuation of bodily parts.[10] In the context of biological individuality, a pluralism based on the presence of multiple epistemic aims has recently been advocated by Beckett Sterner (2015) and by Lidgard and Nyhart (this volume), as well as Love and Brigandt (this volume).

In many instances, bodily parts have blurry boundaries. This is especially the case for activity-functions, including biological processes and molecular systems as studied by systems biology. It is notoriously difficult to break a larger system's activity down into functional components (Krohs and Callebaut 2007). While some approaches in systems biology attempt to understand the functioning of some smaller networks (such as network motifs; Alon 2007) in isolation from their connections to the larger system, other systems biologists argue that a more global investigation of larger systems is required to understand their actual functioning (Huang 2004). The phenomenon of distributed robustness mentioned above, where for instance a gene—ostensibly a functionally relevant part—is not really functionally important in that its deactivation does not change the activity of the overall gene regulatory network (because the activities of other genes in the network are adjusted), likewise shows that some systems do not have clearly delineated components (Brigandt 2013b). Biological mechanisms, despite the connotation of being analogous to a machine, do not ontologically have clear-cut spatial or temporal boundaries; instead, biologists epistemically choose the boundaries of the mechanism of interest and individuate its components, based on pragmatic considerations and idealizations (Bechtel 2015).

Something similar holds for *levels of organismal organization.* My discussion has emphasized part-whole relations and hierarchical organization. However, I do not think that there is a global set of ontological levels (e.g., molecular, developmental, anatomical, behavioral), where all the different entities on a level would share important properties. This is analogous to Linnaean ranks in taxonomy, where two taxa of the same nominal rank may differ dramatically in their phylogenetic age and their species diversity, so that some prefer the use of a rankless taxonomy. In the case of organismal organization, talk about levels is only *locally* meaningful and relative to prior *epistemic* choices. If one entity has been deemed to be a bodily part (based on epistemic considerations), then any entity that is a proper part of it is on a lower level, but this does not commit us to a set of global levels into which all characters of an organism or even all characters across taxa could be so arranged that for any two bodily parts one could unambiguously say whether they are on the same level or on different levels.[11] In the case of a complex system of molecular pathways, one may not be able to assign the various system components to an informative set of levels, especially given feedback and circular causal interactions among the components, so that it is a better strategy to represent the system as a network than as a nested hierarchy.

Recent philosophical discussions of mechanism have emphasized part-whole explanations, in that a mechanism is to be decomposed into components and the mechanism's operation is to be explained in terms of the organization and interaction of the components (Bechtel and Abrahamsen 2005; Bechtel 2010). At the same time, such accounts have shown a problematic bias towards molecular detail, and only recently has the widespread use of abstraction and idealization in the modeling of complex mechanism come to be philosophically addressed (Brigandt 2013b, 2015a; Levy and Bechtel 2013). We yet need a serious philosophical understanding of the various epistemic considerations involved in mathematical modeling as a representational and explanatory practice (O'Malley et al. 2014). Philosophers have also tended to focus on mechanisms' structural components and their qualitative interactions (e.g., binding, activating), while often neglecting quantitative and dynamic aspects. Many parts of a molecular system are *transient*, with biochemical reactions rapidly transforming one molecule into a different kind of molecule and breaking down entities into smaller molecular components. A system's characteristic functioning is not so much due to the stable presence of entities, but to short-lived molecules that are replaced by molecules of the same type (Baetu 2015). Beyond philosophers' focus on the *actual* organization and regular operation of a mechanism, my discussion has mentioned robustness as an important scientific issue, which as the way a system would respond upon perturbation pertains to the system's *modified* organization and operation (Brigandt 2013b, 2015a). Generally, many research questions in systems biology, developmental biology, and other fields studying organismal parts are about transformation and the emergence of novel features, so that while bodily parts have to be contextually individuated, we should not expect the represented parts, their boundaries, and their gathering into organismal subsystems to be stable and unchanging.

Acknowledgments

I thank the participants of the *What Is an Individual?* workshop at the University of Wisconsin–Madison (December 2012) for comments on a previous draft of this essay, and Lynn Nyhart and Scott Lidgard for their detailed written suggestions on several drafts. I am also indebted to Lynn Nyhart and Scott Lidgard for securing financial support to attend the project's authors meeting.

Notes to Chapter Ten

1. A similar dichotomy that has shown up in different controversies is microscopic structures vs. interacting molecules in solution (Gilbert 1987).

2. The philosophical account of molecular mechanisms by Machamer et al. (2000) uses a dual ontology of entities and activities. My account of structures and functions aligns with this, but extends it to higher levels of organismal organization and emphasizes hierarchical part-whole relations among structures and functions.

3. It is common to view a species as an individual, which has organisms as its parts (Love and Brigandt, this volume). A homologue as a lineage across organisms can similarly be considered an individual, which has the particular homologues of organisms as its parts (G. Wagner 2014), though this is not the only way to conceptualize this situation (Brigandt 2009).

4. Reynolds (this volume) discusses cells being transformed by their context and emergent interactions among cells that motivate the label "cell sociology." Rieppel (this volume) lays out Martin Heidenhain's hierarchical view of organismal organization, which emphasized downward causation, i.e., how higher-level traits organize lower-level traits. Herron (this volume) argues that the degree of evolutionary individuality can be subject to short-term changes due to modifications in population structure; and Sterner (this volume) points out that inheritance (as a precondition for individuality) can be due not only to material overlap across generations but also to scaffolding, which involves physical features outside of individuals.

5. While my present focus is on the relevance of function, this also shows that a bodily part as a homologue is individuated in a relational fashion. Whether a particular part can vary independently of others depends not just on the properties of this part, but on some of the other parts, and their relation, including the underlying developmental processes. This ties into discussions of modularity (Schlosser and Wagner 2004), which is often characterized by the causal interactions within a module being of a higher amount or stronger *than* the interactions between modules, which implicates the relation among modules.

6. Activity-functions are in fact processes, which raises the question of whether I endorse a process philosophy. Featuring processes as a category rather than the substances of more traditional metaphysics, so as to highlight becoming and occurring rather than what is, the process philosophy of Alfred North Whitehead also influenced Conrad Hal Waddington's (1975) biological thought. However, I do not endorse the view of some process philosophers that substance (or structure) is less fundamental than or even can be reduced to process (or function).

7. Though my focus is on bodily parts, hierarchy is also an issue above the level of organisms, where even extended hierarchies of biological individuals have been proposed and their evolutionary origin studied (Bouchard and Huneman 2013; Nyhart and Lidgard, this volume; Rieppel, this volume). In my context, while structures and function have structures and functions as their parts, which on yet a lower level are composed of structures and functions, there will be a point where physical parts cannot any longer be deemed to be *biological* structures or functions.

8. Owen defined a homologue as "the same organ in different animals under every variety of form and function" (1843, 379). Alan Love argues that the "function" in Owen's definition is to be understood as use-function, i.e., how it contributes to the surrounding system (causal role) or to the organism's fitness. This paves the way for a corresponding account for the homology of functions: "the same *activity-function* in different animals under every variety of form and *use-function*" (Love 2007, 679, my emphasis). Use-function is the kind of function germane to relations of analogy, so this underscores how homology of activity-function is independent from analogy (see also Love 2013a).

9. While Rudolf Raff once endorsed the notion of process homology (as a co-author of Gilbert et al. 1996), he does not favor this idea any longer, among other things because of the alleged homology-function incompatibility (pers. comm., May 2001).

10. While natural kinds are typically seen as an ontological issue, in Brigandt (2009) I argue that we should include epistemological considerations, such as the epistemic purposes for which natural kind concepts are used in science.

11. In the course of phylogeny the relation among characters on different levels and the very levels themselves can be subject to evolutionary modification, as witnessed by the fact that homologous anatomical structures can be due to non-homologous developmental processes or the activity of non-homologous genes (Brigandt 2007, 2011a).

References for Chapter Ten

Abitua, P. B., E. Wagner, I. A. Navarrete, and M. Levine. 2012. "Identification of a Rudimentary Neural Crest in a Non-vertebrate Chordate." *Nature* 492: 104–7.

Alon, U. 2007. *An Introduction to Systems Biology: Design Principles of Biological Circuits.* Boca Raton: Chapman & Hall / CRC Press.

Amundson, R. 2005. *The Changing Role of the Embryo in Evolutionary Thought: Roots of Evo-Devo.* Cambridge: Cambridge University Press.

Amundson, R., and G. V. Lauder. 1994. "Function without Purpose: The Uses of Causal Role Functions in Evolutionary Biology." *Biology & Philosophy* 9: 443–69.

Anderson, M. L. 2007. "Massive Redeployment, Exaptation, and the Functional Integration of Cognitive Operations." *Synthese* 159: 329–45.

———. 2010. "Neural Reuse: A Fundamental Organizational Principle of the Brain." *Behavioral and Brain Sciences* 33: 245–66.

Appel, T. A. 1987. *The Cuvier-Geoffroy Debate: French Biology in the Decades before Darwin.* Oxford: Oxford University Press.

Baetu, T. 2015. "From Mechanisms to Mathematical Models and Back to Mechanisms: Quantitative Mechanistic Explanations." In *Explanation in Biology: An Enquiry into the Diversity of Explanatory Patterns in the Life Sciences*, edited by P.-A. Braillard and C. Malaterre, 345–63. Dordrecht: Springer.

Bechtel, W. 2010. "The Downs and Ups of Mechanistic Research: Circadian Rhythm Research as an Exemplar." *Erkenntnis* 73: 313–28.

———. 2015. "Can Mechanistic Explanation Be Reconciled with Scale-Free Constitution and Dynamics?" *Studies in History and Philosophy of Biological and Biomedical Sciences* 53: 84–93.

Bechtel, W., and A. Abrahamsen. 2005. "Explanation: A Mechanist Alternative." *Studies in History and Philosophy of Biological and Biomedical Sciences* 36: 421–41.

Bechtel, W., and R. C. Richardson. 1993. *Discovering Complexity: Decomposition and Localization as Strategies in Scientific Research.* Princeton: Princeton University Press.

Boisvert, C. A., E. Mark-Kurik, and P. E. Ahlberg. 2008. "The Pectoral Fin of *Panderichthys* and the Origin of Digits." *Nature* 456: 636–38.

Bouchard, F., and P. Huneman, eds. 2013. *From Groups to Individuals: Evolution and Emerging Individuality.* Cambridge, MA: MIT Press.

Braendle, C., and M.-A. Félix. 2008. "Plasticity and Errors of a Robust Developmental System in Different Environments." *Developmental Cell* 15: 714–24.

Brigandt, I. 2003. "Homology in Comparative, Molecular, and Evolutionary Developmental Biology: The Radiation of a Concept." *Journal of Experimental Zoology Part B: Molecular and Developmental Evolution* 299B: 9–17.

———. 2007. "Typology Now: Homology and Developmental Constraints Explain Evolvability." *Biology & Philosophy* 22: 709–25.

———. 2009. "Natural Kinds in Evolution and Systematics: Metaphysical and Epistemological Considerations." *Acta Biotheoretica* 57: 77–97.

———. 2010. "Beyond Reduction and Pluralism: Toward an Epistemology of Explanatory Integration in Biology." *Erkenntnis* 73: 295–311.

———. 2011a. "Essay: Homology." *The Embryo Project Encyclopedia.* http://embryo.asu.edu/handle/10776/1754.

———. 2011b. "Philosophy of Biology." In *The Bloomsbury Companion to the Philosophy of Science*, edited by S. French and J. Saatsi, 246–67. London: Bloomsbury Academic.

———. 2013a. "Explanation in Biology: Reduction, Pluralism, and Explanatory Aims." *Science & Education* 22: 69–91.

———. 2013b. "Systems Biology and the Integration of Mechanistic Explanation and Mathematical Explanation." *Studies in History and Philosophy of Biological and Biomedical Sciences* 44: 477–92.

———. 2015a. "Evolutionary Developmental Biology and the Limits of Philosophical Accounts of Mechanistic Explanation." In *Explanation in Biology: An Enquiry into the Diversity of Explanatory Patterns in the Life Sciences*, edited by P.-A. Braillard and C. Malaterre, 135–73. Dordrecht: Springer.

———. 2015b. "From Developmental Constraint to Evolvability: How Concepts Figure in Explanation and Disciplinary Identity." In *Conceptual Change in Biology: Scientific and Philosophical Perspectives on Evolution and Development*, edited by A. C. Love, 305–25. Dordrecht: Springer.

Brigandt, I., and A. C. Love. 2010. "Evolutionary Novelty and the Evo-Devo Synthesis: Field Notes." *Evolutionary Biology* 37: 93–99.

———. 2012. "Conceptualizing Evolutionary Novelty: Moving Beyond Definitional Debates." *Journal of Experimental Zoology Part B: Molecular and Developmental Evolution* 318: 417–27.

Clark, J. A. 2010a. "Hubristic and Authentic Pride as Serial Homologues: The Same but Different." *Emotion Review* 2: 397–98.

———. 2010b. "Relations of Homology between Higher Cognitive Emotions and Basic Emotions." *Biology & Philosophy* 25: 75–94.

Coleman, W. 1967. *The Interpretation of Animal Form*. New York: Johnson Reprint Corp.

———. 1971. *Biology in the Nineteenth Century: Problems of Form, Function, and Transformation*. New York: Wiley.

Craver, C. F. 2007. *Explaining the Brain: Mechanisms and the Mosaic Unity of Neuroscience*. Oxford: Oxford University Press.

Ereshefsky, M. 2007. "Psychological Categories as Homologies: Lessons from Ethology." *Biology & Philosophy* 22: 659–74.

Finta, C., and P. G. Zaphiropoulos. 2001. "A Statistical View of Genome Transcription?" *Journal of Molecular Evolution* 53: 160–62.

Gans, C. 1985. "Vertebrate Morphology: Tale of a Phoenix." *American Zoologist* 25: 689–94.

García, C. L. 2010. "Functional Homology and Functional Variation in Evolutionary Cognitive Science." *Biological Theory* 5: 124–35.

Gilbert, S. F. 1987. "In Friendly Disagreement: Wilson, Morgan, and the Embryological Origins of the Gene Theory." *American Zoologist* 27: 797–806.

Gilbert, S. F., and J. A. Bolker. 2001. "Homologies of Process and Modular Elements of Embryonic Construction." *Journal of Experimental Zoology (Molecular and Developmental Evolution)* 291: 1–12.

Gilbert, S. F., J. M. Opitz, and R. A. Raff. 1996. "Resynthesizing Evolutionary and Developmental Biology." *Developmental Biology* 173: 357–72.

Godfrey-Smith, P. 2013. "Darwinian Individuals." In *From Groups to Individuals: Evolution and Emerging Individuality*, edited by F. Bouchard and P. Huneman, 17–36. Cambridge, MA: MIT Press.

Gould, S. J. 2002. *The Structure of Evolutionary Theory*. Cambridge, MA: Harvard University Press.

Griffiths, P. E., and K. Stotz. 2013. *Genetics and Philosophy: An Introduction*. Cambridge: Cambridge University Press.

Hall, B. K. 1998. *Evolutionary Developmental Biology*. 2nd ed. London: Chapman & Hall.

———, ed. 2007. *Fins into Limbs: Evolution, Development and Transformation*. Chicago: University of Chicago Press.

Hall, B. K., and R. Kerney. 2012. "Levels of Biological Organization and the Origin of Novelty." *Journal of Experimental Zoology Part B: Molecular and Developmental Evolution* 318: 428–37.

Huang, S. 2004. "Back to the Biology in Systems Biology: What Can We Learn from Biomolecular Networks?" *Briefings in Functional Genomics and Proteomics* 2: 279–97.

Ihmels, J., S. R. Collins, M. Schuldiner, N. J. Krogan, and J. S. Weissman. 2007. "Backup without Redundancy: Genetic Interactions Reveal the Cost of Duplicate Gene Loss." *Molecular Systems Biology* 3: 86.

Jeffery, W. R. 2007. "Chordate Ancestry of the Neural Crest: New Insights from Ascidians." *Seminars in Cell & Developmental Biology* 18: 481–91.

Johanson, Z., J. Joss, C. A. Boisvert, R. Ericsson, M. Sutija, and P. E. Ahlberg. 2007. "Fish Fingers: Digit Homologues in Sarcopterygian Fish Fins." *Journal of Experimental Zoology Part B: Molecular and Developmental Evolution* 308B: 757–68.

Kirschner, M. W., and J. C. Gerhart. 2005. *The Plausibility of Life: Resolving Darwin's Dilemma*. New Haven: Yale University Press.

Kitano, H. 2004. "Biological Robustness." *Nature Reviews Genetics* 5: 826–37.

Krohs, U., and W. Callebaut. 2007. "Data without Models Merging with Models without Data." In *Systems Biology: Philosophical Foundations*, edited by F. C. Boogerd, F. J. Bruggeman, J.-H. S. Hofmeyr, and H. V. Westerhoff, 181–213. Amsterdam: Elsevier.

Laubichler, M. 2000. "Homology in Development and the Development of the Homology Concept." *American Zoologist* 40: 777–88.

Lauder, G. V. 1994. "Homology, Form, and Function." In *Homology: The Hierarchical Basis of Comparative Biology*, edited by B. K. Hall, 151–96. San Diego: Academic Press.

Levy, A., and W. Bechtel. 2013. "Abstraction and the Organization of Mechanisms." *Philosophy of Science* 80: 241–61.

Leys, S. P., and A. Riesgo. 2012. "Epithelia, an Evolutionary Novelty of Metazoans." *Journal of Experimental Zoology Part B: Molecular and Developmental Evolution* 318: 438–47.

Li, L., and T. Xie. 2005. "Stem Cell Niche: Structure and Function." *Annual Review of Cell and Developmental Biology* 21: 605–31.

Love, A. C. 2003. "Evolutionary Morphology, Innovation, and the Synthesis of Evolutionary and Developmental Biology." *Biology & Philosophy* 18: 309–45.

———. 2006. "Evolutionary Morphology and Evo-Devo: Hierarchy and Novelty." *Theory in Biosciences* 124: 317–33.

———. 2007. "Functional Homology and Homology of Function: Biological Concepts and Philosophical Consequences." *Biology & Philosophy* 22: 691–708.

———. 2009. "Typology Reconfigured: From the Metaphysics of Essentialism to the Epistemology of Representation." *Acta Biotheoretica* 57: 51–57.

———. 2010. "Idealization in Evolutionary Developmental Investigation: A Tension between Phenotypic Plasticity and Normal Stages." *Philosophical Transactions of the Royal Society of London B: Biological Sciences* 365: 679–90.

———. 2013a. "Interdisciplinary Lessons for the Teaching of Biology from the Practice of Evo-Devo." *Science & Education* 22: 255–78.

———. 2013b. "Teaching Evolutionary Developmental Biology: Concepts, Problems, and Controversy." In *Philosophical Issues in Biology Education*, edited by K. Kampourakis, 323–41. Dordrecht: Springer.

Lowe, C. J., L. Issel-Tarver, and G. A. Wray. 2002. "Gene Expression and Larval Evolution: Changing Roles of Distal-Less and Orthodenticle in Echinoderm Larvae." *Evolution & Development* 4: 111–23.

Machamer, P., L. Darden, and C. F. Craver. 2000. "Thinking about Mechanisms." *Philosophy of Science* 67: 1–25.

Manu, S. S., A. V. Spirov, V. V. Gursky, H. Janssens, A.-R. Kim, O. Radulescu, et al. 2009. "Canalization of Gene Expression in the *Drosophila* Blastoderm by Gap Gene Cross Regulation." *PLOS Biology* 7:e1000049.

Masel, J., and M. L. Siegal. 2009. "Robustness: Mechanisms and Consequences." *Trends in Genetics* 25: 395–403.

Milloz, J., F. Duveau, I. Nuez, and M.-A. Félix. 2008. "Intraspecific Evolution of the Intercellular Signaling Network Underlying a Robust Developmental System." *Genes & Development* 22: 3064–75.

Minelli, A. 2003. *The Development of Animal Form: Ontogeny, Morphology, and Evolution.* Cambridge: Cambridge University Press.

Moore, K. A., and I. R. Lemischka. 2006. "Stem Cells and Their Niches." *Science* 311: 1880–85.

Morrison, S. J., and A. C. Spradling. 2008. "Stem Cells and Niches: Mechanisms That Promote Stem Cell Maintenance throughout Life." *Cell* 132: 598–611.

Müller, G. B. 1990. "Developmental Mechanisms at the Origin of Morphological Novelty: A Side-Effect Hypothesis." In *Evolutionary Innovations*, edited by M. H. Nitecki, 99–130. Chicago: University of Chicago Press.

Müller, G. B., and S. A. Newman. 2005. "The Innovation Triad: An EvoDevo Agenda." *Journal of Experimental Zoology Part B: Molecular and Developmental Evolution* 304B: 487–503.

Müller, G. B., and G. P. Wagner. 1991. "Novelty in Evolution: Restructuring the Concept." *Annual Review of Ecology and Systematics* 22: 229–56.

Murphy, T. S. 2012. *Cognitive Homology: Psychological Kinds as Biological Kinds in an Evolutionary Developmental Cognitive Science.* MA thesis, University of Alberta.

Nielsen, C., and P. Martinez. 2003. "Patterns of Gene Expression: Homology or Homocracy?" *Development Genes and Evolution* 213: 149–54.

Nyhart, L. K. 1995. *Biology Takes Form: Animal Morphology and the German Universities, 1800–1900.* Chicago: University of Chicago Press.

O'Malley, M. A., I. Brigandt, A. C. Love, J. W. Crawford, J. A. Gilbert, R. Knight, S. D. Mitchell, and F. Rohwer. 2014. "Multilevel Research Strategies and Biological Systems." *Philosophy of Science* 81: 811–28.

Owen, R. 1843. *Lectures on the Comparative Anatomy and Physiology of the Invertebrate Animals, Delivered at the Royal College of Surgeons in 1843.* London: Longman, Brown, Green, and Longmans.

———. 2007 [1849]. *On the Nature of Limbs: A Discourse.* Edited by R. Amundson. With a preface by B. K. Hall and introductory essays by R. Amundson, K. Padian, M. P. Winsor, and J. Coggon. Chicago: University of Chicago Press.

Palmer, A. R. 2012. "Developmental Plasticity and the Origin of Novel Forms: Unveiling Cryptic Genetic Variation Via 'Use and Disuse.'" *Journal of Experimental Zoology Part B: Molecular and Developmental Evolution* 318: 466–79.

Peterson, T., and G. B. Müller. 2013. "What Is Evolutionary Novelty? Process Versus Character Based Definitions." *Journal of Experimental Zoology Part B: Molecular and Developmental Evolution* 320: 345–50.

Platt, M. L., and E. S. Spelke. 2009. "What Can Developmental and Comparative Cognitive

Neuroscience Tell Us about the Adult Human Brain?" *Current Opinion in Neurobiology* 19: 1–5.

Russell, E. S. 1982 [1916]. *Form and Function: A Contribution to the History of Animal Morphology*. With a new introduction by G. V. Lauder. Chicago: University of Chicago Press.

Schlosser, G., and G. P. Wagner, eds. 2004. *Modularity in Development and Evolution*. Chicago: University of Chicago Press.

Sterner, B. 2015. "Pathways to Pluralism about Biological Individuality." *Biology & Philosophy* 30: 609–28.

Umulis, D., M. B. O'Connor, and H. G. Othmer. 2008. "Robustness of Embryonic Spatial Patterning in *Drosophila melanogaster*." In *Multiscale Modeling of Developmental Systems*, edited by S. Schnell, P. K. Maini, S. A. Newman, and T. J. Newman, 65–111. New York: Academic.

von Dassow, G., and G. M. Odell. 2002. "Design and Constraints of the *Drosophila* Segment Polarity Module: Robust Spatial Patterning Emerges from Intertwined Cell State Switches." *Journal of Experimental Zoology (Molecular and Developmental Evolution)* 294: 179–215.

Waddington, C. H. 1975. *The Evolution of an Evolutionist*. Ithaca: Cornell University Press.

Wagner, A. 2005a. "Distributed Robustness versus Redundancy as Causes of Mutational Robustness." *BioEssays* 27: 176–88.

———. 2005b. *Robustness and Evolvability in Living Systems*. Princeton: Princeton University Press.

Wagner, G. P. 2007. "The Developmental Genetics of Homology." *Nature Review Genetics* 8: 473–79.

———. 2014. *Homology, Genes, and Evolutionary Innovation*. Princeton: Princeton University Press.

Wagner, G. P., and P. F. Stadler. 2003. "Quasi-Independence, Homology and the Unity of Type: A Topological Theory of Characters." *Journal of Theoretical Biology* 220: 505–27.

Wake, D. B. 1982. "Functional and Evolutionary Morphology." *Perspectives in Biology and Medicine* 25: 603–20.

Winther, R. G. 2006. "Parts and Theories in Compositional Biology." *Biology & Philosophy* 21: 471–99.

———. 2011. "Part-Whole Science." *Synthese* 178: 397–427.

Wouters, A. 2003. "Four Notions of Biological Function." *Studies in History and Philosophy of Biological and Biomedical Sciences* 34: 633–68.

Commentaries: Historical, Biological, and Philosophical Perspectives

11

Distrust That Particular Intuition:
Resilient Essentialisms and Empirical Challenges
in the History of Biological Individuality

JAMES ELWICK

Each living creature must be looked at as a microcosm—a little universe, formed of a host of self-propagating organisms, inconceivably minute and as numerous as the stars in heaven.

DARWIN 1868, vol. 2, 404

Stop focusing on biological individuals as entities and start seeing biological individuals as systems or as relationships! Such points have been insisted upon since the 1930s or 1940s, by Ludwig von Bertalanffy or Georges Canguilhem (Gayon 1998, 308), or even the 1870s by Herbert Spencer (Gissis, this volume). All well and good, and many of the papers here are written in support. Indeed Scott Lidgard and Lynn Nyhart have declared that one of the purposes of this volume is to expand the range of possible biological individuality concepts that we can entertain (Lidgard and Nyhart, this volume). But the history of biology shows that the habit of seeing biological individuals as tidy and bounded entities is a difficult one to break. Despite repeated empirical demonstrations that biological individuals often violate our preconceptions, people keep returning to those old definitions. I suggest this is because we intuitively individuate by imaginary essences.

"Essences" and "essentialism" are fraught terms in biology, especially in discussions about species and implications for natural kinds (Winsor 2006; Wilson et al. 2007; Rieppel 2010). They are used here reluctantly because only "essence" seems to convey the ontological intuitions I want to describe, and because only "essentialism" can capture the epistemological intuitions. The truly uncomfortable reader might use "intuitive pattern" in their place. Such difficulties show how words can come to fail us when we try to articulate ourselves about biological individuality.

I suggest one intuitively individuates by imagining three kinds of essences. The first is a spatial/anatomical/formal essence that takes the guise of a particular *shape*, which when physically divided ends the unity of the biological individual. The second is a physiological/energetic essence that behaves as a particular optimal *state*, such as health, and anything detracting from that state can be seen as "foreign" to the biological individual. The third is a developmental/genetic essence, a specific *pattern* persisting over time, which when altered fundamentally changes the individual. For brevity I will refer to these individuating intuitions respectively as *anatomical essentialism, physiological essentialism*, and *developmental essentialism*.

Anatomical, physiological, and developmental essentialism have been repeatedly challenged empirically as new discoveries have undermined our intuitions about biological individuality. Hence anatomical essentialism was challenged in the early 1740s by the discovery that the freshwater polyp *Hydra viridis* could regenerate after sectioning. Physiological essentialism has recently been challenged by findings that human physiological functions are assisted by various commensal and mutualistic microorganisms, not simply hindered by parasitic or pathological microbes. Developmental essentialism was challenged by the discovery in the early 1840s that certain jellyfish and parasites alternated between two distinct morphological forms during development. Yet these essentialisms live on. Why?

One reason is because essentialism is used to imagine something that can undergo change—any living organism, then—yet still maintain an identity or develop hidden potential. Inferring the existence of an underlying essence—especially the importance of "inside" rather than "outside"—is reported to emerge in children as young as four. Its early appearance in shaping how one sees the world may explain why it so difficult to dislodge (Gelman and Hirschfeld 1999, 413, 417). Consider the popular contemporary intuition that each individual's genome is unchanging and unique—a form of developmental essentialism. We know this to be false, not only because genomes do change over time, but because of rare but major exceptions such as genetic chimerism, in which a single person develops out of a fusion of two separate cell lineages that otherwise would have been fraternal twins. Sometimes such people end up questioning whether this chimerism renders their selves multiple (Martin 2007, 221). Even renowned philosophers and bioethicists are guided by essentialism. For instance, Jürgen Habermas's views on the advisability of different forms of medical genetic engineering—between good "repair" and bad "enhancement"—pivot upon developmental essentialism (Habermas 2003), although he should know better (see critique in Barnes and Dupré 2008, 232). Meanwhile governments and courts since the 1980s have used the assumption that individual people have unchanging genomic identities to make rulings. (Martin 2007, 206–7, 212).

Yet these three individuating intuitions also live on because they are often useful. Essentialism is cognitively economical. It works to more easily construe the world by assuming an underlying essence shared by a class of entities, meaning that only exceptions or marginal cases have to be explained away, or when the cause behind some change in a given entity is not yet explained (Gelman and Hirschfeld 1999, 434–38). Individuating by essence is not always incorrect, and nicely describes the world to a certain extent. For instance, physiological essentialism may be a reason why the germ theory strikes such a chord with the general public: it nicely chimes with our intuitions about contamination. The same point might be raised for the hypothesis of the "genetic bottleneck" as an explanation for unicellularity, to be discussed below: it appeals to an intuition of developmental essentialism. Any model or image or metaphor must oversimplify if it is to coordinate action between different people—how else could people communicate with one another if not by abstraction?

A third reason for the resilience of these essentialisms is because they fit with the cherished Western belief that each one of us is a unique being possessing agency and autonomy. This is partly fostered by beliefs in individual rights and responsibilities (Martin 2007, 222). But it also seems to emerge from a tendency most recently pointed out by Barry Barnes and John Dupré—the Durkheimian insight that rather than see agency and autonomy emerging out of a set of relationships that a person has with other people, these properties tend to be situated *within* the individual herself (Barnes and Dupré 2008).[1] I expand upon this point in more detail below.

My task as historical commentator is to put these papers into some sort of larger historical pattern. What follows has two sides. On the one hand the story is something classically and unfashionably progressive. It is a tale of a "Copernican shift": a series of discoveries that gradually undermined our easy intuitions about biological individuality, making individuality relativistic, dependent upon the researcher's perspective. Compare modern-day informatic perspectives on individuality with ancient depictions of a "soul" to see just how much change has occurred. The story of progress is not simply a tale of a move towards a greater array of explanations and perspectives: towards pluralism. It is also a shift towards a less-intuitively appealing world—if such a move can be called a progressive one.

Yet on the other hand there seems to be something more fragile. There is a constant danger of a person reverting back to intuitive conceptions of individuality, even when that person should know better. This danger of reversion is partly because the three essentialisms are easy to think with, and partly because audiences sometimes seem to listen only to points that appeal to intuitive individual essences. If there is an implicit tragedy here, it is that appeals to individual essences will always find large audiences, even if those appeals are oversimplifications. By way of an analogy, consider the many histories of science showing how science is a social process (simply because science happens to be done by groups of people) . . . consistently overshadowed in popular culture by biographies about lone scientific geniuses.[2] The problem is not that there are many brilliant people in science, as there clearly are—it is that this kind of science history is an impoverished view of how science works. The same holds true for thinking about biological individuality: our intuitions are insufficient for a complete understanding of it.

Intuitions of Biological Individuality: Three Essentialisms

There may be cognitive reasons for individuating by essence. Observing cross-cultural similarities, the anthropologist Scott Atran argued that intuitively perceiving living kinds may belong to our ability as humans to "fast-map" the world, an ability we share with other metazoa. When classifying animals, people in widely different cultures usually focus on vertebrates, placing them in different major groups; most invertebrates fall into residual ones by default. Carolus Linnaeus's class *Vermes* and the "puchi" of the Hill Pandaram culture of South India were both residual categories containing everything from insects to crustaceans. Such organisms tend to be lumped together because they lack "phenomenal resolution," usually being small and tending to hide or live in alien environments (Atran 1990, 65, 32–33). Atran's point about classification matters because vertebrates do not tend to challenge intuitions about biological individuality. Members of the puchi and *Vermes* do—but usually they are harder to notice.

Atran also points out that humans tend to furnish the standard against which animals are compared and classified (Atran 1990, 32–33). One reason why this is relevant to biological individuality is because we associate our own individuality with autonomy. From the very beginning of its life each organism capable of awareness is the center of its own solipsistic universe. Every thought, every feeling, every action taken occurs at the very center of one's sensorium. In humans, each one of us is utterly accustomed to being at this center. We also associate individuality with agency: we believe we can change many things in our sensorium. Indeed the issue of command and control ought to make biological individuality a key topic in the neurosciences as well, and the neurosciences a key resource for thinking about biological individuality. Is an organism controlled top-down, by a nervous center that firmly controls the rest of the body, or does its apparent unity of action result from the bottom up, emerging out of the harmonious interaction of its various body parts? Nineteenth-century neuroscientists differed wildly over the answers (Elwick 2003, 39; Elwick 2007).[3]

Emerging from our taken-for-granted sense of agency and autonomy are intuitions about three kinds of biological essences that constitute individuality. I will use seemingly "common-sense" examples to convey this intuitionism. They have a strange dual quality of obviousness and falsity. On the surface these examples verge on being platitudes; with further reflection or the mustering of empirical cases, however, they are open to analytical and empirical challenge.

One intuition is that a biological individual is spatially or anatomically indivisible. If at a cabin in the woods I lose my hand, then the hand is no longer a part of me as an individual. It is no longer physically connected to my body; and I certainly have no further control over that hand. Such an intuition can be called *anatomical essentialism*.

Another intuition is that a biological individual constitutes a persistent yet unique set of functions that yield energy. Its body possesses functions such as metabolism which in combination lead to health. Something interfering with this set of functions can be identified as not belonging to the biological individual. If I am unfortunate enough to get a tapeworm, then I have become a host to a parasite. Such an intuition can be called *physiological essentialism*.

A third intuition is that a biological individual maintains a persistent and unique temporal pattern over its lifetime. This can be anything from an Aristotelian formal cause such as a soul, to genealogical thoughts about "bloodlines," to some modern neo-Darwinian thoughts about genes. My genome is necessary for my particular phenotypic traits to be expressed. If some sort of change is engineered in my underlying genome, then I am changed in fundamental ways as an individual. Such an intuition can be called *developmental essentialism*.

Several associated points follow from these intuitions about essences. For one thing, it's difficult to disentangle these three intuitive individual essences from each other. A developmental essence must somehow be linked with an anatomical essence since it dictates what form that organism will take, or the space it will take up. This is another reason why these individuating intuitions, while shallow, are hard to abandon: the three essences seem to be mutually reinforcing.

Furthermore, each of the essences seems necessary for there to be an individual. Remove one and there seems to be no biological individual. Thus if a body is somehow divided into two, intuitively these entities no longer seem to be a single individual. As a corollary, and to develop a point made by philosopher Ellen Clarke, the more of each of the three essences displayed by an entity, the easier that entity is to individuate. Thus a puppy is easier to see as an individual than a Portuguese man o'war or a grove of aspens (Clarke 2010, 322–23). These essences also seem to affect beliefs about higher and lower organisms: the more of the three essences displayed, the "higher" an organism seems to be.

In a related vein, *purity* seems to be important for an intuition about an essence, either by maintaining a boundary separating the essence from the outside, or through some kind of endogenous power this essence holds. Individuation is marked by degree of purity of essence. Individuation-by-purity also facilitates the reception of certain explanations. To repeat, non-life scientists see as reasonable the explanation that the microscopic entities known as "germs" cause diseases by "invading" or "contaminating" a body's purity. One reason for this may be because germs are seen as contaminating by invisibly changing one essence into something else, as suggested by Susan A. Gelman and Lawrence A. Hirschfeld (1999, 427–28).

However, our thinking about biological individuality has taken a Copernican route as we learned that our intuitions might not be entirely correct. Various challenges have forced us to consider alternatives to these essentialisms. What about plant cuttings? Does a genetic chimera get to vote twice? What does the discovery of hundreds of different bacterial species living inside each person—the "holobiont" (Gilbert et al. 2012; Gilbert, this volume) imply for the germ theory of disease?

Notice how each case depends on empirical examples. Indeed the entire point of the paper by Scott Gilbert, Jan Sapp, and Alfred Tauber is to provide various instances that undermine simple intuitions about biological individuality. To paraphrase them—indeed, they are themselves quoting Lynn Margulis—whether convenient to our beliefs or not, one has to note these examples because they're *there* (Dawkins and Margulis 2009; Gilbert et al. 2012, 336).

But such empirical refutations are made possible only by successive discoveries. It was simply not possible to discuss genomic chimerism in 1911, for instance. This is why the history of biology matters to thinking about biological individuality—because it furnishes examples that undermine our intuitions. Often these examples qualified as discoveries *because* they undermined our intuitions. Yet such an assertion is not necessarily progressivist if these discoveries keep getting forgotten—even by historians of biology. It is to these discoveries that we now turn.

Anatomical Essentialism

Anatomical essentialism, to repeat, is an intuition that a biological individual has a spatially or anatomically indivisible essence. This builds on what Olivier Rieppel discusses at the very beginning of his careful chapter. Indeed Cicero introduced the term *individuum* into Latin to describe not a living being, but instead Democritus's indivisible atom (Lecourt 1998, 218), and presumably it is from Latin that the current meaning of "individual" has derived. Accompanying that belief is that this anatomical essence is protected or bounded by some sort of envelope: the presence of other bodies within that spatial envelope may undermine biological individuality. Indivisibility will be addressed first, then purity.

Empirically, the intuition that an anatomical essence consists of its "indivisibility" is easily undermined. For one thing it's zoocentric: plants such as willow trees can be easily propagated through cuttings. Even in animals it has been known at least since Aristotle that certain animals can live after sectioning. Thus in his treatise *Measure of the Soul,* written around 388 CE, St. Augustine of Hippo observed how after being cut into three, each of the separated parts of a millipede seemed to act as individuals (Trembley 1744, 299).

In retrospect this makes Abraham Trembley's discovery of the regenerative abilities of *Hydra viridis* seem unsurprising, though it became a scientific sensation in its day. In 1740 he gathered *Hydra* polyps from ditches in Sorgvliet, near The Hague, and put them in glass vessels. He then cut them into sections, and observed each section regenerate back into a complete *Hydra.* Trembley's research was partly oriented to one simmering dispute at the time—whether polyps, known as "zoophytes," were more properly described as animals (because they moved, somewhat like inchworms) or as plants (because they could form new *Hydra* from cuttings) (Lenhoff et al. 1986, 180–82).

Because *Hydra's* apparent animality made it closer to humans, discussions about its individuality centered upon anatomical and developmental essences. Trembley's correspondent R. A. F. Réaumur wondered what happened to the polyp's "soul" when it was cut into pieces (Roger 1997, 122); one provocative answer, from J. O. de la Mettrie, was that its soul was material (Hankins 1985, 133; Spary 1996, 181–84). Why did the separated pieces become themselves new and complete *Hydra*? Such questions led to disputes over whether the form of an individual preexisted in ova or sperm (preformationism), or whether it gradually emerged from less-organized material (epigenesis) (Roe 1981, 12). All seem to have agreed that there was some sort of essence necessary for the *Hydra* to be considered an individual: thus John Turberville

Needham and the Comte de Buffon argued that all organisms consisted of "organic molecules" circulating in and out of us, but only when attached to an interior mould dictating the organism's form (Roger 1997, 128–29).

The belief that the *Hydra* was also plant-like reveals the long concern of botanists with biological individuality. Plants not only reproduce from cuttings; they also seem to repeat their parts, with leaves, sepals, and petals of flowers all being somewhat alike. Although often described as beginning with Goethe and the Romantic biologists, this claim of part-repetition was made as long ago as Theophrastus in 300 BCE (Nyhart and Lidgard 2011, 380–81). This long concern with individuality helps us situate the cell theory of Matthias Schleiden and Theodor Schwann when they proclaimed that larger organisms were composed of repetitive "elementary particles" called cells, which also acted as "elementary individuals" (Schwann 1847, 167). Relevant here is the word "elementary," which seems to be derived from contemporary chemistry, where Antoine Lavoisier defined an element as any chemical that could not be further subdivided. Thus a cell could not be further subdivided without losing its anatomical and physiological essence. It was only in later instantiations of the cell theory that cells changed from anatomically indivisible units that emerged in a medium outside or within each cell ("exogeny" and "endogeny" respectively) into units that reproduced by division (Baker 1948; Mendelsohn 2003, 16). If depicted in terms of essences one might depict such a shift in cell theory as a move from anatomical essentialism to developmental essentialism.

The botanist Schleiden, following predecessors like Leibniz and Goethe, articulated a relativistic approach towards biological individuality: individuals could exist at different physical scales.

> Now the individual is no conception, but the mere subjective comprehension of an actual object, presented to us under some given specific conception, and on this latter it alone depends whether the object is or is not an individual. Under the specific conception of the solar system, ours is an individual: in relation to the specific conception of a planetary body, it is an aggregate of many individuals. (Schleiden 1849, 127)

By "subjective comprehension" Schleiden seems to have meant the observer's frame of reference—even an entity as large as the solar system could be seen as an individual, if the background was large enough. What counted as an individual depended upon one's perspective.

Alongside such relativistic views about individuality was the notion of *orders* of individuality: a *matryoshka*-like picture of individuals made out of smaller individuals made out of still smaller individuals. Thus it was not simply cells or indivisible epidermis-covered entities that could be seen as individuals; if seen from a certain perspective, a cluster of them could be seen as an individual too. Schleiden's view influenced two very important evolutionists who also thought a great deal about individuality, Ernst Haeckel and Herbert Spencer.

Haeckel was Schleiden's student and in his 1866 *Generelle Morphologie* Haeckel discussed orders of individuality: what he called "tectology." He proposed six, from the simplest "first order" (such as cells), to the most complex "sixth order" (colonial organisms like a Portuguese man o'war). Higher order individuals were built up of simpler order ones (Haeckel 1866, 269–331, 332–63; Richards 2008). Such a conception was relativistic: an individual in one context became an organ in another. In Nyhart's words, for Haeckel "'Individual' and 'organ' were not absolute concepts but relative ones" (Nyhart 1995, 136). At the same time as Haeckel, Herbert Spencer wrote about biological "compositions" of first, second, third, and fourth order, in which more complex aggregations were built up from simpler units. We know that Spencer was also influenced by Schleiden because Spencer—not fond of acknowledging his intellectual debts—actually used the Schleiden quote above in his *Principles of Biology*[4] (Spencer 1864–67, II: 4–5, I: 202–3). Indeed, in 1868 Haeckel wrote to Spencer and noted they shared the "same ideas," calling particular attention to his own sections in the *Generelle Morphologie* on individuality.[5]

Ultimately it is unimportant to determine who first enunciated the principle of levels of organization and orders of individuality. What is important is that by the late nineteenth century there had appeared a number of different systems of levels of organization. In 1911 the entomologist William Morton Wheeler not only pointed out Spencer's and Haeckel's systems, but also drew attention to August Weismann's. He also noted physicist-philosopher Gustav Fechner's very expansive system, one depicting the entire universe as a single individual organism composed of smaller ones (Wheeler 1911, 308–9).[6] Ultimately such systems made it easier to depict individuals relative to a particular background or environment. More recently the very notion of discrete levels, in which lower levels are necessary for higher ones, has come into question (Lidgard and Nyhart, this volume). Yet orders of individuality still made it possible to see biological individuality relativistically.

Developmental Essentialism

Developmental essentialism is the intuition that a biological individual consists of an essence remaining stable over time. In 1819 Adelbert von Chamisso asserted that many marine polyps took on one form, then a completely different form, then returned to the first form; it was also suspected that certain parasites such as trematode worms also did this. In 1842 the Danish naturalist J. J. Steenstrup confirmed the suspicion, and claimed that this "alternation of generations" was a common pattern in plants and animals (Steenstrup 1845). Nyhart and Lidgard (this volume) discuss how over the next decade the discovery excited many European naturalists, forcing them to think and re-think how they designated relationships between parts and whole.

One reason why the claim of alternation of generations was controversial and exciting was because it challenged the intuition that an individual should be spatially indivisible: anatomical essentialism. How to solve this? One famous response was made by T. H. Huxley, who simply redefined the phenomenon using developmental essentialism. Huxley presented his solution as a way to cut through overly complex details. The individual could be defined as everything emerging from a sexually fertilized ovum. From "phytoid" in botany Huxley created the word "zoöid," using it to denote any entity emerging from an act of sexual fertilization that seemed to simulate an individual (Elwick 2007, 133).

Moving ahead forty years to 1892, not only can Weismann's proposed "germ plasm" be depicted as a persistent and pure developmental essence; its favorable reception owed much to its appeal to popular intuitions about that essence. Weismann proposed a complete separation between the "germ plasm" of the chromosomes and the rest of the body ("soma"): it was the germ plasm that determined the form that the body was to take. The standard account is that in so doing Weismann undermined contemporary "Lamarckian" claims to soft inheritance (Mayr 1982, 699–700). But Weismannism can also be seen as holding a commitment to a temporally persistent essence of biological individuality. That is, the germ plasm was able to persist over time by remaining upstream from any somatic changes: it was powerful because it was kept separate and thus kept pure. Some interesting contemporary challenges—the most famous being Spencer's question of why worker ant larvae could develop into a queen in her absence—were dismissed not only with the "crucial experiment" of five generations of mice tails forcibly removed (Weismann et al. 1891), but also by appealing to an intuitively pure developmental essence. Yet upon reflection, of course the germ changes over time, over generations: this is the only way Darwinian evolution can occur. The germ line cannot be completely stable—Weismann's intuition is useful, but only to a point.

It is telling that challenges to developmental essences are characterized as "contamination." A fairly recent example of this can be found in the case of

Wolbachia. The latest version of developmental essentialism is that each individual, and each species, possesses a distinct and unchanging genome. *Wolbachia* is an intracellular bacterium most famous for its prevalence in arthropods, where it affects their development. But *Wolbachia* is also mutualistic in nematodes: its removal, by antibiotic, often kills or stunts the development of its wormy host. The implication is that *Wolbachia*'s presence is actually necessary for nematodes to develop properly. Yet such mutualism is difficult to recognize if one is a developmental essentialist: in 1999 researchers studying nematode genomes complained of "bacterial sequences contaminating their filarial isolates," apparently ignorant of *Wolbachia*'s presence (Kozek and Rao 2007, 11). That notion of contamination—implying a pure genomic essence—blinded the researchers to the fact that the *Wolbachia*-nematode relationship is a case of a "multi-genome ecosystem" (Bettan 2013).

Besides "contamination," another way to describe violations of pure developmental essence is to describe the result as mongrelism or chimerism. Consider human genomic "chimeras" who possess two genetically distinct cell lines. Known specifically as "tetragametic chimerism," this phenomenon occurs when two fertilized eggs fuse. A 2003 report in *New Scientist* describes one such chimera—a person known only as "Karen"—as "a mixture of two different people." The story continues to use the language of developmentalist essentialism when claiming that "far from being pure-bred individuals composed of a single genetic cell line, our bodies are cellular mongrels" containing cells from our mothers as well as grandparents and siblings. The reporter obviously sees the subject as an individual, but it is significant that she uses the word "mongrelism" to explain the phenomenon to a general audience (Ainsworth 2003; Barnes and Dupré 2008). The sociologist of biomedicine Aryn Martin notes the ways in which popular culture defended individual personhood against threats such as Karen: from a Discovery Network documentary entitled "I Am My Own Twin" to "Bloodlines," the 2004 season finale of *Crime Scene Investigation* (CSI). In that CSI episode the chimeric villain is not initially charged with sexual assault because his semen and blood do not seem to genetically match, though eventually he confesses to the crime (Martin 2007, 215).

That said, although the effects of genomic chimerism are sometimes barely noticeable,[7] they can often be quite harmful. The possible impairment caused by chimerism has been used to explain just why so many organisms in their life cycles go through a unicellular phase (a single pair of gametes), despite the inherent vulnerability of such an adaptation. The unicellular phase has been described as a "reproductive bottleneck" that ensures that only a single genome is present; it is a beneficial adaptation because the benefits of passing through a single-cell stage outweigh the costs (Grosberg and Strathmann 1998). Intuitively such an explanation seems reasonable because we automatically recognize the function of the reproductive bottleneck: as a filter to purify the developmental essence.

One problem with such intuitions is that they may lead us to prematurely close off alternatives—to overlook alternative forms of development that do not involve differentiation from a single cell. This would seem to be a point of Beckett Sterner's paper—that most work on the benefits and costs of multicellularity assume cases in which multicellularity emerges from the division of a single cell. But this is to ignore alternative states of how multicellularity arises, such as by aggregation of multiple separate cells from the same organism or even from different species, as with biofilms (Sterner, this volume). Consider how slime molds develop: rather than differentiate from a single fertilized cell, *Dictyostelium discoideium* goes through a life cycle that starts as a mass of spores, each of which divides into an amoeba; these amoebae then divide into free-swimming independent daughter cells, which then aggregate toward a central point to form a migrating and seemingly unitary "slug," which then turns into a stalk that forms a mass of spores (Sunderland 2011, 510–11). Are there other forms of development that are being overlooked because of our intuitions?

Physiological Essentialism

Physiological essentialism is the intuition that a biological individual often consists of a set of functions producing a unique yet persistent pattern of energy transformation. Metabolism can be seen in this way: converting the different physiological essence of another organism into one's own essence. As Claude Bernard said, the dog "does not get fat on mutton fat, it makes dog fat" (Bernard in Landecker, this volume). Hannah Landecker points out that not only was homeostasis the action of maintaining an internal system that was autonomous from the outer environment; the very activity of metabolism was a form of "transubstantiation" from one substance into another. Indeed it is telling that where Huxley saw in protoplasm the unity of all living substance from cells to microbes to plants and humans, making it interchangeable, a critic saw infinite difference and uniqueness (Landecker, this volume)—infinite different essences, perhaps.

What happens when other physiological essences within our bodies become rivals, ingesting the food we need to maintain our own energy and health? Michael Osborne's paper discusses a historical moment in which the meaning of parasitism changed from something relatively innocuous—

organisms eating together as "messmates"—to something pathological. I suggest that parasitism became pathological when scrutinized by the light of physiological essentialism: the purity of the host's energy pattern (as homeostasis, health, or simple well-being) contaminated by outsiders. In this sense health became defined as freedom from interference by outsiders—fitting our preference for autonomy. We ought instead to prefer definitions of health such as Canguilhem's—the "margin of tolerance with respect to the environment's betrayals" (Canguilhem 1989, 199, 156; Gayon 1998, 313; Lecourt 1998, 221), but such a conception is more complex than a simple intuition of physiological purity.

Interestingly, physiological essentialism seems to mesh well with the definition of individuality as the aligned "fitness interests" of different bodily components. Body parts or functions that prosper at the expense of other components are depicted as "cheating"—hence cancer cells are highly fit as cell lineages, but their spread ends up harming or destroying the body in which they live (Folse and Roughgarden 2010, 448). This seems to be one point of Matthew Herron's chapter—how organisms such as the volvocine algae manage to prevent such "within-organism conflicts." At the same time, however, Herron eschews physiological essentialism by pointing out that there are intermediate degrees of individuality also (Herron, this volume).

To discuss pathologies caused by internal disharmony also brings us to developments such as the germ theory. It would seem that the germ theory is not only instrumentally successful at preventing disease—it is also easy to think with. As noted above, its immense success is helped by capitalizing on our intuitions about our own internal purity and the threat of invasion or contamination. Yet intuitions about purity can only take us so far: the germ theory oversimplifies the role of foreign microbes, depicting them all as pathological. What about "good bacteria"? Gilbert, Sapp, and Tauber (2012) note that certain diseases are actually prevented by the presence of some of our symbiont microbes, sometimes going even so far as to suppress harmful immune responses that lead to conditions like inflammatory bowel disease.[8]

Indeed current research on the microbiome challenges physiological essentialism. It is generally agreed that there are about 10 times the number of microbial (protists, bacteria, fungi) cells relative to "human" cells. While the numbers differ considerably, this ratio stays relatively stable. Thus one 2010 paper in *Nature* estimated 100 trillion to 10 trillion. More importantly, this paper estimated that among these microbial cells, there were at least 160 different bacterial species in each individual human body surveyed, and sometimes far more (Qin et al. 2010, 64). Many of these bacteria fit the popular definition of being "good," since they contribute in different ways to the health of their human host, such as converting (fermenting) sugars to fatty acids and contributing vitamins. Gilbert, Sapp, and Tauber (2012) see this paper as marking a larger shift in how we perceive not only human health but also biological individuality. The existence of a microbiome—a microbial ecosystem—inside the human body would seem to strengthen the claim of their paper's title that "we have never been individuals."

But what Gilbert, Sapp, and Tauber do not note is a reaction appearing almost exactly a year later, also in *Nature*. Another group of scientists claimed that despite this enormous variety of bacteria, there could be found only three stable types across humanity, regardless of national origin. These stable groups the authors call "enterotypes," and suggest these types imply a finite set of well-balanced symbiotic states between microbes and host (Arumugam et al. 2011). The philosopher John Huss is skeptical of this typology, and describes it as part of an emerging dispute over whether the "metagenome" is a genomic unit or a system (Huss 2013). In the context of this paper, the attempt at "enterotyping" might be seen as a way to set out unique yet persistent patterns of energy transformation: a claim there are ultimately only several physiological essences. The concern is not whether this particular typology is useful, or even whether it oversimplifies (which it must do in order to coordinate action between people)—it is that people come to think that there are no alternatives.

Conclusion: Interfaces, Not Individuals

How does one avoid essentialism when it comes to biological individuality? One route is to change perspective, boldly challenging the assumption that perspectives on structure and perspectives on function are antagonistic (Brigandt, this volume). Another strategy is to focus on relationships among biological individuals, whatever those individuals happen to be. To be sure, all the chapters in this volume—and biologists for that matter—do discuss relationships of some sort. Indeed it is probably not possible to be a pure essentialist in biology. Even the most staid descriptive anatomist must discuss how one body part is linked with others, for instance.

Olivier Rieppel's discussion of the enkaptic hierarchies of Martin Heidenhain (Rieppel, this volume) not only talks about nested hierarchical sets and the entities within each set, but also the links between these entities—what is variously called their integration, or the part-whole relation, or the new properties emerging when a set of parts at one level are integrated enough to compose a new whole. Andrew Reynolds focuses not on cells by themselves, or as parts making up greater wholes, but characterizes them as "gregarious social organisms in constant communication with one another by chemical and physical signals" (Reynolds, this volume). Snait Gissis insists that what characterized Herbert Spencer's thought was a focus on the interface between an "individual" at whatever level of organization and its environment: this interface was a "progressing equilibration" (Gissis, this volume). Evolution was a dialectical change resulting from that interaction. "Everywhere the differentiation of outside from inside comes first," he said, such as the simplest cell wall distinguishing the cell contents from their environment (Spencer 1864–67, II: 378). Indeed it is odd that Spencer has become known to history as the key Victorian "individualist." I suspect what is over-simplistically described as Spencer's "Lamarckism" is derived from the intuition that evolution must occur in individuals. But such an intuition impoverishes our view of Spencer's biology and indeed his social theory too.

The same intuition also oversimplifies our understanding of other evolutionists. This chapter began with an epigraph by Charles Darwin stating how each living creature ought to be seen as a microcosm. Although Darwin is well studied, his concerns about biological individuality have rarely been discussed. An interesting alternative history of Darwinism along these lines has periodically been suggested but never brought to fruition. One might move from his early *Beagle* experiments in which he bisected planarian flatworms (*P. tasmaniana*) and watched how in 25 days they formed "two perfect individuals" (Darwin 1844, 244), or how in his notebooks he described trees as "great compound animals" (Darwin 1987, 529). Jonathan Hodge described how in the late 1830s Darwin saw larger entities ("species" or "trees of life") as analogous to individuals while also depicting smaller ones (buds, cells, gemmules, living atoms, monads) as individuals too (Hodge 1985, 209). Phillip Sloan noted how colonial invertebrates allowed Darwin to draw analogies between the species and the individual (Sloan 1986, 421)—in other words, to see the individual relativistically.

Such a history would describe how the particles that Darwin called "pangenes" weren't hypothesized simply to explain heredity, as they tend to be explained today (cf. Bowler and Morus 2005, 195). Darwin also proposed the mechanism of pangenes to solve problems noted above, such as why "lower animals reproduce so many perfect individuals." After all, Darwin saw sexual and asexual generation as fundamentally the same process: he did not see processes such as parthenogenesis as odd. Rather, "the wonder is that it should not oftener occur" (Darwin 1868, II: 358). Such a history would enrich our view of Darwinism, which is generally explained using intuitive biological individuals. This bad habit was begun by Darwin himself, who in the *Origin* tended to use examples of dogs and birds to explain natural selec-

tion and descent with modification. Modern-day Darwinists and nature doc-
umentaries tend to do the same—as far as I am aware there are no breathless
descriptions by David Attenborough of slime mold development. While using obvious biological individuals as examples facilitates explanations of the
neo-Darwinian synthesis and the biological species concept, it is to rely on an
impoverished range.

There is no reason for historians of biology—particularly those involved
in the "Darwin industry"—to be stuck in this rut. For by taking up past questions about biological individuality we can revisit topics which have been so
extensively discussed it seems that there is nothing left to say. That alternative history of Darwinism and biological individuality, in addition to taking up pangenes and compound animals, might be able to say new things
about contemporary scientific opposition to Darwin's proposals. Consider
the zoologist Louis Agassiz, now popularly known for his "creationism" or
his three-fold parallelism.[9] Agassiz leveled a now-forgotten challenge to Darwinism that took up the matter of biological individuality and what he saw as
Darwin's simplistic reading of it.

> Would the supporters of the fanciful theories lately propounded, only ex-
> tend their studies a little beyond the range of domesticated animals,—would
> they investigate the alternate generations of the Acalephs [cnidarians, com-
> monly jellyfish], the extraordinary modes of development of the Helminths
> [flatworms], the reproduction of the Salpae [pelagic colonial tunicates], etc.,
> etc.,—they would soon learn that there are in the world far more astonishing
> phenomena, strictly circumscribed between the natural limits of unvarying
> species, than the slight differences produced by men, among domesticated
> animals; and, perhaps, cease to be so confident, as they seem to be, that these
> differences are trustworthy indications of the variability of species. For my
> own part, I must emphatically declare that I do not know a single fact tending
> to show that species do vary in any way, while it is true that the individuals of
> one and the same species are more or less polymorphous. (Agassiz 1857–62,
> III: 98–99)

Ironically, Agassiz could not have realized that Darwin was just as privately
intrigued about biological individuality as he himself was, since the *Origin*
used only "domesticated animals" for its examples rather than, say, members of the puchi group. And to someone unaware that biological individuality was a highly contested issue in mid-nineteenth-century biology (Nyhart and Lidgard 2011, 374–75; Elwick 2004), Agassiz's own criticism seems
confused—how can an individual be polymorphous? Lacking alternative
terms, Agassiz had no choice but to use the word "individual," a term that
seems to betray him.

An alternative history of Darwinism would take up what was the *real* mystery of mysteries—just what was a biological individual—but would not use the very word whose definition it was probing. What sort of term might it use? One possibility is to follow biologists such as T. H. Huxley and Ernst Haeckel and shift the conceptual ground by coining new words. For instance, "holobiont" is favored by Scott Gilbert (this volume; Gilbert et al. 2012) to denote the complex interacting system of organisms and symbionts.

But would such names be favored in the wider culture, or even among scientists? There will likely be resistance because such words challenge our everyday experience as seemingly autonomous agents, and because scientific explanations that appeal to our own intuitions about essences tend to be more popular. I mentioned that this chapter was partly a story about a "Copernican shift" from an easy intuitive view of biological individuals to one that was more relativistic. But even in astronomy the change was not entirely smooth; our old intuitions die hard. Almost 500 years after Copernicus undermined the view that the sun went around the Earth by suggesting that the Earth instead went around the sun, astronomers still use the words "sunrise" and "sunset." Our own intuitions about individuality will probably be just as difficult to dislodge.

Acknowledgments

I thank Lynn Nyhart and Scott Lidgard for their repeated and detailed comments and suggestions to improve and polish this paper, and to an anonymous referee for recommending Gelman and Hirschfeld's paper. I also am indebted to Pierre-Olivier Méthot, Aryn Martin, Mary Sunderland, and Erica Bettan for suggestions and examples of interesting cases that undermine our intuitions about individuality. The letter from Ernst Haeckel to Herbert Spencer MS.791/73 is used with the permission of the Herbert Spencer Papers, University of London Archives, Senate House.

Notes to Chapter Eleven

1. Indeed, much of this commentary and its focus on essences is deeply indebted to the powerful meditations on essentialism in the final two chapters of Barnes and Dupré (2008).

2. There is even a rueful name for this phenomenon: the "Sobel effect" (Miller 2002).

3. At various points during our conferences it was asked why biological individuality is currently a "hot" topic—one answer might be because issues of authority and control by agencies whose legitimacy used to be unproblematic are now hotly questioned, and new ways of conceiving how systems can be ordered and governed are emerging.

4. Spencer credits Schleiden, but does not give the actual source.

5. Ernst Haeckel to Herbert Spencer, 1 November 1868. Herbert Spencer Papers, University of London Library, MS.791/73.

6. There is also a possibility that the concept of "emergent" properties originated from all of this leveling. It may be a coincidence, but Wheeler credited Spencer's closest friend, G. H. Lewes, with being the first to use the word "emergence" to denote how properties emerge at higher levels (Wheeler 1928, 15).

7. The subject of the *New Scientist* article only discovered she was a tetragametic chimera when she needed a kidney transplant and so underwent blood tests to discover suitable donors.

8. Which is doubly strange—a foreign intervention to prevent the body's overreaction to what it perceives as a foreign intervention, all in the effort to maintain internal "purity."

9. This was his pre-1859 discussion about the apparent correspondence of paleontological sequence, embryological development, and "relative standing" of a species in the animal kingdom (Agassiz 1962 [1857], 29, 84).

References for Chapter Eleven

Agassiz, L. 1857–1862. *Contributions to the Natural History of the United States of America.* Boston: Little Brown.

———. 1962 [1857]. *Essay on Classification.* Cambridge, MA: Belknap.

Ainsworth, C. 2003. "The Stranger Within." *New Scientist* 180: 34–37.

Arumugam, M., J. Raes, E. Pelletier, D. Le Paslier, T. Yamada, D. R. Mende, G. R. Fernandes, et al. 2011. "Enterotypes of the Human Gut Microbiome." *Nature* 473: 174–80.

Atran, S. 1990. *Cognitive Foundations of Natural History : Towards an Anthropology of Science.* Cambridge: Cambridge University Press.

Baker, J. R. 1948. "The Cell Theory: A Restatement, History, and Critique." *Quarterly Journal of Microscopical Science* 89: 103–25.

Barnes, B., and J. Dupré. 2008. *Genomes and What to Make of Them.* Chicago: University of Chicago Press.

Bettan, E. 2013. "The Origin and Development of Research on *Wolbachia*: Methodological, Theoretical and Practical Implications." Unpublished paper, York University, Toronto, Canada.

Bowler, P. J., and I. R. Morus. 2005. *Making Modern Science: A Historical Survey.* Chicago: University of Chicago Press.

Canguilhem, G. 1989. *The Normal and the Pathological.* New York: Zone.

Clarke, E. 2010. "The Problem of Biological Individuality." *Biological Theory* 5: 312–25.

Darwin, C. 1844. "Brief Descriptions of Several Terrestrial *Planariae,* and of Some Remarkable Marine Species, With an Account of Their Habits." *Annals and Magazine of Natural History* 14: 240–51.

———. 1868. *The Variation of Animals and Plants under Domestication.* 2 vols. London: John Murray.

———. 1987. *Notebooks, 1836–1844: Geology, Transmutation of Species, Metaphysical Enquiries.* Edited by Paul H. Barrett, Peter J. Gautrey, Sandra Herbert, David Kohn, and Sydney Smith. Ithaca, NY: Cornell University Press.

Dawkins, R., and L. Margulis. 2009. "Homage to Darwin Debate." http://www.voicesfromoxford.com/homagedarwin_part3.html.

Elwick, J. 2003. "Herbert Spencer and the Disunity of the Social Organism." *History of Science* 41: 35–72.

———. 2004. "Compound Individuality in Victorian Biology, 1830–1872." PhD diss., University of Toronto, 2004.

———. 2007. *Styles of Reasoning in British Life Science: Shared Assumptions, 1820–1858.* London: Pickering and Chatto.

Folse, H. J., III, and J. Roughgarden. 2010. "What Is an Individual Organism? A Multilevel Selection Perspective." *Quarterly Review of Biology* 85: 447–72.

Gayon, J. 1998. "The Concept of Individuality in Canguilhem's Philosophy of Biology." *Journal of the History of Biology* 31: 305–25.

Gelman, S. A., and L. A. Hirschfield. 1999. "How Biological Is Essentialism?" In *Folkbiology*, edited by Douglas L. Medin and Scott Atran, 403–46. Cambridge, MA: MIT Press.

Gilbert, S. F., J. Sapp, and A. I. Tauber. 2012. "A Symbiotic View of Life: We Have Never Been Individuals." *Quarterly Review of Biology* 87: 325–41.

Grosberg, R. K., and R. R. Strathmann. 1998. "One Cell, Two Cell, Red Cell, Blue Cell: The Persistence of a Unicellular Stage in Multicellular Life Histories." *Trends in Ecology and Evolution* 13: 112–16.

Habermas, J. 2003. *The Future of Human Nature.* Cambridge: Polity.

Haeckel, E. H. P. A. 1866. *Generelle Morphologie der Organismen. Allgemeine Grundzüge der Organischen Formen-Wissenschaft, mechanisch begründet durch die von Charles Darwin reformirte Descendenz-Theorie.* 2 vols. Berlin: Reimer.

Hankins, T. L. 1985. *Science and the Enlightenment.* Cambridge: Cambridge University Press.

Hodge, M. J. S. 1985. "Darwin as a Lifelong Generation Theorist." In *The Darwinian Heritage*, edited by D. Kohn, 207–43. Princeton: Princeton University Press.

Huss, J. 2013. "Placing the Accent in Microbiome Research." Paper presented at the International Society for the History, Philosophy and Social Studies of Biology Conference, Montpellier, France.

Kozek, W. J., and R. U. Rao. 2007. "The Discovery of *Wolbachia* in Arthropods and Nematodes: A Historical Perspective." In *Wolbachia: A Bug's Life in Another Bug*, edited by A. Hoerauf and R. U. Rao, 1–14. Basel: Karger.

Lecourt, D. 1998. "Georges Canguilhem on the Question of the Individual." *Economy and Society* 27: 217–24.

Lenhoff, S. G., H. M. Lenhoff, and A. Trembley. 1986. *Hydra and the Birth of Experimental Biology, 1744: Abraham Trembley's Mémoires Concerning the Polyps.* Pacific Grove, CA: Boxwood.

Martin, A. 2007. "The Chimera of Liberal Individualism: How Cells Became Selves in Human Clinical Genetics." *Osiris* 22: 205–22.

Mayr, E. 1982. *The Growth of Biological Thought : Diversity, Evolution, and Inheritance.* Cambridge, MA: Belknap.

Mendelsohn, J. A. 2003. "Lives of the Cell." *Journal of the History of Biology* 36: 1–37.

Miller, D. P. 2002. "The 'Sobel Effect.'" *Metascience* 11: 185–200.

Nyhart, L. K. 1995. *Biology Takes Form: Animal Morphology and the German Universities, 1800–1900.* Chicago: University of Chicago Press.

Nyhart, L. K., and S. Lidgard. 2011. "Individuals at the Center of Biology: Rudolf Leuckart's *Polymorphismus der Individuen* and the Ongoing Narrative of Parts and Wholes." *Journal of the History of Biology* 44: 373–443.

Qin, J., R. Li, J. Raes, M. Arumugam, K. S. Burgdorf, C. Manichanh, T. Nielsen, et al. 2010. "A Human Gut Microbial Gene Catalogue Established by Metagenomic Sequencing." *Nature* 464: 59–65.

Richards, R. J. 2008. *The Tragic Sense of Life: Ernst Haeckel and the Struggle over Evolutionary Thought.* Chicago: University of Chicago Press.

Rieppel, O. 2010. "New Essentialism in Biology." *Philosophy of Science* 77: 662–73.

Roe, S. A. 1981. *Matter, Life, and Generation: Eighteenth-Century Embryology and the Haller-Wolff Debate.* Cambridge: Cambridge University Press.

Roger, J. 1997. *Buffon: A Life in Natural History.* Translated by L. P. Williams. Ithaca, NY: Cornell University Press.

Schleiden, M. J. 1849. *Principles of Scientific Botany: or, Botany as an Inductive Science.* Translated by E. Lankester. London: Longman, Brown, Green and Longmans.

Schwann, T. 1847. *Microscopical Researches into the Accordance in the Structure and Growth of Animals and Plants.* Translated by H. Smith. London: Sydenham Society.

Sloan, P. R. 1986. "Darwin, Vital Matter, and the Transformism of Species." *Journal of the History of Biology* 19: 369–445.

Spary, E. 1996. "Political, Natural and Bodily Economies." In *Cultures of Natural History,* edited by N. Jardine, J. A. Secord, and E. Spary, 178–96. Cambridge: Cambridge University Press.

Spencer, H. 1864–1867. *The Principles of Biology.* London: Williams and Norgate.

Steenstrup, J. J. S. 1845. *On the Alternation of Generations; or the Propagation and Development of Animals through Alternate Generations: A Peculiar Form of Fostering the Young in the Lower Classes of Animals.* London: Ray Society.

Sunderland, M. E. 2011. Morphogenesis, *Dictyostelium,* and the Search for Shared Developmental Processes. *Studies in History and Philosophy of Biological and Biomedical Sciences* 42: 508–17.

Trembley, A. 1744. *Mémoires pour servir à l'histoire d'un genre de polypes d'eau douce.* Leiden: Jean and Herman Verbeek.

Weismann, A. 1892. *Das Keimplasma: Eine Theorie der Vererbung.* Jena: Gustav Fischer Verlag.

Weismann, A., E. B. Poulton, S. Schönland, and A. E. Shipley. 1891. *Essays upon Heredity and Kindred Biological Problems.* 2nd ed. Oxford: Clarendon Press.

Wheeler, W. M. 1911. "The Ant-Colony as Organism." *Journal of Morphology* 22: 307–26.

———.1928. *Emergent Evolution and the Development of Societies.* New York: Norton.

Wilson, R. A., M. J. Barker, and I. Brigandt. 2007. "When Traditional Essentialism Fails: Biological Natural Kinds." *Philosophical Topics* 35: 189–215.

Winsor, M. P. 2006 "The Creation of the Essentialism Story: An Exercise in Metahistory." *History and Philosophy of the Life Sciences* 28: 149–74.

12

Biological Individuality: A Relational Reading

SCOTT F. GILBERT

Wholes and Parts: Composite Individuals

Reading these essays brought me back to a time decades earlier, when I was reading *Zen and the Art of Motorcycle Maintenance* during a particularly hot Baltimore summer. That was where I was introduced to the dissection of wholes into structural and functional, anatomical and physiological, components. That was a time when I was introduced to wholism in my history of biology courses while performing cell culture for my PhD in biology. It was when I was gripped by the fights between Huxley and Owen, and when I was learning how classification could be used as a political tool. All these things proved useful in reading these essays.

That time, the mid-'70s, was a time of great turmoil of parts and wholes. The "sixties" had been a time when masses of people dared to contest the traditional boundaries and functions of parts and wholes. Did the government have the right to coerce individuals to fight in a war they did not find virtuous? Did the government have the right to coerce states and citizens to afford civil rights to blacks when it went against their values and traditions? What was a family, now that divorce was common? Civil rights, women's liberation, and the ecology movements saw blacks, women, and nature as a triad demanding to be an interactive part of the community, respected as agents, not seen merely as resources (Gilbert 1979).

Parts and wholes are similarly being contested now. Economic globalization has turned nation-states into inconvenient boundaries; electronic media have realized science fiction fantasies of instantaneous communication across the planet; artificial reproductive technologies have totally altered the definition of the family; industry has fused together science, medicine, and education; and gender and religion have become matters of choice. It is not an accident that during the past few years there have been so many symposia on part/whole relationships.

Some of the most dramatic reappraisals of part/whole dynamics have been in biology. Twenty-first-century biology threatens to subsume twentieth-century biology into a new paradigmatic framework. The biology of anatomic individualism that has been the basis of genetics, anatomy, physiology, developmental biology, and immunology has been shown to be, at best, a weak first approximation of nature (Gilbert et al. 2012; McFall-Ngai et al. 2013). We are neither anatomic nor physiological individuals. This has been shown consistently not only in invertebrates and in vertebrate model organisms, but also in humans. More than half the cells in the human body are bacterial. Moreover, bacterial products comprise over 30% of our blood metabolites (McFall-Ngai et al. 2013), and they are necessary for our normal physiological maintenance. Kwashiorkor, historically thought of as a protein deficiency disease, has been found to be pathological only if certain bacteria reside in the person's gut (Smith et al. 2013). Pregnancy alters the microbiota of the gut, and these microbes induce some of the characteristic metabolic changes of pregnancy when placed into germ-free mice (Koren et al. 2012). Certain gut inflammatory diseases can be cured by altering the types of bacteria in the intestines or by fecal transplants (Bakken et al. 2011; Chow et al. 2010). So we have come to be considered "holobionts," consortia consisting of the eukaryotic cells plus our persistent bacterial communities (Rosenberg et al. 2007; Gilbert et al. 2012). We are multilineage individuals.

Next, it has also been shown that our bacteria are necessary and expected for our normal development (Gilbert et al. 2012; Gilbert 2013). We are not only the product of the fertilized egg, we are also the product of the bacterial consortia that colonize us. Certain sugars in mother's milk are digested not by baby, but by the baby's bacteria. Moreover, the bacteria have evolved ways of colonizing (in the most literal way) the body. In mice, the blood vessels taking food from the intestine don't form properly without bacteria, nor do the gut-associated lymph nodes. The bacteria accomplish this by secreting factors that induce gene expression in the mammalian intestine (Hooper et al. 2001; Becker et al. 2013). This induction is expected and normal. In some genes, 90% of their expression is induced by bacteria. In zebrafish, bacteria induce the normal division of intestinal stem cells. Without these bacteria, not enough stem cells are made, and the intestine lacks many of its most important cell types (Rawls et al. 2004). The life cycle of an organism requires the life cycles of other organisms (Fig. 12.1). We are not individuals by developmental criteria.

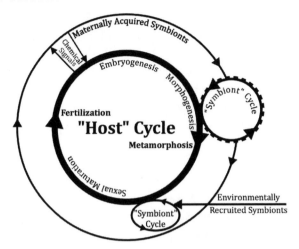

FIGURE 12.1. A holobiont life cycle. The "traditional" life cycle of the animal is shown by the dark black circle. This is supported by the life cycles of symbionts acquired from the egg (thinner circle), and symbionts acquired from the environment. The symbionts can provide chemical signals necessary for early development and protection of the embryo, while the "host" can provide signals to sustain and differentiate these symbionts. Symbionts can be essential for completing host development (as in the mammalian gut) and/or traversing developmental stages such as metamorphosis, as shown by the "gear" form of the symbiont life cycle. They can also provide chemical signals for larval settlement and facultative morphogenesis. (Drawn by David Gilbert after a draft from the 2011 NESCent conference on the origin and evolution of animal-microbe interactions.)

And we are certainly not individuals by genetic criteria. There are about 160 major species of bacteria normally resident on and in our bodies, and the human microbial genome has 150-times more genes than our zygotic genome. In many invertebrates, symbiont genomes are a source of selectable genetic variation. The phenotype of pea aphids—their color, their thermotolerance, and their resistance to parasitoid infections—depends on alleles of their symbionts (Dunbar et al. 2007; Oliver et al. 2009). Many invertebrates receive their symbionts directly from the egg. Indeed, the symbionts are packaged into the egg along with ribosomes, mitochondria, and mRNAs. Mammals and most other vertebrates receive symbionts by infection, primarily from the mother, during pregnancy and parturition (Funkhouser and Bordenstein 2013). We go from one symbiotic system (that of the mother) to another (that of symbionts).[1] Thus, the symbionts must be considered as a third genetic system along with the nucleus and mitochondria (Moran 2007; Gilbert 2011). And birth, which has long been conceived as the creation of a new individual, is actually the continuance of community (Gilbert 2014)! We are not genetic individuals.

In the old paradigm, our genetically pure body was protected against microbial assault by the immune system. Indeed, we lived in a voracious microbial world that would devour us if it were not for the immune system. This is why we received inoculations and booster shots, and this fact was driven home relentlessly and mercilessly by the AIDS epidemic. The discipline of immunology had been called "the science of self/non-self discrimination" (Klein 1982). In this view, the immune system consists of defensive "weaponry," evolved to protect the body against threats from pathogenic microbes.

In a fascinating inversion of this view of life, recent studies have shown that an individual's immune system is in part created by the newly acquired microbiome (see Pradeu 2012). In vertebrates, the gut-associated lymphoid tissue is specified and organized by bacterial symbionts (Rhee et al. 2004; Lanning et al. 2005). When symbiotic microbes are absent in the gut, the immune system fails to function properly and its repertoire is significantly reduced (Lee and Mazmanian 2010; Round et al. 2010). Similarly, Hill et al. (2012) have shown that microbial symbionts provide developmental signals that limit the proliferation of basophil progenitor cells and thereby prevent basophil-induced allergic responses. Lee and Mazmanian (2010) conclude, "multiple populations of intestinal immune cells require the microbiota for their development and function."

The immune system, therefore, appears to be more of a "passport control agent" or even a "bouncer" rather than a defensive army posted to keep the zoological organism "pure." Indeed, the immune system actively recruits the symbionts. The antibodies produced in the intestinal crypts might actually play a "critical role in establishing a sustainable host-microbial relationship" and might be involved in "the creation of an optimal symbiotic environment on the interior of the PPs [Peyer's patches]" (Peterson et al. 2007; Obata et al. 2010). Thus, the immune system, built, in part, under the supervision of microbes, does not merely guard the body against other hostile organisms in the environment. It also mediates the body's participation in a community of "others" that contribute to its welfare (Tauber 2000, 2008).

Ideologically, this signals a huge change. No longer is biology a matter of "us versus them," "eat or be eaten." The existential and Darwinian mode of defining one's self as "being against all others" has been replaced by a more Harawayan and Deleuzian notion of us "becoming with the other."

So if there is no genetic, developmental, immunological, anatomical, or physiological individual, what is "individual selection" in evolution? Can it be that organisms are selected as multigenomic associations? Is the fittest in life's struggle the multispecies consortium, and not an individual of a single species in that group? This possibility has been raised by Bateson (1988), who has argued that "the outcome of the joint action of individuals could become a character in its own right." Since it replicates with selectable variation, the holobiont may be an important level of evolutionary selection.

This moves the biological discussion of symbiotic associations into the arena of "group selection." Most discussions of group selection, however, are not germane here, because they assume that the group in question is composed of members of a single species. A holobiont is a team of different species. However, one important concern is relevant to our discussion of the holobiont: cheaters. The major problem for all group selection theories (and the groups, themselves) are potential "cheaters," those lower-level members of the group that would proclaim their own autonomy and that would multiply at the expense of the others. The problem of "cheaters" then has to be solved in such a way that associates in a symbiotic relationship are under the social control of the whole, the holobiont (Stearns 2007).

This strong socializing and unifying force is found in the immune system (see Burnet and Fenner 1949; Tauber 2000, 2009; Ulvestad 2007; Eberl 2010; Pradeu 2010). If the immune system serves as the integrating system, keeping the animal and microbial cells together, then to obey the immune system is to become a citizen of the holobiont. To escape immune control is to become a pathogen or a cancer. Cheaters are destroyed by the immune system. It is now possible to envision selection as more like team competitions than individual competitions. An American football team may have the best quarterback in the league; but it will not get into the playoffs if its other members can't defend him or catch his passes.

Thus, the symbionts are welcomed into the animal body and are regulated by the immune system. As part of the body, the microbes not only help the body develop and remain physiologically intact, they also provide a secondary system of genetic transmission from parent to offspring. They can provide selectable variation from generation to generation. Moreover, in addition to providing this selectable variation, microbial symbionts may have played, and continue to play, other roles in animal evolution as well (Margulis and Fester 1991). Animal speciation may be mediated, in part, through the ability of microbes to induce reproductive isolation. This can be achieved through symbiont-induced cytoplasmic incompatability between hybrids (Brucker and Bordenstein 2012, 2013) or by symbiont-induced mate selection (Sharon et al. 2010).

Last, we may never have been "pure" animals, innocent of symbionts. We must remember not only that eukaryotic protists were created by endosymbiosis, but that the protist world is full of complex symbioses (Margulis 1981;

Margulis and Fester 1991; Sapp 2009). Some of these microbial symbioses may have led to multicellularity. The choanoflagellates are unicellular protists that are the sister group of the animals. However, these unicellular forms can be converted into multicellular entities—complete with an extracellular matrix and cytoplasmic bridges between cells—by a specific bacterium that often coexists with them. If the protists are cultivated in filtered water, they remain unicellular. If the bacteria are added back, they can form multicellular entities (Dayel et al. 2011; Alegado et al. 2012). Thus, bacteria symbionts may have been important in initiating multicellularity, the ultimate part/whole dialectic in biology.

Science News Magazine's December 28, 2013, issue ranked "microbes ascendant"—the holobiont idea of organisms—as the top story in science for 2013. Its next issue had four stories on how bacterial symbiosis is redefining life. The biology of the twenty-first century will have a different perspective on parts and wholes.

And that's only the beginning. Biology is changing in other ways, too, and these changes are renegotiating the parts/whole distinction. Systems biology sees information flow as a common denominator of all biology, seeking to place all biological disciplines under a common rubric. This, of course, has been a common strategy for dominance in biology. However, for several reasons, the boundaries of the biological disciplines have become remarkably porous. Indeed, it's difficult to justify calling anything "interdisciplinary" or "transdisciplinary," because there are no such disciplines. Biology has become—to use a biological word—syncytial. Just as there are several nuclei within the cytoplasm of a syncytium (such as found in certain muscles and placental cells), there are centers of professional power and training within the common cytoplasm that is biology. So there are good reasons for biologists to speak about parts and wholes. The relationships we had learned are being transformed into something new and very different, and perhaps much richer and more profound.

Relations with these Chapters

The chapters in this volume are the constituent "parts" of this book, but reading them collectively allows them to interact with one another and with this new context I have been discussing. I'll discuss them here, not as a review (that would make this another and very large chapter), but as a conversation. If the chapter is meaningful to a reader, then the "reader's portion" is not passive, but rather an actively engaged interaction. One can even say that the reader and text form a system, and that the text has no meaning outside that system. So I will allow you, the next reader, to selectively eavesdrop on some of the conversations I've been having with these chapters.

And what better place to start than with Scott Lidgard and Lynn Nyhart's discussions in chapter 1 about the contexts framing the part/whole debates. The body politic metaphor acquires new dimensions in our discussions of symbionts. Just as modern immunology was framed during the Franco-Prussian War, where boundaries and defensive weaponry were paramount, symbiosis theory is framed during a time of massive migrations of people across borders. In some instances, the migrations are encouraged (leading, for instance, to Chinese and Indian information technologists coming to northern Finland to work for Nokia); and in other instances, the migrations have been forced by the combination of ecological and political deterioration (leading, for instance, to Islamic populations migrating from Africa and Asia to Europe and the Americas). In such sociopolitical contexts, who is foreign and who is a citizen takes on a particular immediacy. Are symbionts foreigners to a racially pure body (descended from a single cell)? Are the symbionts legal resident aliens, green-card holders whose work visas are checked constantly by an especially sensitive immune system? Or are the symbionts full citizens of a multilineage polity that help generate the body as a normal function? Moreover, the debates about what criteria enable the holobiont to be considered an individual are far from over (Gilbert et al. 2015; Moran and Sloan 2015). The discussion of parts and wholes must be seen as occurring within a sociopolitical context that is evident in our daily lives, in our political rhetoric, and in our front-page headlines.

The chapter on the alternation of generations by Lynn Nyhart and Scott Lidgard (this volume) provides a deeper historical context, showing that those nineteenth-century questions of what constitutes individuality are still with us today. It was Thomas Henry Huxley who (contra Richard Owen, of course) defined the individual as the progeny of the fertilized egg, and it made me wonder if Huxley knew about the work of the embryologist Robert Remak on the origin of cells by division of preexisting cells, being performed at about the same time. Huxley had a penchant for seeing life in its embryological context. He even said that "evolution was not speculation but a fact; and it takes place by epigenesis" (1893), and he instructed Darwin (while the latter was writing the *Origin*) that differences between organisms "result not so much of the development of new parts as of the modification of parts already existing and common to both divergent types." Huxley's relationships of parts to wholes are interesting on many levels, not the least in that they change and remain integrated. This will lead to a notion of homology that is still being hotly debated today.

Once the body was seen as a collection of asexually generated cells, the link could be made to plants not only in terms of cell structure but in terms of reproductive modes. One of the great debates on individuality centered on the alternation of generations, where a sexual generation would produce an asexual generation (often with a different body form), which would produce a new sexual generation. Cnidarian polyps and medusae were discovered to be part of the same life cycle. Here, many biologists considered such an animal not just one distinct "individual," but two, or even more. Steenstrup, Owen, Rudolf Leuckart, Johannes Müller, and others took up the challenge of relating parts to wholes in the complex worlds of colonial plants, jellyfish, and corals, eusocial insects, and parasites. Botanists like Matthias Schleiden, Alexander Braun, and Wilhelm Hofmeister struggled with the nature of plant individuality and alternation of generations. Steenstrup saw that a life cycle comprised multiple generations of "individuals," and Leuckart was particularly adept at making his arguments that no individual of a sexually reproducing species could fulfill all the functions of the species itself (Nyhart and Lidgard 2011). So no animal or plant of that sort was a "perfect" specimen. Leuckart also brought up the notion of "polymorphic" individuals, each with its own task in the division of labor.

Indeed, as we study more organisms and as we see how lineages interact, the questions of individuality are more with us than ever. Take *Mastotermes darwinensis*. Is a worker termite an individual? Or is it the hive, since only the queen is fertile? Or is it even the termite, since it can't digest wood without its bacterial symbiont, *Myxotricha*. But *Myxotricha* is itself a composite of at least five different species, none of which exists except in such an association and in the gut of these termites (Margulis and Sagan 2001). And the "individual" changes as it develops. The notion of the life cycle as the "individual" has returned to evolutionary discussions through the writings of John Tyler Bonner (1995). This has important ramifications if the "life cycle" is that of the holobiont, with its persistent communities of microbes.

In his great synthetic work, *The Cell in Inheritance and Development* (1896), E. B. Wilson wrote, "There is at present no biological question of greater moment than the means by which the individual cell-activities are co-ordinated, and the organic unity of the body maintained; for upon this question hangs not only the problem of transmission of inherited characteristics and the nature of development, but our conception of life itself." The nineteenth-century questions of what constitutes an individual have been recast in twenty-first-century terms and have again moved to the center of biology.

At one end of the symbionts spectrum are the parasites, and Michael Osborne (this volume) has provided a fascinating analysis of the evolution of the parasitic idea in both biology and sociology. The formulation and separation of these ideas took place during the Third French Republic by such eminent scientists as Claude Bernard, Edmond Perrier, and Raphaël Blanchard. Originally, the term did not carry negative connotations and denoted a person who ate alongside a more senior person. (Although Darwin would add his barnacles to the debate, it's difficult to describe him as Captain Fitzroy's parasite.) In biology, the usage became restricted to a member of a different species who lacked something that the host provided. In society, though, a parasite became defined as a member of the same species who "leeched off" his conspecifics. The social context is the Third French Republic, where Solidarism undergirded the social contract, where there was debate concerning the status of the French colonial possessions and peoples, and where biological parasites persisted in preventing colonial expansion.

As these studies of symbiosis were being performed, so was the analysis of metabolism; and often by the same people. Hannah Landecker (this volume) provides an elegant analysis of one of the most fundamental notions of what constitutes an organism: metabolism. Metabolism is nothing less than the ability to preserve the identity of the whole by continually changing its component parts. Because metabolism enables the stabilization of an "individual" by permitting the organism to retain constancy while constantly changing its component parts, "individuals" are not just material things, but always relational processes in time as well (Gilbert 1982). That's what distinguishes us from machines, and that's what our machines look for when they go to Mars to determine if there is "life" there. Metabolism is the paradoxical foundation for existence. Levinas (1969) joyfully celebrates this interconversion of one life into another: "Nourishment as a means of invigoration, is the transmutation of the other into the same, which is the essence of enjoyment: an energy that is other, recognized as other . . . becomes, in enjoyment, my own energy, my strength, me."

Thomas Mann (1969 [1924]) viewed metabolism more pessimistically: "What then was life? It was warmth, the warmth generated by a form-preserving instability, a fever of matter, which accompanied the process of ceaseless decay and repair of albumen molecules that were too impossibly complicated, too impossibly ingenious in structure . . . It was not matter and it was not spirit, but something between the two, a phenomenon conveyed by matter, like the rainbow on the waterfall, and like the flame."

In 2013, a new notion has come into being: "co-metabolism." This is the idea that the metabolic flux that enables us to persist is in fact, a flux between us, our diet, and our symbionts (Smith et al. 2013). Co-metabolism is critical for our physiological homeostasis. Metabolism is now being taken out of the context of "eat or be eaten," where, as Landecker notes, it helped build "commonly held notions of individual organisms as alone in the world." It is presently being placed into a context of messmates. Indeed, the word "commensal" means just that. In the holobiont, we see an incredible play on the historical opposition between life as lineages and life as metabolically self-sustaining wholes (Gilbert 1982; Dupré and O'Malley 2009). The metabolic entity—the holobiont—is made up of several interacting lineages.

Andrew Reynolds's (this volume) sociological perspective on cell-cell communication shows that this idea has become central to the integration of parts and wholes in biology. Indeed, Reynold's chapter on the integration of parts and wholes through communication gives us a vantage point to look at the roles of social metaphors in biology. Endocrine factors are the hormones that act globally (traveling through the blood from one group of cells to the body) to integrate the entirety of the organism/polity. Paracrine factors, influencing nearby cells through the intercellular fluid, act in their local neighborhoods; while juxtacrine factors (on cell membranes) interact only with adjacent cells. Together, these signaling molecules allow cells to act globally or locally to form organs and keep them intact. One of the generalizations that embryologists have made is that cells lead double lives. As "adult" cells, they have a specific function, such as making insulin, pumping fluid, or being transparent to light; but as "young" cells, they are in the construction trades. They produce paracrine factors that build organs from cells. They are both the material and efficient causes of the embryo, and the embryo builds itself from immature cells. In the development of eyes, the cells that induce the lens to form are the cells that will later give rise to the retina.

The embryo as "cell society" has been a critically important metaphor in embryology, and one of the extrapolations of this metaphor concerns who rules this body politic (Gilbert 1988, 1992). Interestingly, the three major types of model organisms in developmental biology reflect the three major models of how the body politic is governed: realist, liberal, and constructivist (see Walt 1998; Copeland 2000). The genetic model systems in developmental biology approximate the "realistic" view of the body politic. The genes are the central authorities running the show. The embryological model systems of developmental biology are like the "liberal" view of the body politic. There are several cellular centers of authority, and interactions between these centers make possible unique and emergent institutions. Last, the evo-devo/eco-devo model systems of developmental biology model are like the "constructivist" model of the body politic. Here, the body is generated not only by internal factors but also by the interactions of that body with its biotic, cultural, historical, and environmental milieu.

I am particularly drawn to Ingo Brigandt's chapter (this volume) that argues that processes are bodily parts, just as structures are. I think this is a critical point to make, and one that brings the structure/function question squarely into the part/whole controversy. This new perspective on considering functions as parts allows Brigandt to see the intimate relationships between structure and function. It also allows us a new perspective on homology through evolutionary time. I think that one of the reasons we have been distracted from this view is the way that processes are represented graphically: all those arrows. But the arrows, denoting temporal sequence and causation, are actually showing stereocomplementary structural fit (Gilbert and Greenberg 1984). Just like bones fitting together, just like the lungs having a space on the left-hand side for the heart to fit into, the proteins of a signal transduction pathway must bind together in a stereocomplementary pattern. Brigandt is able to relate "activity" and "function" in ways that allow us to see the homology of related activities, independently from their functions.

What constitutes the natural part also becomes a question. The construction units of the body need not be the same as the adult anatomical units. Flies have parasegments, and vertebrates have rhombomeres and medial rib modules. These construction units are not seen in the adult. It may be that what the embryo considers as a natural unit is what the enhancer "perceives." There is an enhancer for gene expression in the medial rib and in each hindbrain rhombomere. And exaptation is the rule. Brigant's view of a homologue as a "unit of morphological evolvability" is a great point for discussion, because it leads to considering exaptations and their roles in evolution.[2]

And like the questions brought up by other chapters in this book, this philosophical chapter is intensely relevant to today's biology. For those of us who work on the origin of the turtle shell, homology is a huge issue, and questions of connectivity, embryological origin, and altered function that Brigant addresses are at the center of these investigations (Lyson et al. 2013; Cebra-Thomas et al. 2013).

And this problem of turtle bone homology has been one that has been discussed brilliantly by Olivier Rieppel (2012), but not here. In this volume, Rieppel discusses a critically important, but much neglected, historical episode in the debates over parts and wholes and their extrapolation from biological entities to social entities. Focusing on Martin Heidenhain's biological notion of nested structural hierarchies, enkapsis, Rieppel analyses the strange history of wholism in Nazi Germany. This was an area where wholism and

Goethe's notion of multiplicity in unity were used both by Jewish mandarin scientists and by Nazis constructing a new German biology for the Third Reich. (*Bildung und Kultur* were the watchwords of the Reform Jewish movement, and Jews swelled the ranks of the Goethe Society [Mosse 1997].)

The use of wholism by the Nazis and the destruction of the scientific infrastructure of central Europe during World War II (including the Prater Vivarium in Vienna) effectively wiped out a tradition of wholistic biology that attempted to counter the reductionist tendencies of Anglophone science. This paper tries to show how important this tradition was to the biology of central Europe. In America, where genetics, cell biology, and an engineering approach to the science predominated, this perspective was seen as a minor variant favored by Nazis. Wholism is a war casualty; and like most refugee communities that fled the Nazi-occupied areas, it never had the same vitality or centrality that it had before "Blut und Boden" became standard-bearers for lineage and environment.

The biological idea that the whole preceded its parts can be found in Kant's Third Critique, and it became a normative part of embryology, which was, after all, a predominantly German discipline. Charles Otis Whitman (who received his PhD in Leipzig) was one of the most influential American embryologists in the early 1900s, and he made this a fundamental principle of the developmental biology he taught at the University of Chicago and at the Marine Biology Laboratories. The notion that the parts form the whole simultaneously with the whole defining the function of the part has been a major paradigm of animal development, and can be seen as a major concept in the work of Hans Spemann and Paul Weiss. This was a doctrine that did not have to be National Socialist, even if extrapolated into society, which Whitman did, on a more republican basis.

But extrapolated into the German *Volksgemeinschaft* by people like August Thienemann, it meant that an individual person would have the moral obligation to sacrifice his life for the greater good of the whole. This was the morality of Nazi Germany, but it can also be said to be the morality of any country during war. It is the role of governments to remind their citizens that their country persists thanks to those who made "the supreme sacrifice." The part/whole dichotomy is at its most existential summit here. E. O. Wilson (2012) recently looked at this part/whole controversy in evolutionary biology. He noted the paradox that within groups, selfish individuals will outcompete altruistic individuals, but that between groups, those groups with altruistic individuals will be favored over those whose individuals do not cooperate. There will be selection, he postulates, on genes promoting both behaviors.

Therefore, "the victory can never be complete; the balance of selection pressures cannot move to either extreme. If individual selection were to dominate, societies would dissolve. If group selection were to dominate, human groups would come to resemble ant colonies." It seems we inherit, as the Talmud declares, a Yetzer Ra impelling us to selfish competitive acts as well as a Yetzer Tov, propelling us to social cooperation. There will always be tension between individuals and society, between love and duty. So we can expect great suffering as well as great literature.

But the genome does not always give the same orders. The directives from the nucleus are modified by information coming from the environment. The genome is not a text to be decoded, but a composition to be interpreted (Gilbert and Bard 2014). Every organism is a performance of the genome, and each performance is a new interpretation. This perspective is particularly appropriate in light of recent research on Predictive Adaptive Responses (PAR). According to the PAR view (Gluckman and Hansen 2004), the genome produces receptors that enable it to monitor the environment. The organism has developmental plasticity such that the phenotype produced is responsive to the environmental signal. However, the signal may or may not be a true signal of the actual environmental change. So a photoperiod getting smaller each day may tell an organism that it should change its pelage from brown to white. Winter is coming. But if it doesn't snow, the white pelage is dangerously sharp against the rocks. Similarly, the mammalian fetus receives signals from maternal nutrition concerning the caloric content of its environment. If provided a poor diet in utero, gene expression in the liver becomes that making a protein suite that stores calories. If there is a "mismatch" and such an infant is born into a well-nourishing world, that infant has a much greater risk of developing obesity, diabetes, and congestive heart failure due to the faulty prediction made in utero.

The notion that predictive testing and planning, from the subatomic to the social level, are the driving forces of sustained individuality is also interesting from the view that humans are planning animals and that fantasy is an important force in human social evolution. Humans can plan strategies by imaging scenarios that never happened and may never happen. We can imagine alternative possibilities and plan to maximize our continuity in the different environments. Our brain can even fool itself by having the body physiologically react to imaginary conditions (Gilbert 2003). This is the basis of entertainment and certainly of sexual fantasy. Humans are self-consciously planning animals. Information theory leaves open the mechanisms by which the modular interactions make possible such long-term exploring, and the mechanisms would be expected to be different at each level of organization, going from atomic-level constraints (see Deacon 2011) to social planning and campaign strategies.

Beckett Sterner's paper on the mechanisms of cell type inheritance continues the discussion of environmental and internal mechanisms of hereditary control. Here, he introduces the concept of the "demarcator," either a material agent or a causal process that is responsible for integrating the parts or the organism and its life cycle together. Biological entities should be able to be distinguished based on their possession of such demarcators. This concept, Sterner asserts, is still being developed, and it extends the notions of overlap and scaffolding used by Wimsatt and Griesemer (2007). Thus, inheritance could be achieved either by pushes from within (*material overlap* between generations) or influences from without (*scaffolding* directing the phenotypic inheritance). Examples from unicellular organisms—both prokaryotic and eukaryotic—bring together these two modes of inheritance in one scheme. Indeed, it is reminiscent of the "alternation of generations" project that sought to unite asexual and sexual inheritance schemes into a common mode. Indeed, the life cycle is crucial in these discussions of inheritance.

It seems that symbionts use both the scaffolding and material overlap means of propagating phenotype, and that these are not mutually exclusive (Chiu and Gilbert 2015). Many arthropods receive their symbionts primarily through their mother's oocyte, where it has been sequestered. Vertebrates and many invertebrates usually acquire symbionts through infection. This infection can be at the moment of egg laying or parturition; but it is usually from the mother's cache of symbionts, but with contributions from the entire community (Funkhouser and Bordenstein 2013). As mentioned above, symbionts can be the source of hereditable variation as well as reproductive isolation. It will be interesting to see how the demarcator idea plays out in complex symbiotic life cycles where individuality is lost or gained (such as in salamander kleptogenesis [Bogart et al. 2007] and angler fish fusions [Pietsch 2005]).

Some of the most interesting scaffolds, though, involve the inheritance of behavioral phenotypes. Meaney's laboratory has shown that anxiety in rats can be inherited; but only through a complex interaction of what could be interpreted as both material and scaffold. Weaver and his colleagues (2004, 2007) showed that anxious rats had high levels of corticosterone. This was due to the absence of the gluocortocoid receptor in the brains of these rats. This receptor mediates the negative feedback loop, downregulating corticosterone production. The receptor wasn't there because the enhancer regulating that gene's expression in the brain was methylated. And methylation was permitted in that enhancer only in rats that did not receive adequate maternal care. In the rats that had received maternal care, this same region of DNA was unmethylated, allowing the glucocorticoid receptor gene to be expressed in the brain. So rats without adequate maternal care become more anxious. And what do female rats having this condition do? They give their pups less maternal care. Cross-fostering rats between anxious and non-anxious mothers changes the methylation and the anxious phenotype. Here, behavior controls gene expression, and gene expression helps generate that behavior. The behavior makes the scaffolding.

A similar behavioral-epigenetic scaffolding has been seen in rat sexual behaviors (Champagne et al. 2006; Cameron et al. 2008). Here, high levels of maternal care also cause the demethylation of the regulatory regions of the estrogen receptor genes, enabling their expression in the MPOA region of the brain that is associated with sex-specific behaviors. Those female rats with low estrogen receptors in the MPOA region of the brain have a more receptive sexual phenotype than the rats who had been licked and groomed thoroughly when young. They also do not lick and groom their offspring, thus continuing the inheritance of the trait. These differences, moreover, are not "good and bad," "normal and pathological." Rather, they are variations that may become advantageous in different environments. This is not a pathology, but a norm of reaction. But what is important is that the behaviors are hereditary despite there being no mutational difference in the DNA between the variants. Maternal behavior can create a scaffolding that allows inheritance.

Matthew Herron's paper (this volume) deals directly with one of the major evolutionary transitions, from unicellularity to multicellularity, as shown by the Volvocales. His contribution lays out that there are transitions within the major transitions, and that the progression from one level of individual to another is not a simple binary step. Rather, there are multiple steps, and both genetic and physical parameters appear to be regulating the transitions. In the volvocine algae individuality appears to be partitioned along three levels of the biological hierarchy: cells, colonies, and clones. In each, one sees the principle put forth by Queller and Strassman (2009) that the new individual is characterized by high levels of internal cooperation and low levels of internal conflict. In addition to genetic homogeneity and the single-cell bottleneck of a zygote, *Volvox* "clones" also become individuals through division of labor into distinct soma and germ cells, retention of cytoplasmic bridges between cells, the formation of a common extracellular matrix holding the cells together, and the establishment of organismal polarity through the rotation (on the cellular level) of the basal bodies that produce the flagellum.

In some species, there is even a gastrulation-like movement in which all the cells participate. Clones in different groups of Volvocales have one or more of these marks of individuality.

So is a cancer a "zooid," Huxley's term for parts of colonies that are like individuals but not fully so? There have been some interesting speculations on cancer as an atavistic return to a colonial stage of individuality. Weinberg (2007), for instance, claims that the genes responsible for cellular coopera- tion during the origins of multicellularity are those whose malfunction causes cancer. Davies and Lineweaver (2011) explicitly claim that cancer is an atavis- tic condition when genetic or epigenetic instructions "re-establish the domi- nance of an earlier layer of genes that controlled loose-knit colonies of only partially differentiated cells." Since tumors are often clonal, but develop ge- netically or epigenetically distinct subclones (for instance, cancer stem cells), it would be interesting to think about a metastasizing cancer as ramets, each having a similar, but distinct genetic or epigenetic identity.

With Snait Gissis' paper (this volume), we come to the nineteenth- century social philosopher (and sometime biological theorist) Herbert Spen- cer and the explicit analogy of the society being a body politic. But what I am impressed with is what kind of body politic it is. Spencer modeled his society on the embryo. Specifically, he based his views on those of Karl Ernst von Baer, who described development as the change from homogeneity to specialization. "The development of a society as well as the development of man and the development of life generally," said Spencer (1851, 319), "may be described as a tendency to individuate—to become a thing." At the close of *Social Statics*, Spencer wrote, "Yet this phrase of von Baer, expressing the law of individual development, awakened my attention to the fact that the law which holds of the ascending stages of each individual organism is also the law which holds of the ascending grades of organisms of all kinds." And in an 1864 letter to George Lewes, Spencer (1906) claimed, "if anyone says that had von Baer never written, I should not be doing that which I am now, I have nothing to say to the contrary."

I think Spencer prescient in this respect. His people have dual functions— they are parts that make a whole, and they are defined by the whole. Thus, as Gissis notes, Spencer had a hybrid view of individuality, which could be viewed either as "collective individuality" or as a "collectivity of individu- als." So it is with the embryo, and the fates of embryonic cells come by their lineage, their interactions with other cells, and their interactions with the en- vironment. The cells make the embryo, and the embryo, in relationship with the environment, makes the cells. Parts and wholes are in relationship from atoms through societies. If societies are like organisms, they are not adults, they are embryos. If Gaia exists, she's not an adult, either. Without doubt, she is an embryo.

The parts contribute to the whole and the whole determines the parts. The chapters here will interact with each other and with the mind of the beholder in ongoing dialogue. There will be selection and discernment, and there will be the "becoming with," the growth of the whole into a new whole through the interaction with something new. It's been a pleasure and a privilege to be one of the first people who interact with these chapters.

Acknowledgments

I wish to thank Lynn Nyhart and Scott Lidgard for their wonderful organization of the symposia in Philadelphia and Madison. Also, I want to thank Carin Berkowitz for the hours of excellent conversation during our travels. SFG is funded by the Swarthmore College faculty research and by the Academy of Finland.

Notes to Chapter Twelve

1. I wish to thank Martin Jacobs for this observation.

2. So I must tell the story of my favorite exaptation. A year or so ago, I asked the chief science librarian, Meg Spencer, "Why do we have all these rows of *Biological Abstracts*, going back to the 1920s. I'm sure nobody's looked at them since we got internet service." She looked at these rows of unused heavy tomes and said, "Soundproofing. It would take $500 worth of curtains to do what these books do." All the information catalogued in those volumes is now worthless. Rather, the physical properties of the paper have become critical for their preservation in the library. Same object, different pathway of relationships.

References for Chapter Twelve

Alegado, R. A., L. W. Brown, S. Cao, R. K. Dermenjian, R. Zuzow, S. R Fairclough, J. Clardy, and N. King. 2012. "A Bacterial Sulfonolipid Triggers Multicellular Development in the Closest Living Relatives of Animals." *Elife* 1:e00013. doi:10.7554/eLife.00013.

Bakken, J. S., T. Borody, L. J. Brandt, J. V. Brill, D. C. Demarco, M. A. Franzos, C. Kelly, et al. 2011. "Treating *Clostridium Difficile* Infection with Fecal Microbiota Transplantation." *Clinical Gastroenterology and Hepatology* 9: 1044–49.

Bateson, P. 1988. "The Biological Evolution of Cooperation and Trust." In *Trust: Making and Breaking Cooperative Relations*, edited by Diego Gambetta, 14–30. Oxford: Blackwell.

Becker, S., T. A. Oelschlaeger, A. Wullaert, K. Pasparakis, J. Wehkamp, E. F. Stange, and M. Gersemann. 2013. "Bacteria Regulate Intestinal Epithelial Cell Differentiation Factors Both *in Vitro* and *in Vivo*." *PLOS ONE* 8 (2): e55620. doi:10.1371/journal.pone.0055620.

Bogart, J. P., K. Bi, J. Fu, D. W. Noble, and J. Niedzwiecki. 2007. "Unisexual Salamanders (Genus *Ambystoma*) Present a New Reproductive Mode for Eukaryotes." *Genome* 50 (2): 119–36.

Bonner, J. T. 1995. *Life Cycles.* Princeton, NJ: Princeton University Press.

Brucker, R. M., and S. R. Bordenstein. 2012. "Speciation by Symbiosis." *Trends in Ecology and Evolution* 27: 443–51.

———. 2013. "The Hologenomic Basis of Speciation: Gut Bacteria Cause Hybrid Lethality in the Genus *Nasonia.*" *Science* 341: 667–69.

Burnet, F. M., and F. Fenner. 1949. *The Production of Antibodies.* 2nd ed. Melbourne, Australia: Macmillan.

Cameron, N. M., D. Shahrokh, A. Del Corpo, S. K. Dhir, M. Szyf, F. A. Champagne, and M. J. Meaney. 2008. "Epigenetic Programming of Phenotypic Variations in Reproductive Strategies in the Rat through Maternal Care." *Journal of Neuroendocrinology* 20 (6): 795–801.

Cebra-Thomas, J. A., A. Terrell, K. Branyan, S. Shah, R. Rice, L. Gyi, M. Yin, et al. 2013. "Late-Emigrating Trunk Neural Crest Cells in Turtle Embryos Generate an Osteogenic Ectomesenchyme in the Plastron." *Developmental Dynamics* 242 (11): 1223–35.

Champagne, F. A., I. C. Weaver, J. Diorio, S. Dymov, M. Szyf, and M. J. Meaney. 2006. "Maternal Care Associated with Methylation of the Estrogen Receptor-Alpha1b Promoter and Estrogen Receptor-Alpha Expression in the Medial Preoptic Area of Female Offspring." *Endocrinology* 147 (6): 2909–15.

Chiu, L., and S. F. Gilbert. 2015. "The Birth of the Holobiont: Multi-species Birthing through Mutual Scaffolding and Niche Construction." *Biosemiotics* 8 (2): 191–210.

Chow, J., S. M. Lee, Y. Shen, A. Khosravi, and S. K. Mazmanian. 2010. "Host Bacterial Symbiosis in Health and Disease." *Advances in Immunology* 107: 243–74.

Copeland, D. C. 2000. "The Constructivist Challenge to Structural Realism: A Review Essay." *International Security* 25: 187–212.

Davies, P. C., and C. H. Lineweaver. 2011. "Cancer Tumors as Metazoa 1.0: Tapping Genes of Ancient Ancestors." *Physical Biology* 8 (1): 015001.

Dayel, M. J., R. A. Alegado, S. R. Fairclough, T. C. Levin, S. A. Nichols, K. McDonald, and N. King. 2011. "Cell Differentiation and Morphogenesis in the Colony-Forming Choanoflagellate *Salpingoeca rosetta.*" *Developmental Biology* 357 (1): 73–82.

Deacon, T. 2011. *Incomplete Nature: How Mind Emerged from Matter.* New York: Norton.

Dunbar, H. E., A. C. Wilson, N. R. Ferguson, and N. A. Moran. 2007. "Aphid Thermal Tolerance Is Governed by a Point Mutation in Bacterial Symbionts." *PLOS Biology* 5 (5): e96. doi:10.1371/journal.pbio.0050096.

Dupré, J., and M. A. O'Malley. 2009. "Varieties of Living Things: Life at the Intersection of Lineage and Metabolism." *Philosophy & Theory in Biology* 1: 1–25.

Eberl, G. 2010. "A New Vision of Immunity: Homeostasis of the Superorganism." *Mucosal Immunology* 3: 450–60.

Funkhouser, L. J., and S. R. Bordenstein. 2013. "Mom Knows Best: The Universality of Maternal Microbial Transmission." *PLOS Biology* 11 (8): e1001631.

Gilbert, S. F. 1979. "The Metaphorical Structuring of Social Perceptions." *Soundings* 62: 166–86.

———. 1982. "Intellectual Traditions in the Life Sciences: Molecular Biology and Biochemistry." *Perspectives in Biology and Medicine* 26 (1): 151–62.

———. 1988. "Cellular Politics: Ernest Everett Just, Richard B. Goldschmidt, and the Attempt to Reconcile Embryology and Genetics." In *The American Development of Biology*, edited by R. Rainger, K. Benson, and J. Maienschein, 311–46. Philadelphia: University of Pennsylvania Press.

———. 1992. "Cells in Search of Community: Critiques of Weismannism and Selectable Units in Ontogeny." *Biology & Philosophy* 7 (4): 473–87.

———. 2003. "The Role of Predator-Induced Polyphenism in the Evolution of Cognition: A Baldwinian Speculation." In *Evolution and Learning: The Baldwin Effect Reconsidered*, edited by B. H. Weber and D. J. Depew, 235–52. Cambridge: MIT Press.

———. 2011. "Genetic and Epigenetic Sources of Selectable Variation." In *Transformations of Lamarckism: From Subtle Fluids to Molecular Biology*, edited by E. Jablonka and S. Gissis, 283–93. Cambridge: MIT Press.

———. 2013. "Symbiosis as a Way of Eukaryotic Life: The Dependent Co-origination of the Body." *Journal of Biosciences* 39: 1–9.

———. 2014. "A Holobiont Birth Narrative: Epigenetic Transmission of the Human Microbiome." *Frontiers in Genetics* 5: 282. doi:10.3389/fgene.2014.00282.

Gilbert, S. F., and J. Greenberg. 1984. "Intellectual Traditions in the Life Sciences: II. Stereocomplementarity." *Perspectives in Biology and Medicine* 28 (1): 18–34.

Gilbert, S. F., and J. Bard, 2014. "Formalizing Theories of Development: A Fugue on the Orderliness of Nature." In *Towards a Theory of Development*, edited by A. Minelli and T. Pradeu, 129–43. Oxford University Press.

Gilbert, S. F., J. Sapp, and A. I. Tauber. 2012. "A Symbiotic View of Life: We Have Never Been Individuals." *Quarterly Review of Biology* 87 (4): 325–41.

Gilbert S. F., T. C. Bosch, and C. Ledón-Rettig. 2015. "Eco-Evo-Devo: Developmental Symbiosis and Developmental Plasticity as Evolutionary Agents." *Nature Reviews Genetics* 16 (10): 611–22.

Gluckman, P. D., and M. A. Hanson. 2004. "Living with the Past: Evolution, Development, and Patterns of Disease." *Science* 305 (5691): 1733–36.

Hill, D. A., M. C. Siracusa, M. C. Abt, B. S. Kim, D. Kobuley, M. Kubo, T. Kambayashi, et al. 2012. "Commensal Bacteria-Derived Signals Regulate Basophil Hematopoiesis and Allergic Inflammation." *Nature Medicine* 18 (4): 538–46.

Hooper, L. V., M. H. Wong, A. Thelin, L. Hansson, P. G. Falk, and J. I. Gordon. 2001. "Molecular Analysis of Commensal Host-Microbial Relationships in the Intestine." *Science* 291 (5505): 881–84.

Huxley, T. H. 1893. *Darwiniana: Collected Essays, Vol. II.* London: Macmillan.

Klein, J. 1982. *Immunology: The Science of Self-Nonself Discrimination.* New York: John Wiley & Sons.

Koren, O., J. K. Goodrich, T. C. Cullender, A. Spor, K. Laitinen, H. K. Bäckhed, A. Gonzalez, et al. 2012. "Host Remodeling of the Gut Microbiome and Metabolic Changes during Pregnancy." *Cell* 150 (3): 470–80.

Lanning, D. K., K. J. Rhee, and K. L. Knight. 2005. "Intestinal Bacteria and Development of the B-Lymphocyte Repertoire." *Trends in Immunology* 26 (8): 419–25.

Lee, Y. K., and S. K. Mazmanian. 2010. "Has the Microbiota Played a Critical Role in the Evolution of the Adaptive Immune System?" *Science* 330 (6012): 1768–73.

Levinas, E. 1969. *Totality and Infinity.* Dusquene: Duquesne University Press.

Lyson, T. R., B. A. Bhullar, G. S. Bever, W. G. Joyce, K. de Queiroz, A. Abzhanov, and J. A. Gauthier. 2013. "Homology of the Enigmatic Nuchal Bone Reveals Novel Reorganization of the Shoulder Girdle in the Evolution of the Turtle Shell." *Evolution & Development* 15 (5): 317–25.

Mann, T. 1969 [1924]. *The Magic Mountain.* Translated by H. T. Lowe-Porter. New York: Vintage.

Margulis, L. 1981. *Symbiosis in Cell Evolution: Life and Its Environment on the Early Earth.* New York: W. H. Freeman.

Margulis, L., and R. Fester. 1991. *Symbiosis as a Source of Evolutionary Innovation.* Cambridge: MIT Press.

Margulis, L., and D. Sagan. 2001. "The Beast with Five Genomes." *Natural History* 110 (5): 38.

McFall-Ngai, M., M. G. Hadfield, T. C. Bosch, H. V. Carey, T. Domazet-Loso, A. E. Douglas, N. Dubilier, et al. 2013. "Animals in a Bacterial World: A New Imperative for the Life Sciences." *Proceedings of the National Academy of Sciences USA* 110 (9): 3229–36.

Moran, N. A. 2007. "Symbiosis as an Adaptive Process and Source of Phenotypic Complexity." *Proceedings of the National Academy of Sciences USA* 104 (Supplement 1): 8627–33.

Moran, N. A., and D. B. Sloan. 2015. "The Hologenome Concept: Helpful or Hollow?" *PLOS Biology* 13 (12): e1002311.

Mosse, G. 1997. *German Jews beyond Judaism.* Cincinnati: Hebrew Union College Press.

Nyhart, L. K., and S. Lidgard. 2011. "Individuals at the Center of Biology: Rudolf Leuckart's *Polymorphismus der Individuen* and the Ongoing Narrative of Parts and Wholes. With an Annotated Translation." *Journal of the History of Biology* 44: 373–443.

Obata, T., Y. Goto, J. Kunisawa, S. Sato, M. Sakamoto, H. Setoyama, T. Matsuki, et al. 2010. "Indigenous Opportunistic Bacteria Inhabit Mammalian Gut-Associated Lymphoid Tissues and Share a Mucosal Antibody-Mediated Symbiosis." *Proceedings of the National Academy of Sciences USA* 107 (16): 7419–24.

Oliver K. M., P. H. Degnan, M. S. Hunter, and N. A. Moran. 2009. "Bacteriophages Encode Factors Required for Protection in a Symbiotic Mutualism." *Science* 325 (5943): 992–94.

Peterson, D. A., N. P. McNulty, J. L. Guruge, and J. I. Gordon. 2007. "IgA Response to Symbiotic Bacteria as a Mediator of Gut Homeostasis." *Cell Host and Microbe* 2 (5): 328–39.

Pietsch, T. W. 2005. "Dimorphism, Parasitism, and Sex Revisited: Modes of Reproduction among Deep-Sea Ceratioid Anglerfishes (Teleostei: Lophiiformes)." *Ichthyological Research* 52 (3): 207–36.

Pradeu, T. 2010. "What Is an Organism? An Immunological Answer." *History and Philosophy of the Life Sciences* 32 (2–3): 247–67.

———. 2012. *The Limits of the Self: Immunology and Biological Identity.* Translated by E. Vitanza. Oxford: Oxford University Press.

Queller, D. C., and J. E. Strassmann. 2009. "Beyond Society: The Evolution of Organismality." *Philosophical Transactions of the Royal Society of London. Series B, Biological Sciences* 364 (1533): 3143–55.

Rawls, J. F., B. S. Samuel, and J. I. Gordon. 2004. "Gnotobiotic Zebrafish Reveal Evolutionarily Conserved Responses to the Gut Microbiota." *Proceedings of the National Academy of Sciences USA* 101 (13): 4596–601.

Rhee, K. J., P. Sethupathi, A. Driks, D. K. Lanning, and K. L. Knight. 2004. "Role of Commensal Bacteria in Development of Gut-Associated Lymphoid Tissue and Preimmune Antibody Repertoire." *Journal of Immunology* 172: 1118–24.

Rieppel, O. 2012. "The Evolution of the Turtle Shell." In *Morphology and Evolution of Turtles: Origin and Early Diversification.* Edited by Donald B. Brinkman, Patricia A. Holroyd, James D. Gardner, 51–61. New York: Springer.

Rosenberg, E., O. Koren, L. Reshef, R. Efrony, and I. Zilber-Rosenberg. 2007. "The Role of Microorganisms in Coral Health, Disease and Evolution. *Nature Reviews Microbiology* 5: 355–62.

Round, J. L., R. M. O'Connell, and S. K. Mazmanian. 2010. "Coordination of Tolerogenic Immune Responses by the Commensal Microbiota." *Journal of Autoimmunity* 34 (3): J220–25.

Sapp, J. 2009. *The New Foundations of Evolution: On the Tree of Life*. New York: Oxford University Press.

Sharon, G., D. Segal, J. M. Ringo, A. Hefetz, I. Zilber-Rosenberg, and E. Rosenberg. 2010. "Commensal Bacteria Play a Role in Mating Preference of *Drosophila Melanogaster*." *Proceedings of the National Academy of Sciences USA* 107 (46): 20051–56.

Smith, M. I., T. Yatsunenko, M. J. Manary, I. Trehan, R. Mkakosya, J. Cheng, A. L. Kau, et al. 2013. "Gut Microbiomes of Malawian Twin Pairs Discordant for Kwashiorkor." *Science* 339 (6119): 548–54.

Spencer, H. 1851. *Social Statics: or, The Conditions Essential to Happiness Specified, and the First of Them Developed*. London: John Chapman.

———.1906 [1864]. "Letter to G. H. Lewes." Quoted in *Herbert Spencer*, by J. Arthur Thomson, 139–40. London: J. M. Dent. http://www.gutenberg.org/files/39002/39002-h/39002-h.htm.

Stearns, S. C. 2007. "Are We Stalled Part Way through a Major Evolutionary Transition from Individual to Group?" *Evolution* 61: 2275–80.

Tauber, A. I. 2000. "Moving beyond the Immune Self? *Seminars in Immunology* 12: 241–48.

———. 2008. "Expanding Immunology: Defense Versus Ecological Perspectives." *Perspectives in Biology and Medicine* 51 (2): 270–84.

———. 2009. "The Biological Notion of Self and Non-self." *Stanford Encyclopedia of Philosophy*. http://plato.stanford.edu/entries/biologyself/.

Ulvestad, E. 2007. *Defending Life: The Nature of Host-Parasite Relations*. Dordrecht: Springer.

Walt, S. M. 1998. "International Relations: One World, Many Theories." *Foreign Policy* 110: 29–46.

Weaver, I. C. 2007. "Epigenetic Programming by Maternal Behavior and Pharmacological Intervention. Nature Versus Nurture: Let's Call the Whole Thing Off." *Epigenetics* 2 (1): 22–28.

Weaver, I. C., N. Cervoni, F. A. Champagne, A. C. D'Alessio, S. Sharma, J. R. Seckl, S. Dymov, M. Szyf, and M. J. Meaney. 2004. "Epigenetic Programming by Maternal Behavior." *Nature Neuroscience* 7: 847–54.

Weinberg, R. A. 2007. *The Biology of Cancer*. New York: Garland Science.

Wilson, E. B. 1896. *The Cell in Development and Inheritance*. New York: Columbia University Press.

Wilson, E. O. 2012. *The Social Conquest of Earth*. New York: Norton.

Wimsatt, W. C., and J. R. Griesemer. 2007. "Reproducing Entrenchments to Scaffold Culture: The Central Role of Development in Cultural Evolution." In *Integrating Evolution and Development: From Theory to Practice*, edited by R. Sansom and R. N. Brandon, 227–324. Cambridge, MA: MIT Press.

13

Philosophical Dimensions of Individuality

ALAN C. LOVE AND INGO BRIGANDT

Introduction

Natural philosophers have long been interested in individuality and the relationship between parts and wholes. A key source for this interest has been empirical examples from animals and plants where intuitive notions of individuality seem to break down, such as in cases of colonial marine invertebrates or insect metamorphosis. In the early modern period, Leibniz offered a novel view of nested individuality in which genuine individuals could be compositional elements or parts of larger individuals ad infinitum (Smith 2011). This view was inspired by microscopical discoveries of Leeuwenhoek and his contemporaries and provided strong persistence conditions for individuals through major life history transformations, such as metamorphosis and even death. Increased attention to complex life cycles of terrestrial and marine invertebrates in the nineteenth century led to an explosion of competing perspectives on how to conceptualize individuality (Elwick 2007). For example, Rudolf Leuckart tried to articulate general laws about biological individuals and their part-whole relations by interpreting the alternation of generations as an instantiation of the division of labor, where the different parts of a colonial organism were individuals in their own right (Nyhart and Lidgard 2011).

More recently, individuality has been of primary interest to contemporary philosophers because of its crucial role in different aspects of evolutionary biology. These debates include whether species are individuals or classes, what counts as a unit of selection, and how transitions in individuality occur evolutionarily. These discussions often rely on prior accounts of individuality to determine *whether* the concept is applicable, such as in the case of species, and *when* the concept is applicable, such as in cases where there is a question about whether a group can exhibit properties that are not simply an aggregate sum of its individual member organisms. For example, Ellen Clarke reviews thirteen distinct conceptions of individuality and argues that the situation demands a solution: "there is a real problem of biological individuality, and an urgent need to arbitrate among the current plethora of solutions to it . . . there is a choice to be made about which definition, or how many definitions to accept" (Clarke 2010, 312 and 314). One reason for the urgency derives from the needs of evolutionary theory, which must be able to identify individuals in order to measure reproductive fitness: "counting the units enables us to predict and explain how the traits of such units are changing over time, under the action of natural selection" (Clarke 2013, 429). Clarke offers "a monistic account of organismality" to solve the problem of individuality (2013, 429); her unified account says what a biological individual is in all circumstances.

We will return to Clarke's particular proposal below (in the subsection "Evolutionary Individuality"), but here it serves as an introductory illustration of how philosophical analyses are often conducted in terms of metaphysics ("what is an individual?"), rather than epistemology ("how can and do researchers conceptualize individuals to address different scientific goals?").[1] As Clarke reminds us, a "plethora of solutions" are on offer—different ways that researchers conceptualize individuals. These include displaying the capacity of reproduction, having a single-cell bottleneck during the life cycle, or exhibiting a separation of germ and soma. A metaphysical framing of the issue suggests that one property or a combination of several can be used to univocally answer the question of what an individual is. This metaphysical orientation often takes the shape of fundamental theorizing. Many philosophers of science assume that our best fundamental theories inform us about the basic furniture of the world. The necessary and sufficient conditions for individuality are ascertained from abstract theorizing that is fundamental to all of biology; in this case, evolutionary theory adopts the mantle of fundamentality.[2] Once formulated, the fundamental theory of what an individual is governs scientific practice—that is, it tells scientists what to count when measuring fitness and drawing evolutionary inferences. The metaphysics of what an individual is determines how biologists do their epistemology or go about the practice of individuation.

In this chapter we review and characterize several philosophical distinctions pertinent to individuality, such as metaphysics versus epistemology, individuals versus classes, and monism versus pluralism, in light of the diverse contributions to the volume. We pay special attention to the way metaphysical assumptions have animated controversies, past and present. Both biological and philosophical researchers have frequently assumed that they were engaged in fundamental theorizing to determine what an individual is, but this assumption is often unnecessary and unwarranted. Indeed, our aim is to shift attention in discussions of individuality from metaphysics to epistemology. We argue that some of these controversies involve epistemological differences rather than metaphysical disagreement. In addition to shedding light on several cases explored in the contributions, this reorientation implies that a pluralist stance about biological individuality is warranted. The epistemological conclusion resulting from this reorientation yields consequences for metaphysics because the pluralism arises out of different scientific interests that produce distinct approaches to the complexity of living phenomena in the world.

Individuality and the Return of Metaphysics

Despite a predisposition against metaphysics in early and mid-twentieth-century philosophy, where epistemology and the analysis of language were favored in the study of the sciences, the past twenty-five years have been marked by a revival of metaphysical theorizing as a central area of analytic philosophy of science.[3] This renaissance of metaphysical theorizing has run in parallel with analyses from a variety of philosophers interested in questions about the identity of objects, parts and wholes, the persistence of objects through time and across change, and accidental versus essential properties of individuals. For example, mereology—the logic of parts and wholes—has been pursued as an alternative to set theory, where parts and wholes are concrete objects unlike a set, which is an abstract entity even when having concrete objects as its members (Varzi 2014). However, mereological theory does not put empirical constraints on what objects can count as a whole (e.g., a *biological* individual), and instead offers a general logical characterization of the relationship between a part and the whole or among different parts of a whole.

One prominent metaphysical task is to articulate a coherent framework of change that recovers how an object can be the same entity at two different time points. To this end, some metaphysicians conceptualize a material object as a four-dimensional entity composed of three spatial dimensions and one temporal dimension or extended region of time during which it exists. On this type of account, objects have temporal parts in addition to spatial parts (Hawley 2010).[4] One paradox of this view is that an object cannot (technically speaking) change because two of the four-dimensional object's temporal parts—no matter how different they are—are still two existing parts of one overall object. Finally, metaphysicians often distinguish between an individual's accidental and essential properties (Robertson and Atkins 2013).

An accidental property is one an individual happens to have but which it could lack. Thus, an individual could change any of its accidental properties and still remain the same individual. In contrast, an individual must possess its essential properties because they are what it is to be this individual. If any of these essential properties were absent then the object would not exist (any longer). This is about the defining properties of a *particular* individual, but it does not address the properties that define the category "individual," in particular which objects qualify as biological individuals (e.g., organisms).

These debates about mereology, temporal parthood, and essential properties in analytic metaphysics are disconnected from the sciences and biology in particular. They are conducted in full abstraction from concrete details and pertain to questions that are not specifically biological (e.g., the very possibility of change, in general). As we observed, philosophical discussions were not always disconnected from empirical cases and philosophers' reflections on individuality have been motivated by intriguing biological and other material examples. In contrast to contemporary analytic metaphysicians, philosophers of biology have addressed some of these examples when treating biological individuality, such as whether physiologically linked and genetically identical stands of quaking aspen are a group of individual trees or a single individual. However, many of these philosophers of biology share a methodological assumption with analytic metaphysicians; namely, that there is a single correct account of what an individual is or how we should understand individuality. Just as an account of temporal parthood in analytic metaphysics is intended to cover all cases of parts through time, so also a fundamental theory of individuality in philosophy of biology is usually intended to cover all cases of biological individuals. This monist impulse—there is a single correct account—derives from a type of metaphysical orientation that assumes a univocal parsing of the world into individuals that can be counted and non-individuals that must be treated otherwise.[5] Although we favor an emphasis on biological examples that provoke questions about what counts as an individual, we resist the monist impulse by reorienting analyses of individuality epistemologically. In order to achieve this reorientation, we first need to see the contours of some discussions in contemporary philosophy of biology.

Philosophical Interest in Biological Individuality

SPECIES AS INDIVIDUALS

A classic debate in biology concerns how to conceptualize species, especially given their mutability. Biologists and philosophers jointly effected a major transformation in this debate with the idea that a species is an individual. Proposed by Michael Ghiselin (1974) and elaborated by David Hull (1978), this conceptualization challenged a traditional and predominant idea that a particular species is a natural kind. On the natural kind view, organisms are members of a species; on the individual view, organisms are parts of a species-individual, just like cells are parts of an individual organism. Implicit in the natural kind view is a commitment to some form of similarity among species members, possibly even essential properties shared by all members. An individuality thesis has no such commitment; cells can be quite different and still be parts of one individual (e.g., mammalian blood cells that lack a nucleus). Likewise, the variation among organisms within a species-individual need not be circumscribed by a morphological similarity metric or any presumption about shared genetic composition.

More generally, the species-as-individuals thesis was meant to accommodate the fact that a species taxon: (i) is denoted by a proper name; (ii) is a particular object that occupies a certain region of space and exists during some period of time; and, (iii) exhibits variation at any time and can be subject to significant (evolutionary) change across time, while still being the same species. An individual has these three basic properties, whereas a natural kind is often understood not to have them, especially spatiotemporal boundedness and mutability. This same line of argument has been extended to higher taxa and homologues. Thus, the higher taxon "mammals" is considered an individual, and homologous structures in different organisms, such as kidneys, are not members of a natural kind but parts of a homologue-individual (Ereshefsky 2009; Wagner 2014; see also Brigandt, this volume).[6]

The notion of "individual" at work in these arguments is relatively generic because it was typically assumed that the important difference is marked by distinct ontological categories: individuals versus natural kinds. This notion does not provide more specific conditions on individuality, which might be desirable for distinguishing particular species of microorganisms as individuals from an individual organism containing interacting microorganisms from different taxa. This highlights a lacuna in discussions of species as individuals: even though there have to be criteria for determining which organisms constitute a species-individual (and which do not), most proponents of the species-as-individuals thesis have not explicitly addressed those criteria. A key reason for this lacuna is that species-as-individuals proponents associate such criteria with the membership conditions for natural kinds (Ereshefsky 2010).

Although the species-as-individuals thesis has become near orthodoxy among both biologists and philosophers, Richard Boyd (1999a) introduced a revised conceptualization of natural kinds—the homeostatic property cluster (HPC) account—that does not involve traditional assumptions (e.g., natural kinds are spatiotemporally unrestricted), and therefore permits species and other biological entities to be natural kinds. A key element of the HPC account is that a whole cluster of properties, which are merely correlated, can characterize a kind. As a consequence, a particular member of a kind need not have all of these properties, and any one property need not be found in all kind members (e.g., members of a species). While this element makes room for diversity within a kind (i.e., for a species taxon to exhibit variation), another element severs the commitment to these characteristic properties of a kind being intrinsic (e.g., genetic composition or morphological similarity). Instead, relational properties, such as "having the same ancestor as [another organism]" or "being able to interbreed with [another organism]," which explicitly include criteria used in species concepts, are part of the homeostatic cluster of properties characterizing the kind. These types of relational properties are fully compatible with phenotypic diversity and evolutionary change.[7]

One positive feature of the HPC approach is that it is a general account of kinds in biology and other special sciences. An HPC approach goes beyond a narrow focus on species and attempts to capture other kinds, such as stem cells and genes, which exhibit considerable internal diversity (Wilson et al. 2007).[8] Another significant feature, especially in the present context, is that the HPC approach introduces epistemic considerations into a discussion about species as individuals that has tended to focus exclusively on ontological issues about the nature of species. By asking what classificatory, explanatory, and other scientific purposes are addressed by grouping different objects together, regardless of whether they are viewed as forming a kind or an individual (Brigandt 2009; Boyd 1999a), the strategy of the HPC account moves away from the metaphysically framed question—what are species—to epistemologically framed questions—how and why are biologists grouping organisms into species.

EVOLUTIONARY INDIVIDUALITY

Recent conceptual reflection has concentrated on the issue of biological individuality in evolutionary theory, where there has been renewed attention to how individuality originates evolutionarily (e.g., see the contributions in Bouchard and Huneman 2013). Philosophical accounts of individuality often try to capture two kinds of challenging phenomena. The first set of phenomena involves cases where individuality is difficult to assess across the diversity of life forms. An aspen tree, for instance, is connected underground to other trees within a whole grove of aspen. These interlinked aspen trees are genetically identical so that the spatially circumscribed boundedness and genetic uniqueness often used to define paradigmatic individuals do not hold. There are many different examples of colonies of organisms with highly integrated causal linkages and specialized roles among their constituent organisms. Sometimes, as in the case of an ant colony, most organisms cannot even reproduce—resulting in something analogous to the separation of germ-line cells from somatic cells and a bottleneck across generations, which are observed in many (though not all) metazoans. Such a "superorganism" colony raises the question of whether it is a biological individual. The constituent organisms of a Portuguese man o'war (*Physalia physalis*) exhibit a functional division of labor, have spatial contiguity, and form a spatially bounded whole—like a paradigmatic multicellular individual. Since these constituent organisms cannot survive in isolation, one could argue that they are not individuals, unlike the whole Portuguese man o'war, even though the latter is often conceptualized as a colony *composed of* multiple individuals.

The widespread phenomenon of symbiosis complicates these issues further, given that a putative individual can include organisms from very different taxa. Many of these seemingly strange cases involve multicellular organisms, but philosophers have increasingly scrutinized the realm of microbes (O'Malley 2014). Microbial communities can involve the same kind of close interactions and functional specialization observed in paradigmatic individuals, but they also exhibit lateral gene transfer, all of which can provide evolutionary coherence without genetic identity (Ereshefsky and Pedroso 2013, 2015; Clarke 2016).

Theories of individuality typically seek necessary and sufficient conditions for ascertaining when biological objects are individuals (e.g., Clarke 2013) or offer dimensional analyses that return judgments in terms of degrees of individuality (e.g., Godfrey-Smith 2009). The properties often associated with individuality go together in many metazoans: being internally contiguous and having a spatial boundary, having specialized and physiologically integrated parts, being able to reproduce, bearing adaptations, and having mechanisms that reduce internal evolutionary conflict, such as germ-soma separation (Clarke 2010). However, the various nonstandard cases demonstrate that across all taxa, different criteria of individuality do not always align.[9] One philosophical response is to use an HPC approach (introduced above in the subsection on Species as Individuals) because it is intended to make room for diversity within a complex kind. Rob Wilson and Matt Barker (2013) argue that the ontological category of "biological individual" is characterized by several different properties. These properties are correlated, but only imper-

fectly, so that many organisms do not possess all of the characteristic properties and some non-organisms turn out to be biological individuals.

A more nuanced strategy is to treat some branches of the tree of life differently. For each taxon or lineage, the goal would be to offer some precise
criteria of individuality that should be met in this circumscribed context.
Ellen Clarke (2012) adopts this approach by looking for criteria in plants that
have biological effects analogous to situations known from animals. For example, instead of using the animal-specific separation of germ and soma as
a criterion for plants, Clarke encourages us to concentrate on mechanisms
that contribute to the effect of producing heritable variance in fitness. Even
though plants do not have the reproductive division of labor found in animals, they exhibit other mechanisms that have the same effect, such as apo-
mictic reproduction or meristem stratification. Thus, while such criteria were
originally derived from abstract considerations about individuality (i.e., having the effect of producing heritable variance in fitness), the particular account is plant-specific. More recently, Clarke (2013) has offered an account
that relies on the concept of multiple realization. The individuating mechanisms that underlie individuality can be instantiated in many ways; the particular way in which a mechanism is instantiated in a taxon is not as crucial.
The two abstract mechanisms on this account are (a) policing mechanisms
that prevent an object's constituents from being subject to differential selection, and, (b) demarcating mechanisms that facilitate an object's integrity so
that it can undergo selection. There are many ways to police or demarcate,
hence multiple realization, but the mechanisms must be present. According to Clarke, a biological individual is any object that exhibits both types of
mechanisms simultaneously.

The second set of phenomena that philosophical accounts of individuality attempt to capture is major evolutionary transitions and the evolution of
individuality. How did unicellular organisms give rise to multicellular organisms? How did some multicellular animals come to form superorganisms?
In addition to providing a characterization of what a biological individual
is, these questions require an account of how individuals emerge at new levels of organization. For this second set of phenomena, the explanatory focus
is on natural selection and the concomitant notions of fitness, conflict, and
cooperation (Sterner, this volume). Once a "genuine" individual at a higher
level has arisen, mechanisms must be in place to eliminate (or minimize)
fitness differences among its constituent parts. For example, germ-soma separation can eliminate evolutionary conflict among constituent cells so that
the multicellular animal is the level at which selection operates. Under these
conditions, any fitness difference and selection at the level of constituent

cells—though potentially beneficial (e.g., clonal selection in lymphocytes) or harmful (e.g., cancer)—would be evolutionarily inert. This encourages a theoretical orientation that highlights the capacity to eliminate fitness differences among constituent parts as a critical prerequisite to stabilizing new types of individuality. By implication, mechanisms that suppress conflict at sub-organismal levels become criteria for individuality, and entities that do not exhibit these mechanisms can be considered suspect as individuals.

In addition to the elimination of conflict and facilitation of cooperation after an individuality transition, the transition process itself must be explained. During a transition in individuality, there are still potential evolutionary conflicts among the lower-level individuals; some have the opportunity to cheat on and exploit the cooperation within the colony. To the extent that natural selection favors cooperation in the transition process, it is an instance of group selection because the colony is still a group of lower-level individuals and not yet a higher-level individual. For this reason, philosophers invoke multilevel selection theory (Okasha 2006). Moreover, even in highly derived organisms, the formation of chimeras that have genetically heterogeneous cellular constituents regularly occurs. Slime molds (dictyostelids and myxomycetes) have a life cycle where unicellular conspecifics aggregate to form a multicellular "organism." In many sponges, cnidarians, bryozoans, and ascidians, two or more multicellular organisms—which are conspecific, yet genetically distinct—fuse to form a chimera ("intergenotypic fusion"). Apart from creating problems for criteria of individuality that insist on genetic homogeneity, genetic chimeras raise questions for selection-based explanations given that some cheaters can be horizontally transmitted (Grosberg and Strathmann 2007).

Inspired by the role that individuals play in evolutionary theory and the difficult questions raised by the evolution of individuality, most contemporary philosophical accounts of individuality are evolutionary in orientation (Clarke 2012). These accounts construe a biological individual as the bearer of fitness and the entity on which natural selection operates. The notion of fitness presupposes a conception of how many offspring a parent has, so one must define which biological object qualifies as a parent individual, and which one counts as a separate offspring individual. Peter Godfrey-Smith (2009) approaches these questions by starting from populations (rather than individuals), which he terms "Darwinian populations" if they are able to undergo evolution by natural selection. Any member of such a population is, derivatively, a "Darwinian individual," which can be an entity from genes to superorganisms. Godfrey-Smith conceptualizes individuality using five quantitative properties that describe how Darwinian populations can differ:

(1) heritability (i.e., the degree of parent-offspring similarity), (2) the abundance of variation, (3) the degree of competition within the population (i.e., the extent to which a fitness gain in one individual lowers the fitness of others), (4) the smoothness of the fitness landscape (i.e., the extent to which a small change in an individual's traits results in a small change in fitness), and (5) the degree to which reproductive fitness is determined by an individual's internal character (as opposed to the influences from external features). This yields a five-dimensional space in which populations from different taxa occupy different positions. If all values for the properties are high, cumulative selection is possible and we are dealing with a paradigmatic Darwinian population. Other populations can exhibit different combinations of values for these properties and thereby exhibit different degrees of individuality for its members. An interesting feature of this account is that one individual may score higher than another individual in one dimension, but score lower (i.e., be less like a paradigmatic individual) with respect to a different dimension.[10]

In the history of biology, discussions of individuality tended to focus on physiological features, especially prior to the advent of evolutionary theory. Diverging from the philosophical trend of focusing on evolutionary construals of individuality, Thomas Pradeu (2012) reinvigorates the earlier physiological perspective.[11] Pradeu's account of biological individuality and identity concentrates on having an immune system, which establishes and maintains an individual's boundaries. He claims that all known organisms have an immune system, including invertebrates, plants, and prokaryotes, and thereby possess the required basis for individuality. In addition to highlighting the physiological properties of the immune system, Pradeu argues that his approach provides a criterion of individuality relevant to evolutionary accounts: "It is necessary, in fact, to closely observe the physiological mechanisms of organisms, especially immunological mechanisms, in order to determine what is or is not an evolutionary individual" (2012, 259). Apart from preventing fitness conflicts at lower levels by mechanisms such as a separation between the germ-line and somatic cells, there is a need to eliminate new variants by policing mechanisms, and the immune system is instrumental in this task.

Shifting Attention from Metaphysics to Epistemology

Whether it is species construed as individuals or HPC kinds, or biological individuals defined evolutionarily or physiologically, these discussions are frequently conducted in terms of metaphysics ("What are species? What is an individual?"), rather than epistemology ("How can and do researchers conceptualize species or individuals to address different scientific goals?").

This is especially noticeable in the widespread assumption that historical construals of individuality in terms of physiology are false or subsidiary because biological individuals are primarily entities on which natural selection acts. From a metaphysical standpoint, it is not sufficient to provide characterizations of different criteria for individuality in the context of scientific practice; the situation calls for adjudicative action: "there is an urgent need for the concept to be cleaned up" (Clarke 2010, 323). Thus it is not surprising that Ellen Clarke offers "a monistic account of organismality" (2013, 429). This urgent need derives from an assumption that there is a single correct or monistic account of what species or individuals are (metaphysics), regardless of the diverse and incompatible ways that scientists designate species and individuals (epistemology). We recommend reversing the orientation and starting from the vantage point of epistemology, then (if desired) proceeding to metaphysics.

How would one effect this reorientation? A first step is to explore answers to questions surrounding investigations into or reliant on a notion of biological individuality: How does a biologist decide to count individuals? How does a biologist decompose an individual into (meaningful) parts? What criteria are used and should be used depends on the underlying goals of inquiry (see also Lidgard and Nyhart, this volume). This leads to a second step that explores answers to questions about epistemic goals: Why does a biologist use a particular conceptualization of individuality? What methodological or explanatory purpose do certain decomposition criteria serve? In evolutionary inquiry, measurements of fitness are important for explaining changes in populations. With respect to this aim, a well-defined fitness value is a criterion of adequacy for an account of what an individual is. Similarly, some considerations for decomposing individuals into parts are motivated and constrained by the need to account for evolutionary transitions, which require the suppression of selective dynamics among constituent parts. In developmental inquiry, in contrast, spatial boundaries of parts are important for understanding changes during ontogeny. Therefore, ways of counting individuals or decomposing individuals into parts and tracking them through time have spatial separation as a criterion of adequacy (Love forthcoming). In physiological inquiry, functional interconnections are central to tracking activity in different systems. In such an epistemic context, counting individuals may require taking into account mutualistic relationships and decomposing individuals into parts that transgress spatial boundaries and diverge from what is relevant to selective dynamics. In systematic inquiry, a robust operationalization of characters and character states, including the absence of features, is a criterion of adequacy for classification and building phylogenies. Therefore, criteria for counting individuals and decomposing them into parts can be heterogeneous, including both structural and functional features, as long as reliable and robust character codings are achieved.

This coarse-grained way of distinguishing how different goals constrain the epistemology of individuality or individuation can be traced out further to show that within evolutionary, developmental, physiological, or systematic inquiry (inter alia), there are additional constraints that operate during scientific inquiry. Although there are thematic differences in the conceptualization of individuals and their parts across disciplines, a more fine-grained tracing of the goals of inquiry within a discipline can reveal additional constraints and suggests that there are not simply independent concepts of individuality across disciplines (e.g., developmental individuals, physiological individuals, evolutionary individuals). While one cannot exclude a priori the possibility of there being a unique definition of individuality that meets all relevant epistemic goals in all disciplines, the complexity of the biological world generally means that all-purpose concepts will not be useful even if they can be formulated. Given this, the diversity of goals in scientific research generates an expectation that there will always be many extant approaches to individuation and decomposition in biological inquiry, even within the same discipline. Importantly, when the presence of these different investigative and explanatory goals is explicitly acknowledged and their details are characterized, we can gain philosophical insights into how different approaches to what counts as individuals and their meaningful parts fruitfully coexist in biological sciences.

Philosophers have drawn attention to this feature of biological epistemology for more than four decades (Kauffman 1971; Wimsatt 1974; Winther 2011). This raises the question of how most philosophical accounts of individuality could not have such a central element of scientific reasoning in view. Fundamental theorizing is one answer. Particular accounts advance specific properties as defining of individuality by treating one area of inquiry as the most fundamental. This then governs how individuals are understood in all areas of inquiry and is commensurate with a metaphysical orientation aiming to determine the single correct (monistic) account of what individuals are. The strategy of fundamental theorizing is most frequently observed with respect to individuality and evolutionary processes (see the above subsection "Evolutionary Individuality"). A common justification is that the connection between individuality and fitness considerations is somehow primary or basic:

> It is in an evolutionary context that the notion of the individual really does a lot of work. . . . The notion of the biological individual is inextricably bound up with the notion of fitness. (Clarke 2010, 313)

Clarke's claim that biological individuals are "inextricably bound up" with concerns about survival and reproduction implies that the evolutionary perspective is somehow fundamental for all of biology when evaluating the nature of individuality. Although fundamental theorizing can be reductionist in character, focusing on lower levels of organization (e.g., molecular biology, chemistry, or physics), the present situation shows that "fundamental" is not always the lowest level of size and spatial parthood. In Clarke's claim, the notion of fitness from evolutionary biology is deemed to be most basic. Peter Godfrey-Smith holds a similar view: "The link between 'individuality' and reproduction is in some ways inevitable. Reproduction involves the creation of a new entity, and this will be a countable individual" (Godfrey-Smith 2009, 86). Although his notion of individuality is more multifaceted than the one offered by Clarke, the underlying methodology of fundamental theorizing is shared. Even Thomas Pradeu (2012), who focuses on a physiological, immune system construal of biological individuality, views his account as picking out the most fundamental feature of individuality because he maintains that evolutionary aspects of individuality rely on the immunological properties he singles out: "It is still necessary to examine the physiological processes produced in the organism to arrive at a precise definition of what, in each case, counts as an evolutionary individual" (2012, 260).

The strategy of fundamental theorizing requires discounting individuation practices from areas of inquiry that lead to conflicts with the resulting account of individuality. Divergent characterizations used by biologists are seen as something to be eliminated or reinterpreted because the metaphysical account of individuality is meant to govern how biologists do their theorizing—it says what individuals really are. At best, different individuation practices can be subsumed within the fundamental account; at worst, they are mere tools that aid inquiry but lack correspondence to the structure of the world. While developmental or physiological inquiry speaks of individuals, these are "real" only if they map onto the account of individuality drawn from fundamental theorizing in the context of evolution.

Although the strategy of fundamental theorizing about evolutionary individuality has some benefits, its commitment to a monistic account and discounting of successful individuation practices that diverge from it are problematic and leave opaque the forms of reasoning we find among biologists engaged in individuation and decomposition. If we shift from metaphysics to epistemology in discussions of individuality by foregrounding the diverse goals of inquiry that shape and constrain what counts as individuals and their meaningful parts, then we are in a position to achieve at least three significant philosophical aims (see also Brigandt 2013; Sterner 2015; Lidgard and Nyhart, this volume). First, we are able to comprehend why diverse individuation practices within disciplines successfully function in the sense of leading to fecund investigative approaches and increasing explanatory depth. If we can make explicit how definitions of individuality match the underlying epistemic goals of biologists, then we better comprehend why these practices work (and conversely, whether there is some form of mismatch). Second, we gain tools for identifying sources of conflict within and across disciplines, such as the existence of different criteria of adequacy (e.g., functional versus structural) due to different goals of inquiry. When conflict arises, scrutinizing these criteria and underlying goals can help in evaluation without assuming that one approach to individuality is wrong. Third, we can discriminate more finely among philosophical approaches to individuality by explicitly distinguishing between definitions of individuality and epistemic goals of inquiry.[12] For example, the HPC account endorsed by Wilson and Barker (2013), which views many imperfectly correlated properties as constitutive of individuality, does not inherently favor any particular property or biological domain as fundamental. At the same time, their approach is monistic given its unique HPC definition of individuality, as opposed to acknowledging that there are multiple concepts of individuality, each answering to different epistemic goals of inquiry (something only acknowledged by some other HPC approaches).

In the next section we explore how contributions to this volume can be understood to fulfill these three aims: comprehending successful matching of characterizations of individuality and epistemic goals, diagnosing conflicts between characterizations of individuality resulting from different goals, and discriminating more precisely among philosophical perspectives on individuality.

Illuminating Case Studies of Individuality in Biological Inquiry

The historical, philosophical, and scientific analyses in this volume offer a treasure trove of materials for illustrating the value of shifting from metaphysics to epistemology in the context of discussions of biological individuality. This value has at least two dimensions in the context of comprehending distinct but successful practices of individuation, diagnosing conflicts about individuality, and discriminating among philosophical perspectives:

(1) interpretive, providing new and helpful perspectives on historical and contemporary scientific situations; and (2) explanatory, offering accounts of why biologists reason the way they do in different contexts of inquiry where distinct aspects of complex biological phenomena are in view. We highlight examples from both dimensions drawn from the chapters herein and take up their implications for formulating novel perspectives on the metaphysics of individuality in the final section. Our aim is not to be exhaustive but illustrative. This means individual chapters may be germane to both dimensions, but we concentrate attention on only one. The hope is that others would see the various benefits of an epistemological reorientation, which investigates the different epistemic goals underlying accounts of individuality as well as their metaphysical implications, and follow suit.

INTERPRETIVE

Michael Osborne (this volume) takes us back to the context of parasitology in relation to social theory in late nineteenth- and early twentieth-century France. This was a fertile period for cross-pollination between approaches because parasitology, as a labeled field, was coalescing as a medical specialization in schools of tropical medicine that might supersede bacteriological pathology. What ensued was an interweaving of different understandings of individuality, which our reorientation interprets as arising from distinct epistemic goals of inquiry. For example, if the aim was to understand the origins of human disease, then individuality was conceptualized immunologically as a relationship of host to pathogen, where the pathogen was of a different species than the host and the causal dynamics of significance pertained to the host's functional integrity. But if the aim was to understand the parasite and its life cycle, then individuality was understood physiologically in terms of whether the parasite was able to live on its own apart from one or more hosts (i.e., functional autonomy). Whether the parasite was autonomous from the host is bracketed when studying medical pathology; whether the host's functional integrity was compromised in a particular way was relegated to the background when studying the natural history of the parasite. Some researchers adhered to both epistemic aims as they moved between natural history and medical pathology, even though these aims generated distinct conceptualizations of individuality. This engendered unexpected friction as the matching relationship between aim and conceptualization sometimes exhibited slippage.

In addition to this interweaving of individuality notions germane to investigating parasitism, socioeconomic relations between France and its African colonies nurtured the application of biological concepts like parasitism to describe societal relationships. As a consequence, different conceptualizations of individuality, which were not always explicitly articulated, ended up being invested with commitments of social ideology. The inclusion of social-political aims added a layer of complexity to the multiple scientific goals in play, which in this case actually encouraged the scientists involved to understand their conceptual differences monistically—society could not be organized in more than one way at a time. If the focus was on an African colony drawing a benefit from France (another individual), which bore the cost, then understanding the colony as a distinct individual, negatively valenced (i.e., as a parasite), seemed intuitive. If the focus was on the health of France in light of its component parts, then the autonomy of the colony was less salient than the integrity of the nation's physiological functioning. Thus, the distinct conceptualizations of individuality, motivated by different epistemic goals of inquiry, became reified as competitors as they trafficked through social discourse in a way that did not occur when moving between the contexts of natural history and medical pathology. At the same time, the movement through social discourse led to reciprocal effects, such as using a supposed developmental trajectory for a category of humans to model a common evolutionary trajectory for parasitic species.

A different historical episode surrounding biological individuality comes to us in early twentieth-century German idealistic morphology through the work of Martin Heidenhain and his notions of "enkapsis" and enkaptic hierarchy, which Olivier Rieppel (this volume) investigates. These notions were meant to capture an integrated structural and functional organization of organisms composed of multiple levels and could be extrapolated to other levels (e.g., ecosystems and human societies). This complex, integrated understanding of individuality grew out of an organicist perspective, which rejected approaches to living systems that treated them as nothing but physicochemical parts. Thus, we can note that a particular epistemic goal constrained what counted as an individual and as meaningful parts; individuality had to be conceptualized in a way that supported a reciprocal dynamic of causal relations between the components acting to produce the complex whole and the complex whole influencing the behavior of the parts. But the epistemic goal also contributed to a monistic orientation because the enkaptic hierarchy literally encompasses all of the causal relations relevant to individuals as such.

Heidenhain was engaged in a form of fundamental theorizing and this is evident in the fact that enkapsis extended far beyond the organism. As noted earlier, "fundamental" does not necessarily mean lowest level of spatial parthood, so Heidenhain could eschew reduction to the physicochemical level and at the same time engage in fundamental theorizing by construing en-kapsis as the basis for various biological as well as societal phenomena. En-kapsis applied to ecology as much as evolution and moved across boundaries between zoology and social theory where it was used to justify aspects of Nazi ideology (e.g., "the individual will sacrifice his life if this is required for the survival of the whole," August Thienemann, cited in Rieppel, this volume). Apart from this export of biological ideas into political ideology, it is intriguing to hypothesize that Heidenhain's monistic interpretation of enkapsis potentially contributed to its methodological sterility. If enkapsis was understood in the context of investigating a specific type of individual, such as a metazoan organism, then it could be more successful in engender-ing a fruitful line of research about relationships between parts and wholes, especially if seen as one strategy among many for dissecting the causal archi-tecture of complex living systems. Rieppel sees value of this kind in applying the idea of enkapsis within contemporary biological investigations, and oth-ers have argued that it is distinctively helpful in going beyond standard part-whole relationships when conceptualizing hierarchically organized biological individuals (Zylstra 1992).

That monism might be associated with methodological sterility—that the complexity of the biological world generally means that all-purpose concepts will not be useful even if they can be formulated—is further supported in the work of Herbert Spencer. Snait Gissis (this volume) explicitly recognizes this: "perceiving any living entity as an individual for some purposes and as a collectivity for others would have greatly simplified his work." But Spencer was undaunted and advocated the methodological principle that the simple should be understood in terms of the complex and not vice versa. A hybrid notion of "collective individuality" was needed to capture the totality of in-tricate interplay and fundamental entanglement of organism-environment couplings. The distinctive epistemic goal of inquiring into the properties of interaction was so paramount that Spencer emphasized incessant relations occurring between individuals conceptualized as a joint array or system, not how to individuate them from one another or decompose them into mean-ingful parts. Individuality is not constitutive of interactive processes but the result or outcome of those processes. These processes accomplish indi-viduation by finding temporary but stable equilibrium points of individual-environment interactions.

It is difficult to operationalize Spencer's fundamental theorizing that took complex interactions and relations, such as organism-environment cou-plings, as the basis for understanding everything else. Identifying the single correct view of individuality involves treating relationships between parts, wholes, and their environments at a high degree of abstraction. This is a key reason why general all-purpose concepts are less useful in the investigation of complex biological phenomena. While it is in some sense true that nature exhibits "spatiotemporal bounding," biologists require more specific conceptualizations of individuality to meet epistemic goals of inquiry (e.g., spatiotemporal bounding that seals off internal physiological activities, as illustrated by an epithelium occluding the passage of ions). Both Heidenhain and Spencer recognized the complexity of biological individuality (multilevel and interacting with external conditions), but lost their grip on it by trying to capture individuality with a single concept that lacked operational traction. An important lesson is that the most fruitful matchings of individuality conceptions and epistemic goals typically derive from more circumscribed epistemic goals that carve off a portion of biological complexity rather than embracing it in full.

The alternation of generations is a striking biological phenomenon that puts direct pressure on conceptualizations of individuality, as Lynn Nyhart and Scott Lidgard (this volume) indicate. Organisms sexually reproduce to beget offspring that develop into adults that do not resemble their parents and reproduce asexually to beget offspring that develop differently but resemble the sexually reproducing, initial generation. Steenstrup concentrated on the radically different morphologies exhibited in these reproductive transitions and interpreted them as shifts in the nature of individuality ("something more than a metamorphosis is concerned," cited in Nyhart and Lidgard, this volume). Others tried to find a single conception of the individual that would account for the material continuity of an alternating, successive pattern across generations. Depending on the epistemic goal—whether to account for a change in the strategy of sexual reproduction (since reproduction was often taken as defining of individuality), or to account for the stable succession of alternating patterns of sexual reproduction across generations— different conceptualizations of individuality seemed better suited to the task. And when Steenstrup focused his attention across generations, he treated sexual reproduction as the higher form of individuality, defined in terms of increasing autonomy over time, which progressively succeeds the asexual form of individuality.

More generally, the phenomenon of a life cycle (illustrated dramatically by the alternation of generations) shows that goals of inquiry involve commitments to temporality that affect suitable characterizations of individuality. Richard Owen's epistemic goal of providing an explanation of diverse types of development in terms of a causal law of nature encouraged an emphasis on processes where hierarchical levels of individuals are cumulatively built up and propagated asexually and sexually in different ways. Huxley, in contrast, focused his epistemic attention on the "independent existence" or autonomy seen in vertebrates, and thus anchored individuality across time in sexual reproduction from a single egg in one generation to a single egg in the next generation. Asexual budding and other forms of propagation were renamed and relegated as subsidiaries, transformations of parts rather than changes in what counted as the whole. Plant phenomena introduced distinctive wrinkles since there was no paradigmatic part to fix the temporal origination or achievement of "autonomous" individuality, and yet these accented novel developmental processes in hierarchically structured systems relevant to the stability and propagation of organization characteristic of individuals. Both Owen and Huxley understood their accounts of individuality monistically, but from an epistemological perspective we can interpret their value in terms of distinctive goals of inquiry with different criteria of adequacy that are helpful in particular contexts for investigating and explaining the temporal origins and maintenance of biological individuality.

EXPLANATORY

Several of the interpretations above are suggestive of the explanatory dimension of our epistemic approach to individuality, which consists in investigating why biologists reason the way they do in different contexts of inquiry. Treating cells as individuals is one locus where we gain such traction. Biologists routinely shift their perspective on cellular individuality as a means to investigate biological systems. Conceptualizing cells in relationships of reciprocal regulation stresses their parthood, but focusing—as Andrew Reynolds (this volume) does—on cell-cell communication or cell sociology stresses their role as wholes or individuals.[13] The switch in focus makes additional conceptual resources accessible, such as reaching for a notion of the division of labor in modern nation states as a model for cell-organism relations when cells are understood as components, or resisting a machine metaphor for describing biological systems (e.g., cells as building blocks) by accenting joint cellular agency; appealing to cell sociology flags this because machine parts do not exist in community.

The ability to switch between different conceptualizations of individuality for cells (parts versus a whole) provides researchers with increased reasoning capacity to achieve different goals of inquiry surrounding individuality. If the aim is to comprehend the cellular level as functionally autonomous, then characterizing cells as individuals is warranted. For example, cellular sociobiology, in contrast to cell sociology, literally treats cells as independent agents. If the aim is to comprehend aggregate properties, such as are observed during development, then cellular independence is replaced by how a cell is affected and transformed by its "social" environment, and the agency of individual cells is muted to emphasize their joint contribution to a greater whole. These "group level" and genuinely social effects are better captured through definitions of individuality above the cellular level; and structural relationships, such as adhesion, emphasize the contribution of the cell to a larger whole. In each of these cases, what counts as the internal and external environment changes, thereby drawing attention to different causal relations that would otherwise be obscured. Conceptualizing cells as parts and wholes in different contexts of inquiry is a powerful analytical tool for dissecting complex biological phenomena that would be unavailable if individuality was fixed at a particular level.

Switching perspectives between cells as individuals and cells as parts of individuals is on full display for transitions of individuality (Herron, this volume). In these cases biologists need to understand why a particular direction of change has taken place, which means tracking the fitness costs and benefits related to different arrangements of and interactions among cells. Thus, individuality is typically conceptualized in terms of genetic and evolutionary criteria (see the above subsection "Evolutionary Individuality"). Matthew Herron uses the volvocine algae as a model taxon because of the diversity exhibited for between-cell organization and evolutionary trajectories towards higher-level individuality. He shows the epistemological value of analyzing evolutionary individuality from multiple vantage points by zooming in on developmental and ecological changes during the life cycle. This offers insight into what sequence of steps is most plausible for the transition from cells as individuals to cells as components of a multicellular individual. By implication, Herron is open to the evolutionary reversibility of different between-cell organizational features, which goes beyond the unidirectional explanatory focus taken by standard studies of transitions in individuality. This epistemic perspective makes it possible to experimentally explore changes in individuality that occur due to short-term ecological changes affecting population organization and thereby enrich our understanding of long-term evolutionary modification.

As opposed to answering whether a given *Volvox* spheroid is an individual, based on a monistic definition of individuality, Herron's epistemic goal of inquiry is to investigate how much individuality is present at each of the different levels: cells, colonies, and clones (genets). He lays out thirteen derived traits relevant to multicellularity, such as the conversion of the cell wall into extracellular matrix, an incomplete cytokinesis, and the retention of cytoplasmic bridges in adults. This sounds similar to Wilson and Barker's (2013) HPC construal of individuality, which views a cluster of properties as metaphysically defining of individuality even though not every individual needs to possess all of these properties. Yet Herron is not aiming for *the* definition of individuality; instead, he is trying to tease apart which derived traits are present in a certain taxon of the volvocine algae, and what sequence of steps occurred in a particular lineage. This empirical approach makes it possible to mechanistically dissect transitions in a way that relying only on evolutionary individuality at an abstract level, as we have seen in some instances of fundamental theorizing, does not.

Beckett Sterner (this volume) can be understood as making a similar point about the explanatory potential of more concrete perspectives on individuality. He argues that the standard view of an evolutionary individual deriving from fundamental theorizing with fitness as the most basic feature is inadequate to meet the epistemic goal of mechanistically explaining evolutionary transitions in individuality. Instead, he introduces a new concept— the "demarcator"—to refer to both developmental and ecological structures or processes occurring during a life cycle that are necessary (though not sufficient) to circumscribe individuals as such. This strategy involves examining concrete mechanisms that fulfill the role of demarcator, specifically through the control of inheritance, and thereby provide conditions of individuation in biological systems. Including concrete examples of scaffolding, these mechanisms are instantiations of the abstract properties appealed to when characterizing individuality (e.g., establishing a boundary by policing), and add what is missing when only fitness is utilized to mark out individuals. These mechanisms provide the conditions for individuality through change and multiplication during reproduction and criteria for what counts as a genuine part of the whole. Working out the detail of these concrete mechanisms also provides insight into why particular events, such as a unicellular bottleneck, occur *when* they do within a life cycle and why those events exhibit specific ranges of variation.

Although the concept of a demarcator needs to be fleshed out more thoroughly in empirical research, we can see that it could assist in fulfilling the aim of explaining evolutionary transitions more adequately, as well as aiding the investigation of how much individuality is present at different levels of organization and what sequences of steps are most plausible for transitions in different multicellular lineages. Here we have the explanatory dimension in potentia: if biologists aim to mechanistically explain evolutionary transitions in individuality, then additional conceptualizations of individuality beyond those based on an abstract construal of fitness in multilevel selection theory are required. Additionally, other epistemic goals, such as explaining why particular events occur within a life cycle, can be pursued with a characterization of individuality in terms of demarcators, which are otherwise opaque on the standard view of an evolutionary individual. Overall, these possibilities offer a rationale for why biologists could or should reason about biological individuality with different definitions, rather than a single all-purpose conception, in different contexts of inquiry given their specific epistemic aims (see also Sterner 2015).

Cells are not the only or primary meaningful conceptualization of parts in organisms. Biologists have a variety of reasons for decomposing an individual into parts (Winther 2011). Ingo Brigandt (this volume) focuses his attention on organismal parts in the context of one of the most perennial axes of debate: structure versus function. Each of these is manifest in different suites of epistemic goals, reflected in venerable terminology: anatomy versus physiology, comparative morphology versus functional morphology. Brigandt's central claim is that structure and function are on par in the context of decomposing individuals into parts; neither is more fundamental than the other. The key move in his argument is the isolation of a particular sense of function (function as *activity*), which permits the conceptualization of a functional body part that has equal standing to structural body parts by abstracting away from the contribution that an activity-function makes to the larger system or whole organism (Wouters 2003). His hierarchical approach, according to which a bodily part (structure as well as activity-function) is composed of both lower-level structures and activity-functions, heightens the parity between structure and function by using them in combination. Conceptual abstraction also permits biologists to use structure and function separately as principles of decomposition for the parts of biological wholes in different contexts of inquiry, thereby accounting for why biologists reason the way they do in different contexts of inquiry.

Although Brigandt focuses on how structure and function can work together to generate a more robust ontology of parts, he is fully aware that functional and structural characterizations of parts match different epistemic goals. Many structures have their identity by virtue of some activity, which might play multiple roles in different biological contexts. The practices of individuation for parts adopted by biologists are geared toward apprehending the different types of individuals that result from distinct contextual situations. The epistemological reorientation we advocate makes more salient those research programs that look specifically at what structures do and how that changes through time, such as during development. By analogy, the ability to work with multiple concepts of individuality not only serves different epistemic goals of inquiry but also makes possible their *joint* use to serve the *same* epistemic goal. This option is not visible to fundamental theorizing premised on monism and motivated by the metaphysical question of what a part is. And the value of our reorientation extends beyond the parts of organisms on which Brigandt's chapter focuses, as it holds from the molecular to the supraorganismal level, and thereby highlights why keeping multiple characterizations of individuality in the toolbox of scientific reasoning contributes to the achievement of diverse epistemic goals.

Even though it has receded to the background in most philosophical analyses, Hannah Landecker (this volume) reminds us in her contribution of the aim of explaining physiological autonomy and thereby considerably broadens the array of diverse epistemic goals. Her emphasis on conversions of agency in biological systems indicates that there is a different kind of transition in individuality that is of interest to biologists—metabolic transitions. If having the capacity to metabolize is one way of characterizing individuality physiologically, how do we understand the dynamics of causal agency in situations where individuals are consuming and transforming other individuals (a "logic of conversion")? Akin to Sterner's emphasis on concrete mechanisms controlling features of inheritance that are a necessary condition for individuality, Landecker's historical argument traces the emergence of a concept of metabolism that is a necessary condition for individuality: "metabolism is intrinsic to concepts of the process of individuation: how an organism becomes an individual, set off from the world, that becomes the unit of analysis for those who study locomotion, perception, and desire—physiology, neurology, psychology, etc." (Landecker, this volume).

Here it is clear that characterizations of individuality in terms of fundamental theorizing from evolutionary biology are not in the foreground. Those studying physiology, neurology, and psychology simply have different epistemic aims and came to them through different historical paths. Landecker highlights this tension in two ways. First, she notes that Hans Jonas exhibited a monist impulse to fundamentally theorize that metabolism is the sina qua non of individuality: "Jonas, as is the case with most philosophers, was trying to generate a universally applicable analysis of what is." The complicated historical stage of actors conceptualizing and investigating metabolism that Landecker details, alongside of contemporary research, destabilizes this monism dramatically. Second, she mentions that definitions of life have often been cast as a choice between two fundamental viewpoints: the capacity to form lineages and the capacity to be metabolically self-sustaining (Dupré and O'Malley 2009). Our epistemological reorientation encourages a rejection of this dichotomy because it is insensitive to the different epistemic aims at play in biological investigation. If the goal is to explain the persistent agency for an individual that has been consumed, then the capacity to form a lineage (of some sort) may be extremely useful given that the metabolic sustaining capacity is gone. If the goal is to account for a logic of conversion where the capacity to form a lineage is destroyed (in the eaten) in the service of preserving or maintaining another individual (the eater), then a capacity for self-sustaining metabolism necessarily moves to the foreground.

Implications and Conclusions

The epistemological reorientation for discussions of biological individuality that we have argued for does not require returning to metaphysical questions, but we believe the implications are worth exploring, if only briefly. As we have seen, most recent analyses of biological individuality have assumed that there is one correct account of the nature of individuality. For many philosophers, this monism is motivated by what we labeled "fundamental theorizing," which involves singling out one or more features as the most basic for being an individual (e.g., being a bearer of fitness). Competing accounts of individuality based on different features or combinations thereof (e.g., spatio-temporal boundedness or physiological autonomy) must either be rejected or demonstrated to derive from the fundamental feature. In contrast, both the interpretive and explanatory dimensions of our epistemological reorientation described in the previous section provide fodder for formulating novel perspectives on the metaphysics of individuality. One of these themes is that multiple concepts of individuality matched to different epistemic goals give us a better understanding of the complexity of biological individuality. Our analysis of this complexity goes beyond previous acknowledgements by some philosophers that different criteria of individuality may have to be used for different taxa (Clarke 2012, 2013; Wilson and Barker 2013). First, it includes and articulates scientists' epistemic goals in addition to biological entities and phenomena (e.g., taxa and lineages). Second, it makes room for the possibility that different construals of individuality may be needed for the same biological entity (or taxon), especially when biologists differ with respect to their epistemic goals (see also Lidgard and Nyhart, this volume).

If individuality is pursued monistically within a highly abstract framework intended to cover all levels and contexts where individuality is present, as observed in the historical cases of Heidenhain and Spencer (Rieppel, this volume; Gissis, this volume), then the project seems more resistant to concrete empirical characterization. This is problematic given that one philosophical task is to discern metaphysical implications of biological phenomena, scrutinized and explained scientifically. Part of this resistance may derive from a decoupling of concepts of individuality and epistemic goals. Using the abstract framework of information theory, David Krakauer and colleagues (2014) offer a complete theory of individuality for identifying biologically significant individuals and their meaningful parts at any level of organization—but with little to no discussion of what epistemic aims this would fulfill. Their claim that biologists are using the informational concept of an individual despite being unaware of it does not illuminate scientists' epistemic motivations either.[14] And even if a specific epistemic goal underlying the approach by Krakauer et al. was articulated (e.g., elucidating common features of stable configurations of individuality in different biological systems), then the value of their information-theoretic conception of individuality would be keyed to that goal. Rather than providing a monistic definition, their account would be another conceptualization of individuality in the toolkit of biologists available for pursuing different goals of inquiry.

What then might we learn metaphysically from our epistemological reorientation in contexts where biological individuality is in view? Negatively, the biological complexity investigated with multiple conceptions of individuality in conjunction with diverse epistemic goals is not susceptible to a monist characterization. In direct opposition to the fundamental theorist who seeks to identify a single, complete, and comprehensive account of individuality, a pluralist stance towards the metaphysics of individuality seems warranted (see also Kellert et al. 2006; Brigandt 2013).[15] There are multiple correct, empirically substantiated answers to the metaphysical question of what an individual is. Pluralism, in this context, is not only epistemological but also metaphysical. That there is more than one correct way to account for individuality means that there is more than one way to be an individual. The successful pairing of different characterizations of individuality with the diverse investigative and explanatory aims of biologists is an indicator of the structure of reality qua biological individuals (see also Dupré 1993; Boyd 1999b).

Once a pluralist stance has been adopted, deriving metaphysical implications is less of an exercise in sorting out competing conceptions of individuality and more about identifying successful matches of particular definitions of individuality with particular epistemic goals. Examining these scientific successes, which are related to how a specific definition's use has value vis-à-vis some epistemic goal, tells us something about the way the world is. Conceptualizing evolutionary individuals in terms of fitness considerations is well suited to the epistemic aim of explaining changes in the relative frequency of individuals exhibiting differing traits in populations over time because there are evolutionary individuals of this type in nature. If a mechanistic account of reproduction and differential persistence of individuals is sought, additional empirical facts beyond a mathematically abstract construal of individuality have to be adduced (Sterner, this volume). Conceptualizing physiological individuals in terms of metabolic autonomy is well suited to the epistemic aim of explaining how an individual maintains its distinctness over the course of a lifetime because there are physiological individuals of this type in nature. The physical features underlying the "autonomy" and possibly "agency" seen in these contexts (Landecker, this volume; Reynolds, this volume) are not captured by an evolutionary characterization. While our epistemic reorientation does not preclude the possibility that we will subsequently identify commonalities in matches between definitions and epistemic goals that permit us to combine them into more abstract pairings, there is no prerequisite that it occur. Many types of definition-goal matches are possible. In some cases, there may be a single best conception of individuality for a particular epistemic goal. Sometimes one conception of individuality may be conducive to multiple epistemic goals. In other cases, multiple conceptions of individuality may be required to address particular epistemic goals, as seems to be the case in explaining evolutionary transitions of individuality.

The metaphysical picture that emerges from our epistemological reorientation is messy but has a solid grounding: successful science. That there are many types of biological individuals is something that we have empirically discovered about the biological world; this complexity is not susceptible to a unified characterization of individuality (see also Mitchell 2003). The warrant for this is empirical and derived from the successful practices of individuation and decomposition observed in actual scientific inquiry (Kellert et al. 2006). In our view, the only way to recover a systematic, monist account of biological individuality is to neglect, discount, or actively reject the manifold ways that scientists successfully mark out individuals and carve them up into meaningful parts. The rationale for this rejection would have to come from another source, such as a distinctive commitment to unification not derived from scientific practice: "to pursue the task of seeing how everything hangs together" (Godfrey-Smith 2014, 4). The pluralist stance starts with the task of seeing whether everything hangs together. With respect to whether there is an adequate, all-purpose concept of biological individuality, the answer appears to be "no" (see also Lidgard and Nyhart, this volume).

In this introductory overview we have reviewed and characterized a variety of philosophical distinctions that are relevant to discussions of biological individuality (e.g., individuals versus classes) with the ultimate aim of advocating for an epistemological reorientation of these discussions. Metaphysical assumptions have animated controversies about individuality, both in the past of biology and in the present for philosophy. We have encouraged a rejection of fundamental theorizing to determine what an individual really is as unnecessary and unwarranted. Our argument for a shift in attention from metaphysics to epistemology is motivated by successful biological practices of individuation and decomposition in the context of different epistemic goals, and is amply illustrated in the contributions to this volume. The resulting metaphysical project of elucidating the implications of a pluralist perspective on biological individuality diverges from many approaches in contemporary philosophy of biology but has the advantage of taking seriously the array of empirical successes that biologists have had in investigating the complex nature of individuality in living phenomena.

Acknowledgments

We are grateful to Lynn Nyhart and Scott Lidgard for the invitation to participate in this project, securing financial support to attend the project workshop, and helpful feedback on different portions of this chapter throughout its development. Lynn's and Scott's unfailing editorial patience and perseverance was an undeserved gift. Thanks are due to all of the original workshop participants and chapter contributors whose conversational input and written arguments influenced our analysis. Alan Love is supported in part by grants from the John Templeton Foundation (Integrating Generic and Genetic Explanations of Biological Phenomena, ID 46919; From Biological Practice to Scientific Metaphysics, ID 50191).

Notes to Chapter Thirteen

1. This same point can be illustrated with discussions about the metaphysical nature of species, in particular whether they are individuals rather than classes (covered below in the subsection "Species as Individuals"; see also Brigandt 2009; Love 2009).

2. "Fundamental theorizing" should not be confused with reduction to the molecular or microphysical level. Although the two can sometimes coincide, in this particular case they do not.

3. A prominent exemplification is the Society for the Metaphysics of Science (https://sites .google.com/site/socmetsci/).

4. An analogous conception in the context of organisms as biological individuals is Willi Hennig's account of the relation between a semaphoront, which is his notion for a time-slice of an organism during a very short period, and an individual across its entire lifetime (see Havstad et al. 2015).

5. Our use of the term "monism" does not refer to the philosophical position—much discussed in late nineteenth- and early twentieth-century philosophy—that matter and mind are both manifestations of a united, underlying substance (Weir 2012). Instead, in our analysis, mo-

nism refers to a rejection of the possibility that there is more than one legitimate definition of what a biological individual is.

6. Commenting on the question of whether homologues are natural kinds or individuals, Havstad et al. (2015) extend Hennig's notion of a semaphoront from species to homologues by distinguishing between a bodily part across an organism's whole lifetime, which undergoes substantial ontogenetic change, and an ontogenetic stage of this bodily part.

7. Although nearly all proponents of the HPC account of kinds are philosophers, a few biologists agree that species can be seen as HPC kinds (Rieppel 2007, 2009, 2013; Assis 2011).

8. Whether it should capture all or only scientific kinds are additional questions. Some philosophers have worried that an HPC account does not hold for some classifications of particular successful sciences and includes classifications that are explicitly rejected by some sciences (Ereshefsky and Reydon 2015).

9. This might be expected given the number and diversity of criteria that have been put forward: (1) displaying the capacity of reproduction, (2) exhibiting a life cycle, (3) having a unique genotype, (4) engaging in sex, (5) having a single-cell bottleneck during the life cycle, (6) exhibiting a separation of germ and soma, (7) displaying mechanisms that police for cheaters, (8) having identifiable spatial boundaries and contiguity or physiological integrity, (9) exhibiting histocompatibility or an immune self identity, (10) being under selection as a unit of fitness maximization, (11) having cooperation among but not conflict between parts, (12) exhibiting a common evolutionary fate or co-dispersal, and (13) being an entity that bears adaptations (Clarke 2010). Clarke and Okasha (2013) discuss the diversity of criteria for individuality in comparison to the plurality of species concepts.

10. Though it is generally assumed that evolutionary theory requires individuals to be the entities bearing fitness, Frédéric Bouchard (2008) has argued that the notion of an individual (and an individual's reproductive fitness) may be dispensable. On his account, evolutionary explanations require only a concept of lineage and the differential persistence of lineages; these concepts apply to the full range of difficult cases, from colonial species to multi-species symbioses, which is a major motivation for his approach.

11. Godfrey-Smith (2013) recognizes a difference between the evolutionary and physiological approach by distinguishing between his "Darwinian individual" and an "organism."

12. This is analogous to distinguishing a definition of "evolutionary novelty" and the epistemic goal of explaining the evolutionary origin of novelty (Brigandt and Love 2012). For other examples illustrating the benefits of paying attention to the epistemic goals of inquiry in addition to the definitions of concepts, see Brigandt (2012).

13. Similar themes emerge when talking of "molecular ecosystems" (Nathan 2014).

14. "It is our belief that when biologists speak of individuals they are often invoking informational individuals without always making this assumption explicit" (Krakauer et al. 2014, 7).

15. Lidgard and Nyhart (this volume) articulate the diverse uses of biological individuality concepts in terms of changing "problem spaces," and explicitly view epistemic goals to be ingredients of such problem spaces. Given that a particular construal of biological individuality may serve some epistemic goals while excluding features that would be relevant for others, Lidgard and Nyhart see the need for multiple concepts of individuality. Beckett Sterner (2015) likewise argues for a pluralism about biological individuality on the basis of the presence of multiple epistemic goals (which he calls "epistemic roles" that the concept of individuality must satisfy). He claims that even when evolutionary goals are at stake, the monistic evolutionary accounts of individuality by Clarke and Godfrey-Smith are insufficient.

References for Chapter Thirteen

Assis, L. C. S. 2011. "Individuals, Kinds, Phylogeny and Taxonomy." *Cladistics* 27: 1–3.

Bouchard, F. 2008. "Causal Processes, Fitness, and the Differential Persistence of Lineages." *Philosophy of Science* 75: 560–70.

Bouchard, F., and P. Huneman, eds. 2013. *From Groups to Individuals: Evolution and Emerging Individuality*. Cambridge, MA: MIT Press.

Boyd, R. 1999a. "Homeostasis, Species, and Higher Taxa." In *Species: New Interdisciplinary Essays*, edited by R. A. Wilson, 141–85. Cambridge, MA: MIT Press.

———. 1999b. "Kinds as the 'Workmanship of Men': Realism, Constructivism, and Natural Kinds." In *Rationality, Realism, Revision: Proceedings of the 3rd International Congress of the Society for Analytic Philosophy*, edited by J. Nida-Rümelin, 52–89. Berlin: de Gruyter.

Brigandt, I. 2009. "Natural Kinds in Evolution and Systematics: Metaphysical and Epistemological Considerations." *Acta Biotheoretica* 57: 77–97.

———. 2012. "The Dynamics of Scientific Concepts: The Relevance of Epistemic Aims and Values." In *Scientific Concepts and Investigative Practice*, edited by U. Feest and F. Steinle, 75–103. Berlin: de Gruyter.

———. 2013. "Explanation in Biology: Reduction, Pluralism, and Explanatory Aims." *Science & Education* 22: 69–91.

Brigandt, I., and A. C. Love. 2012. "Conceptualizing Evolutionary Novelty: Moving Beyond Definitional Debates." *Journal of Experimental Zoology Part B: Molecular and Developmental Evolution* 318: 417–27.

Clarke, E. 2010. "The Problem of Biological Individuality." *Biological Theory* 5: 312–25.

———. 2012. "Plant Individuality: A Solution to the Demographer's Dilemma." *Biology & Philosophy* 27: 321–61.

———. 2013. "The Multiple Realizability of Biological Individuals." *Journal of Philosophy* 110: 413–35.

———. 2016. "Levels of Selection in Biofilms: Multispecies Biofilms Are Not Evolutionary Individuals." *Biology & Philosophy* 31: 191–212.

Clarke, E., and S. Okasha. 2013. "Species and Organisms: What Are the Problems?" In *From Groups to Individuals: Evolution and Emerging Individuality*, edited by F. Bouchard and P. Huneman, 55–75. Cambridge, MA: MIT Press.

Dupré, J. 1993. *The Disorder of Things: Metaphysical Foundations of the Disunity of Science*. Cambridge, MA: Harvard University Press.

Dupré, J., and M. A. O'Malley. 2009. "Varieties of Living Things: Life at the Intersection of Lineage and Metabolism." *Philosophy and Theory in Biology* 1: e003.

Elwick, J. 2007. "Styles of Reasoning in Early to Mid-Victorian Life Research: Analysis:Synthesis and Palaetiology." *Journal of the History of Biology* 40: 35–69.

Ereshefsky, M. 2009. "Homology: Integrating Phylogeny and Development." *Biological Theory* 4: 225–29.

———. 2010. "What's Wrong with the New Biological Essentialism." *Philosophy of Science* 77: 674–85.

Ereshefsky, M., and M. Pedroso. 2013. "Biological Individuality: The Case of Biofilms." *Biology & Philosophy* 28: 331–49.

———. 2015. "Rethinking Evolutionary Individuality." *Proceedings of the National Academy of Sciences USA* 112: 10126–32.

Ereshefsky, M., and T. A. C. Reydon. 2015. "Scientific Kinds." *Philosophical Studies* 172: 969–86.

Ghiselin, M. T. 1974. "A Radical Solution to the Species Problem." *Systematic Zoology* 23: 536–44.

Godfrey-Smith, P. 2009. *Darwinian Populations and Natural Selection*. Oxford: Oxford University Press.

———. 2013. "Darwinian Individuals." In *From Groups to Individuals: Evolution and Emerging Individuality*, edited by F. Bouchard and P. Huneman, 17–36. Cambridge, MA: MIT Press.

———. 2014. *Philosophy of Biology*. Princeton: Princeton University Press.

Grosberg, R. K., and R. R. Strathmann. 2007. "The Evolution of Multicellularity: A Minor Major Transition?" *Annual Review of Ecology, Evolution, and Systematics* 38: 621–54.

Havstad, J. C., L. C. S. Assis, and O. Rieppel. 2015. "The Semaphoronic View of Homology." *Journal of Experimental Zoology Part B: Molecular and Developmental Evolution* 324: 578–87.

Hawley, K. 2010. "Temporal Parts." *Stanford Encyclopedia of Philosophy*. http://plato.stanford .edu /entries /temporal-parts.

Hull, D. L. 1978. "A Matter of Individuality." *Philosophy of Science* 45: 335–60.

Kauffman, S. A. 1971. "Articulation of Parts Explanations in Biology and the Rational Search for Them." *PSA: Proceedings of the Biennial Meeting of the Philosophy of Science Association* 1970: 257–72.

Kellert, S. H., H. E. Longino, and C. K. Waters. 2006. "Introduction: The Pluralist Stance." In *Scientific Pluralism*, edited by S. H. Kellert, H. E. Longino, and C. K. Waters, vii–xxvix. Minneapolis: University of Minnesota Press.

Krakauer, D., N. Bertschinger, E. Olbrich, N. Ay, and J. C. Flack. 2014. "The Information Theory of Individuality." *arXiv e-Print Archive* arXiv:1412.2447v1 [q-bio.PE]. http://arxiv.org/pdf/ 1412.2447v1.pdf.

Love, A. C. 2009. "Typology Reconfigured: From the Metaphysics of Essentialism to the Epistemology of Representation." *Acta Biotheoretica* 57: 51–57.

———. Forthcoming. "Individuation, Individuality, and Experimental Practice in Developmental Biology." In *Individuation across Experimental and Theoretical Sciences*, edited by O. Bueno, R.-L. Chen, and M. B. Fagan. Oxford: Oxford University Press.

Mitchell, S. D. 2003. *Biological Complexity and Integrative Pluralism*. Cambridge: Cambridge University Press.

Nathan, M. D. 2014. "Molecular Ecosystems." *Biology & Philosophy* 29: 101–22.

Nyhart, L. K., and S. Lidgard. 2011. "Individuals at the Center of Biology: Rudolf Leuckart's *Polymorphismus der Individuen* and the Ongoing Narrative of Parts and Wholes. With an Annotated Translation." *Journal of the History of Biology* 44: 373–443.

O'Malley, M. A. 2014. *Philosophy of Microbiology*. Cambridge: Cambridge University Press.

Okasha, S. 2006. *Evolution and the Levels of Selection*. Oxford: Oxford University Press.

Pradeu, T. 2012. *The Limits of the Self: Immunology and Biological Identity*. Oxford: Oxford University Press.

Rieppel, O. 2007. "Species: Kinds of Individuals or Individuals of a Kind." *Cladistics* 23: 373–84.

———. 2009. "Species as a Process." *Acta Biotheoretica* 57: 33–49.

———. 2013. "Biological Individuals and Natural Kinds." *Biological Theory* 7: 162–69.

Robertson, T., and P. Atkins. 2013. "Essential vs. Accidental Properties." *Stanford Encyclopedia of Philosophy*. http://plato.stanford.edu /entries /essential-accidental.

Smith, J. E. H. 2011. *Divine Machines: Leibniz and the Sciences of Life*. Princeton: Princeton University Press.

Sterner, B. 2015. "Pathways to Pluralism about Biological Individuality." *Biology & Philosophy* 30: 609–28.

Varzi, A. 2014. "Mereology." *Stanford Encyclopedia of Philosophy.* http://plato.stanford.edu/entries/mereology.

Wagner, G. P. 2014. *Homology, Genes, and Evolutionary Innovation.* Princeton: Princeton University Press.

Weir, T. H., ed. 2012. *Monism: Science, Philosophy, Religion, and the History of a Worldview.* New York: Palgrave Macmillan.

Wilson, R. A., and M. Barker. 2013. "The Biological Notion of Individual." *Stanford Encyclopedia of Philosophy.* http://plato.stanford.edu/entries/biology-individual.

Wilson, R. A., M. J. Barker, and I. Brigandt. 2007. "When Traditional Essentialism Fails: Biological Natural Kinds." *Philosophical Topics* 35: 189–215.

Wimsatt, W. C. 1974. "Complexity and Organization." *PSA: Proceedings of the Biennial Meeting of the Philosophy of Science Association* 1972: 67–86.

Winther, R. G. 2011. "Part-Whole Science." *Synthese* 178: 397–427.

Wouters, A. 2003. "Four Notions of Biological Function." *Studies in History and Philosophy of Biological and Biomedical Sciences* 34: 633–68.

Zylstra, U. 1992. "Living Things as Hierarchically Organized Structures." *Synthese* 91: 111–33.

Acknowledgments

This book began as a workshop in Philadelphia in May 2012, generously funded and hosted by the Chemical Heritage Foundation. We thank the CHF for this opportunity, and for the associated subvention that has been crucial to the publication of the present volume. We especially wish to thank Carin Berkowitz, Director of the Beckman Center for the History of Chemistry at the CHF, for her outstanding management of the many details associated with running a successful conference. A follow-up authors' workshop took place at the University of Wisconsin-Madison in December 2012, with funding from the UW-Madison's Anonymous Fund, a John Templeton Foundation grant on Principles of Complexity awarded to David Krakauer and Jessica Flack, the Robert F. and Jean E. Holtz Center for Science and Technology Studies, and the Department of the History of Science. Judith Kaplan, Daniel Liu, Melissa Charenko, and Eileen Ward provided logistical and technical support for these workshops. Even though they did not end up contributing to the volume, the workshop engagement of invited participants William (Ned) Friedman, Elihu Gerson, Peter Godfrey-Smith, Richard Grosberg, David Hughes, David Krakauer, Phillip Sloan, Günter Wagner, and William Wimsatt contributed significantly to the intellectual project.

We note that while the disciplinary backgrounds of the editors differ, we have shepherded this project as a kind of composite functional individual, learning from one another as we have gone along. The name order of the co-editors is alphabetical.

Finally, we wish to dedicate this volume to our teachers and mentors, especially Jeremy B. C. Jackson and Alan H. Cheetham, and Mark B. Adams and Robert E. Kohler, who collectively showed us how to combine meticulous care with a broad vision of scholarship.

Contributors

Ingo Brigandt
Department of Philosophy
University of Alberta
2-40 Assiniboia Hall
Edmonton, AB T6G 2E7
Canada

James Elwick
Science and Technology Studies
    Department
York University
313 Bethune College
4700 Keele St.
Toronto M3J1P3
Canada

Scott F. Gilbert
Department of Biology
Swarthmore College
500 College Avenue
Swarthmore, PA 19081
United States

Snait B. Gissis
Cohn Institute for the History and
    Philosophy of Science and Ideas
Tel Aviv University

Haim Levanon Str. Ramat Aviv
Tel Aviv
Israel

Matthew D. Herron
Georgia Institute of Technology
North Avenue
Atlanta, GA 30332
United States

Hannah Landecker
UCLA Institute for Society and Genetics
and
UCLA Department of Sociology
264 Haines Hall
375 Portola Plaza
Angeles, CA 90095-1551
United States

Scott Lidgard
Integrative Research Center
Field Museum of Natural History
1400 S. Lake Shore Drive
Chicago, IL 60605
United States

Alan C. Love
Minnesota Center for Philosophy
    of Science
and
Department of Philosophy
University of Minnesota–Twin Cities
831 Heller Hall
271 19th Ave S
Minneapolis, MN 55455–0310
United States

Lynn K. Nyhart
Department of History
University of Wisconsin-Madison
3211 Mosse Humanities Bldg.
455 N. Park St.
Madison, WI 53706–1483
United States

Michael A. Osborne
School of History, Philosophy, and
    Religion
322 Milam Hall
2520 SW Campus Way
Oregon State University
Corvallis, OR 97331-6202
United States

Andrew S. Reynolds
Department of Philosophy and
    Religious Studies
Cape Breton University
Sydney, Nova Scotia B1P 6L2
Canada

Olivier Rieppel
Integrative Research Center
Field Museum of Natural History
1400 S. Lake Shore Drive
Chicago, IL 60605
United States

Beckett Sterner
School of Life Sciences
Center for Biology and Society
Arizona State University
PO Box 873301
Tempe, AZ 85287–3301
United States

Index

Page numbers for illustrations are in boldface.